THE CHINESE COMPUTER

Studies of the Weatherhead East Asian Institute, Columbia University

The Studies of the Weatherhead East Asian Institute of Columbia University were inaugurated in 1962 to bring to a wider public the results of significant new research on modern and contemporary East Asia.

THE CHINESE COMPUTER

A GLOBAL HISTORY OF THE INFORMATION AGE

THOMAS S. MULLANEY

THE MIT PRESS   CAMBRIDGE, MASSACHUSETTS   LONDON, ENGLAND

The MIT Press would like to thank the anonymous peer reviewers who provided comments on drafts of this book. The generous work of academic experts is essential for establishing the authority and quality of our publications. We acknowledge with gratitude the contributions of these otherwise uncredited readers.

This book was set in ITC Stone and Avenir by New Best-set Typesetters Ltd. Printed and bound in the United States of America.

Library of Congress Cataloging-in-Publication Data

Names: Mullaney, Thomas S. (Thomas Shawn), author.
Title: The Chinese computer : a global history of the information age / Thomas S. Mullaney.
Description: Cambridge, Massachusetts : The MIT Press, 2024. | Series: Studies of the Weatherhead East Asian Institute, Columbia University | Includes bibliographical references and index.
Identifiers: LCCN 2023027999 (print) | LCCN 2023028000 (ebook) | ISBN 9780262047517 (hardcover) | ISBN 9780262372435 (epub) | ISBN 9780262372428 (pdf)
Subjects: LCSH: Chinese language—Data processing—History. | Chinese character sets (Data processing) | Chinese language—Written Chinese.
Classification: LCC PL1074.5 .M85 2024 (print) | LCC PL1074.5 (ebook) | DDC 495.10285—dc23/eng/20231206
LC record available at https://lccn.loc.gov/2023027999
LC ebook record available at https://lccn.loc.gov/2023028000

10 9 8 7 6 5 4 3 2 1

This book is dedicated to Thomas (1945–2017), Orfeo (2018–),
Merri (1945–2021), and Arthur (2021–). I wish you all could have met.

CONTENTS

ACKNOWLEDGMENTS

I missed my brother's call at 8:49 a.m. on Saturday morning, in late January 2021. Arthur was in the back seat, not two days old, and Chiara and I were driving him to his first doctor's appointment. Orfeo, our oldest, was at daycare, taking his first steps on the wondrous and confounding path called brotherhood. Scott called back at 8:56. I answered this time.

Hi, Scott. This isn't going to be good news, is it?

No, I'm afraid it's not. Mom died this morning.

I was as prepared as I could've been for the news. It began months earlier. *Mom's doctor found something.* That was the first call. *The cancer has spread,* came next. Finally: *Mom is starting hospice.*

After that third call, my flesh began to wage rebellion. It started in my sleep, when I ground my back molar so hard it broke open. I'd never peered inside my own bones before, but there it was: bright, purple pulp, clear as day in the morning mirror.

Two days later, rebellion spread to my waking hours. On our way back from a two-beer lunch, I pissed my pants on the car ride home. At the last stop light before our intersection, when I realized the struggle was futile, Chiara watched in disbelief as my blue jeans turned five shades darker.

COVID was raging. Chiara was pregnant with our second child. *Do I fly to her? Do I say goodbye in person? What if I bring the disease home?* This was when everything was unknown.

Finally, *yes*: Saturday, December 19, 2020. SFO-PHX. Gate: Not Yet Assigned. Boarding Begins: 7:50 a.m. Seat 1E. P19SR3. UA 794. Chiara would stay with Orfeo. I would go and say goodbye.

I drove myself to the airport. I wiped down every conceivable surface. I gave wide berth to the fool with his mask protestingly beneath his nostrils. I declined water. I held it in. I was up and off that plane the moment the doors unlocked.

We ate take-out, my mom, my brother, my sister-in-law, and I. We talked. We watched TV. A few days later I said goodbye. That was it.

This book is dedicated to four people who I wish could have met each other, but never did: my father, my mother, and my two sons. Dad died months before Orfeo was born. Mom died less than 48 hours after Arthur's arrival. This will always bother me.

In a larger sense, though, I dedicate this to all the people I met over the course of writing this book—some fifteen years—who died along the way. People like Lois Lew, Chan Yeh, Louis Rosenblum, Rolf Heinen, and others who opened up to me and taught me so much. I regularly come across their emails in my inbox, and their text messages, and I still have to stop myself from drafting a reply. With apologies to all the many people I should be thanking by name—and certainly would, were these like other acknowledgments I've written in my career—I've chosen to express thanks to one group of people in particular: those who welcomed me, however briefly, in their lives and the lives of their families, and who are no longer with us.

Fresh news of death came even as I was finishing the page proofs for this book, in fact. First, I had to change all of Lois Lew's verbs to the past tense. And then the same for Halcyon Lawrence and Bruce Rosenblum. Lois Lew lived a remarkably long and rich life, a fact that helped dampen the blow somewhat. News of Halcyon's death was different. It burnt everyone who knew her—even mere acquaintances and professional colleagues like myself—to cinders. That such a brilliant and immense soul should be taken away so young was and is utterly inconceivable.

Finally, news came of Bruce. His passing was not unexpected. He had confided in me his ALS diagnosis years ago. But the news was crushing all the same. I've never met anyone who faced death like Bruce did: steadfast, optimistic, without complaint, even as this cruel disease robbed him of so much life. How a person could be this strong, I may never understand.

Thank you to everyone who opened up to me. I wish you all could have met each other, too.

INTRODUCTION: CHINESE IN THE DIGITAL AGE

One billion Chinese speakers have been laid low by a strange new cognitive disorder.

They're forgetting how to write Chinese.

That's the rumor, at least.

Reports began circulating in the early 2000s, each following an eerily similar narrative arc. One moment, a person was competent, accomplished, often highly educated. A scientist. An entrepreneur. An author. The very next, they were schoolchildren, struggling to recall even the most basic Chinese characters.

They'd "lift the brush, but forget the character": *tibi wangzi*, the expression went.[1]

Aphasia, some called it—a debilitating condition resulting in an inability to speak. *Dysgraphia*, others chimed in—aphasia's cousin, for writing rather than speaking. A "strange new form of illiteracy," still others suggested.[2] No one could make sense of it, this epidemic whose pathology was so at odds with established medicine, its onset so sudden, that the story seemed stolen straight from the pages of science fiction.

"Character amnesia"—this is the label that stuck.[3]

Rumors gave way to sobering statistics: 98.8 percent of respondents in a survey from 2013 reported experiencing *tibi wangzi*, many on a daily basis.[4] The entire country, it seemed, was in the grips of a bizarre "Chinese character crisis" (*Hanzi weiji*).[5]

The titanic scale of "character amnesia" was made only more worrisome by its perplexing behavior. It didn't afflict people living at the fringes of society—the marginalized and impoverished, the way so many public health crises do. This one was stalking China's elite, ruthlessly. The wealthier and more urbanized a person was, the more vulnerable. The greater their expendable income, the greater their risk of losing the ability to write.

A culprit finally revealed itself: digital writing. "Character amnesia" was most prevalent among those who used computers, smartphones, and tablets—any electronic device used to write Chinese with the aid of a QWERTY keyboard or a trackpad. One minute, a person entered steady streams of characters on a laptop or a mobile device. As soon as they powered down, though, it was as if their minds powered down, too.

How do we make sense of these astonishing accounts? Is it another case of moral panic in the digital age—whether concerns over "Textspeak," emoticons, the decline of handwriting, or other matters of "language hygiene?" Or could it be that twenty-first century China is home to hundreds of millions of newly illiterate aphasics, or dysgraphic amnesiacs?[6] If so, why don't we find evidence of this crisis everywhere we look? A cratering economy? The collapse of higher education, perhaps? How, then, can China be one of the world's most vibrant and wealthiest digital economies? How is it possible, moreover, that the Chinese-language internet is boiling over with activity, with an estimated 900 million internet users in mainland China alone, engaged in a frenetic, nonstop traffic in Chinese-language content?[7] If China's most connected, tech-savviest individuals are "incapable of writing" (a baseline definition of dysgraphia), who exactly is doing all this Chinese writing?

The Chinese Computer is the first history in any language about Chinese in the digital age. Grounded in more than fifteen years of research, it charts out the inception of electronic Chinese in the immediate wake of World War II, through its efflorescence in the present day. Based on oral histories, material artifacts, and archives drawn from dozens of collections across Asia, Europe, and North America, it's a tale of eccentric and often brilliant personalities drawn from the ranks of IBM, the Central News Agency of China, RCA, MIT, the CIA, the US Air Force, the US Army, the Pentagon, the RAND Corporation, the British telecommunications giant Cable and Wireless, Silicon Valley, the Taiwanese military, Japanese industrial circles, and the highest rungs of mainland Chinese intellectual, industrial, and military establishments.

But this book does more than present a colorful cast of underdogs. It strives to explain six core dimensions of Chinese computing—six axioms that one must grasp in order to understand Chinese in the digital age.

To explore these axioms, we begin our journey in a winter-chilled auditorium in China's Henan province, where a group of fifty-five talented digerati gathered in December 2013 during the very height of the "character amnesia" crisis. They came together, not to lament *tibi wangzi*, but to best their opponents in head-to-head competition: to take first place in a typing contest and secure bragging rights as the fastest computer keyboardist in China—and perhaps the world.

THE SIX AXIOMS OF CHINESE COMPUTING

ymiw2
   klt4
      pwyy1
        wdy6
          o1
            dfb2
            wdv2
              fypw3
              uet5
             dm2
              dlu1 . . .

A young Chinese man sat down at his QWERTY keyboard and rattled off an enigmatic string of letters and numbers.

Was it code? Child's play? Confusion?

It was Chinese.

The beginning of Chinese, at least. These forty-four keystrokes marked the first steps in a process known as "input" or *shuru*: the act of getting Chinese characters to appear on a computer monitor or other digital device using a QWERTY keyboard or trackpad (figure 0.1).

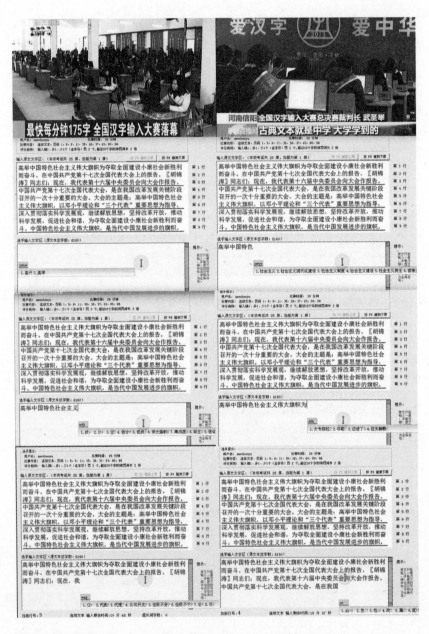

Figure 0.1   Stills taken from a 2013 Chinese input competition screencast

Across all computational and digital media, Chinese text entry relies on software programs known as "Input Method Editors"—better known as "IMEs" or simply "input methods" (*shurufa*). IMEs are a form of "middleware," so-named because they operate in between the hardware of the user's device and the software of its program or application. Whether a person is composing a Chinese document in Microsoft Word, searching the web, sending text messages, or otherwise, an IME is always at work, intercepting all of the user's keystrokes and trying to figure out which Chinese characters the user wants to produce. Input, simply put, is the way *ymiw2klt4pwyy* . . . becomes a string of Chinese characters.[8]

IMEs are restless creatures. From the moment a key is depressed, or a stroke swiped, they set off on a dynamic, iterative process, snatching up user-inputted data and searching computer memory for potential Chinese character matches. The most popular IMEs these days are based on Chinese phonetics—that is, they use the letters of the Latin alphabet to describe the sound of Chinese characters, with mainland Chinese operators using the country's official Romanization system, Hanyu pinyin. (Pinyin phonetic input has not always been the most popular approach to Chinese input, however, a point we will return to shortly.)

With the first key depression—"C," perhaps—IMEs such as Sogou pinyin, QQ pinyin, and Google pinyin begin to present the user with options. These "candidate" characters, appearing in a pop-up menu that follows along at the margins of the monitor, are ones whose phonetic values begin with "C," such as *chi* (吃 "to eat"), *cai* (才 "only then," among other meanings), or hundreds of other possibilities.

When the user depresses a second key—"H," let's say—the IME adjusts this list of candidates. It begins to present only those Chinese characters whose pronunciations begin with "CH" (eliminating the earlier possibility of *cai* but maintaining the possibility of *chi*). Once a user sees their desired character in the pop-up menu, one final keystroke—either the space bar, enter, or a number key—is all that's needed to select characters and add them to the main composition window (perhaps the user's desired term is *chaoxi* 抄袭, or "plagiarism") (figure 0.2). Keystroke by keystroke, this is how Input Method Editors enable the use of a QWERTY keyboard to produce Chinese characters by means of alphanumeric symbols.[9]

Figure 0.2 Example of Chinese Input Method Editor pop-up menu (抄袭 / "plagiarism")

This young man's name was Huang Zhenyu (also known by his nom de guerre, Yu Shi). He was one of around sixty contestants that day, each wearing a bright red shoulder sash—like a tickertape parade of old, or a beauty pageant.[10] "Love Chinese Characters" (*Ai Hanzi*) was emblazoned in vivid, golden yellow on a poster at the front of the hall. The contestants' task was to transcribe a speech by outgoing Chinese president Hu Jintao, as quickly and as accurately as they could. "Hold High the Great Banner of Socialism with Chinese Characteristics," it began, or in the original: 高举中国特色社会主义伟大旗帜为夺取全面建设小康社会新胜利而奋斗.[11] Huang's QWERTY keyboard did not permit him to enter these characters directly, however, and so he entered the quasi-gibberish string of letters and numbers instead: *ymiw2klt4pwyy1wdy6*. . . .

With these four-dozen keystrokes, Huang was well on his way, not only to winning the 2013 National Chinese Characters Typing Competition, but also to clock one of the fastest typing speeds ever recorded, anywhere in the world.[12]

In chapter 1, we confront the most basic, but also most profound, axiom of Chinese computing: *ymiw2klt4pwyy1wdy6* . . . is not the same as 高举中国特色社会主义 . . . ; the keys that Huang *actually* depressed on his QWERTY keyboard—his "primary transcript," as we could call it—were completely different than the symbols that ultimately appeared on his computer screen, namely the "secondary transcript" of Hu Jintao's speech. This is true for every one of the world's billion-plus Sinophone computer users. In Chinese computing, what you type is *never* what you get.

For readers accustomed to English-language word processing and computing, this should come as a surprise. For example, were you to compare the paragraph you're reading right now against a key log showing *exactly* which buttons I depressed to produce it, the exercise would be unenlightening (to put it mildly). "F-o-r-_-r-e-a-d-e-r-s-_-a-c-c-u-s-t-o-m-e-d-_-t-o-_-E-n-g-l-i-s-h . . . ," it would read (forgiving any typos or edits). In English-language typewriting and computer input, a typist's primary and secondary transcripts are, in principle, identical. The symbols on the keys and the symbols on the screen are the same.

Not so for Chinese computing. When inputting Chinese, the symbols a person sees on their QWERTY keyboard are always different from the symbols that ultimately appear on the monitor or on paper. Every single computer and new media user in the Sinophone world—no matter if they are blazing-fast or molasses-slow—uses their device in exactly the same way as Huang Zhenyu, constantly engaged in this iterative process of *criteria-candidacy-confirmation*, using one IME or another. Not *some* Chinese-speaking users, mind you, but *all*. This is the first and most basic feature of Chinese computing: Chinese human-computer interaction (HCI) requires users to operate entirely in code *all the time*.[13]

How is this possible? With tens of thousands of Chinese characters, and a unique alphanumeric code for many if not most of them, how could Huang—or, more importantly, hundreds of millions of other Chinese computer users—be expected to memorize and deploy them all in real time? To answer this question, I venture back in the first chapter not to the 2010s, but to the 1940s and the earliest effort to build an electro-automatic Chinese writing machine: the IBM Electric Chinese Typewriter, invented by the Chinese engineer Chung-Chin Kao and prototyped by the International Business Machines corporation. On the keyboard of this unprecedented machine, as we will see, there were no Chinese characters at all. Instead, typists needed to input Chinese characters by means of a separate "primary transcript"—just like Huang Zhenyu—only in this case by using a series of unique, four-digit ciphers, one cipher for each of the more than 6,000 characters the machine was capable of printing. In this chapter, we will meet one of the first individuals to master this device: a remarkable woman named Lois Lew who, after a nearly ten-year-long quest to learn her story, I had the great fortune of interviewing.

If Huang Zhenyu's mastery of a complex alphanumeric code weren't impressive enough, consider the staggering speed of his performance. He transcribed the first 31 Chinese characters of Hu Jintao's speech in roughly 5 seconds, for an extrapolated speed of 372 Chinese characters per minute. By the close of the grueling 20-minute contest, one extending over thousands of characters, he crossed the finish line with an almost unbelievable speed of 221.9 characters per minute.

That's 3.7 Chinese characters *every second*.

In the context of English, Huang's opening 5 seconds would have been the equivalent of around 375 English words-per-minute, with his overall competition speed easily surpassing 200 WPM—a blistering pace unmatched by anyone in the Anglophone world (using QWERTY, at least).[14] In 1985, Barbara Blackburn achieved a *Guinness Book of World Records*–verified performance of 170 English words-per-minute (on a typewriter, no less). Speed demon Sean Wrona later bested Blackburn's score with a performance of 174 WPM (on a computer keyboard, it should be noted).[15] As impressive as these milestones are, the fact remains: had Huang's performance taken place in the Anglophone world, it would be his name enshrined in the *Guinness Book of World Records* as the new benchmark to beat.[16]

Huang's speed carried special historical significance as well.

For a person living between the years 1850 and 1950—the period examined in the prequel to this book, *The Chinese Typewriter*—the idea of producing Chinese by mechanical means at a rate of over two hundred characters per minute would have been virtually unimaginable. Throughout the history of Chinese telegraphy, dating back to the 1870s, operators maxed out at perhaps a few dozen characters per minute. In the heyday of mechanical Chinese typewriting, from the 1920s to the 1970s, the fastest speeds on record were just shy of eighty characters per minute (with the majority of typists operating at far slower rates). When it came to modern information technologies, that is to say, Chinese was consistently one of the slowest writing systems in the world.[17]

What changed? How did a script so long disparaged as cumbersome and helplessly complex suddenly rival—exceed, even—computational typing speeds clocked in other parts of the world? Even if we accept that Chinese computer users are somehow able to engage in "real time" coding, shouldn't Chinese IMEs result in a lower overall "ceiling" for Chinese text processing as compared to English? Chinese computer users have to jump through so many more hoops, after all, over the course of a cumbersome, multistep process: the IME has to intercept a user's keystrokes, search in memory for a match, present potential candidates, and wait for the user's confirmation. Meanwhile, English-language computer users need only depress whichever key they wish to see printed on screen. What could be simpler than the "immediacy" of "Q equals Q," "W equals W," and so on?

In chapter 2, I answer this question by exploring a second axiom of Chinese computing: Even though Chinese human-computer interaction relies upon forms of mediation unseen in mainstream Anglophone computing, these additional layers of mediation can result in speeds that equal or surpass those of the seemingly "unmediated" world of what-you-type-is-what-you-get. Counterintuitively, the addition of mediation can lead to the *subtraction* of time.

To unravel this seeming paradox, we will examine the first Chinese computer ever designed: the Sinotype, also known as the Ideographic Composing Machine. Debuted in 1959 by MIT professor Samuel Hawks Caldwell and the Graphic Arts Research Foundation, this machine featured a QWERTY keyboard, which the operator used to input—not the phonetic values of Chinese characters—but the brushstrokes out of which Chinese characters are composed. The objective of Sinotype was not to "build up" Chinese characters on the page, though, the way a user builds up English words through the successive addition of letters. Instead, each stroke "spelling" served as an electronic address that Sinotype's logical circuit used to retrieve a Chinese character from memory. In other words, the first Chinese computer in history was premised on the same kind of "additional steps" as seen in Huang Zhenyu's prizewinning 2013 performance.

During Caldwell's research, as we will see, he discovered unexpected benefits of all these additional steps—benefits entirely unheard-of in the context of Anglophone human-machine interaction at that time. The Sinotype, he found, needed far fewer keystrokes to *find* a Chinese character in memory than to *compose* one through conventional means of inscription. By way of analogy, to "spell" a nine-letter word like "crocodile" (c-r-o-c-o-d-i-l-e) took far more time than to *retrieve* that same word from memory ("c-r-o-c-o-d" would be enough for a computer to make an unambiguous match, after all, given the absence of other words with similar or identical spellings). Caldwell called his discovery "minimum spelling," making it a core part of the first Chinese computer ever built. Today, we know this technique by a different name: "autocompletion," a strategy of human-computer interaction in which additional layers of mediation result in faster textual input than the "unmediated" act of typing. Decades before its rediscovery in the Anglophone world, then, autocompletion was first invented in the arena of Chinese computing.

Why was Huang Zhenyu using a QWERTY keyboard in the first place? Given the profound challenge of "fitting" tens of thousands of Chinese characters onto a QWERTY keyboard, why didn't computer engineers simply abandon QWERTY altogether and concentrate on designing a more uniquely "Chinese" interface—something that might have enabled

Chinese computer users to bypass input altogether, and enjoy the same "unmediated" human-computer interaction as their Anglophone counterparts?

As we will see in chapter 3, this is exactly what engineers in the late 1960s and early-1970s tried to do—and for a time, they succeeded. The era's rapid advances in computer processing opened new vistas for engineers across Asia, the United States, and the UK. Rather than attempting to build upon the initial promise of Sinotype and the QWERTY-based approach, they abandoned QWERTY entirely in a quest for what might be termed "immediate Chinese."

The systems designed in this period were astonishing in their diversity. One of the custom-built interfaces we'll encounter offered users 120 levels of SHIFT, as compared to the two found on standard QWERTY devices (i.e., lowercase and uppercase). Another interface featured a set of 256 keys, rather than the typical range of approximately 80 to 100 on Western-built machines. Another from this period offered upward of 2,000 keys. Still another employed a stylus and touch-sensitive tablet with which users could select characters directly from the input surface. Yet another dispensed with flat input surfaces altogether, opting instead to wrap a matrix of Chinese characters around a revolving, cylindric interface.

In addition to charting out this pivotal and poorly understood part of the timeline of Chinese computing, this chapter also serves as an important reminder: the IME has never been a revered technology, no matter its potential or proven track record. From its inception to the present day, input has tended to be understood as an inherently *compensatory* technology, one meant to assist or "work around" the challenges confronted in the computational processing of Chinese text. Despite the many achievements we will chart out in this history—up to and including blazing speed and efficiency—we must be cautious never to imagine that IMEs were somehow heralded as an "alternative modernity" or "competitor" to Western-style human-computer interaction. To the contrary, we will encounter throughout this time period—at times subtly, at times explicitly—the longing for a form of Chinese human-computer interaction that matched or approximated the revered "immediacy" of the English-language world.[18] Input has almost never been anyone's "first choice."

From the late 1970s onward, the time period examined in chapter 4, custom-built Chinese interfaces steadily disappeared from marketplaces and laboratories alike, displaced by wave upon wave of Western-built personal computers crashing on the shores of Reform Era China (1978–1989). This resurgence of QWERTY—and with it, input—was not a return to the status quo ante, however. If the dawn of Chinese computing featured just a pair of input systems (Kao's Four-Digit Code and Caldwell's Sinotype code), the late 1970s and 1980s witnessed an explosion of competing methods—dozens, hundreds, and ultimately more than a thousand different IMEs. Huang Zhenyu and his competitors in 2013 were the great-grandchildren of this era, born into a world when it was second nature to have a wide array of IMEs at one's fingertips. Indeed, even as Huang entered that cryptic alphanumeric sequence *ymiw2klt4pwyy . . .* into his computer, the key logs of the five dozen other competitors would have looked different.

This brings us to a fourth axiom of Chinese computing, and perhaps the most difficult to grasp: Chinese input is infinite. For every single Chinese character, there is an infinite number of input sequences that could be used to produce it—at least in theory.

To understand this bizarre "infinity-to-one" relationship, we need to return to Huang's seemingly nonsensical alphanumeric sequence *ymiw2klt4pwyy . . .* , not just to grasp how it works but, more importantly, to recognize that this sequence was only one of an uncountable number he could have entered on his QWERTY keyboard to achieve the exact same text output. An infinite number of pathways could have all led to the same speech by Hu Jintao.

To delve in, "y-m-i-w-2" corresponds to the characters 高举 (*gaoju*) because, within the specific IME Huang was using—an IME known as "Five Stroke" or Wubi input—the keys "Y," "M," "I," and "W" each correspond to a specific set of graphical shapes or pieces of Chinese characters. Within Wubi input, the key "Y" is assigned to eleven shapes in particular, one of them (亠) being the top-most portion of Huang's intended character 高 (*gao*). By depressing the key "Y," then, Huang was in effect telling the Wubi IME that he was in pursuit of a Chinese character containing that specific shape. As soon as he depressed "Y," therefore, the IME began to present potential character matches in the pop-up menu.

Because the letter "Y" corresponds to ten other shapes as well, however, this initial set of character matches would have contained a great deal of noise. To disambiguate further, Huang then depressed the key "M"—which itself corresponds to a *different* set of fourteen shapes (including the shape 冂, also found within Huang's intended character of 高, this time in the bottom half). By this point, then, the IME already had a great deal of information to go on, thus limiting the potential set of matches to only those Chinese characters containing *both* one of the eleven shapes corresponding to "Y" *and* one of the fourteen shapes corresponding to "M." As the number of possibilities decreased dramatically, the IME refreshed the pop-up menu, and already started to "suggest" the character 高 to Huang, awaiting confirmation. Huang then continued the same process for the rest of the contest (figure 0.3).

For readers taken aback at the seeming complexity of Huang's input string, prepare yourself for the real shock: Huang could have achieved the exact same text output by using a completely different alphanumeric input sequence. Within Wubi input, there is more than just one way to arrive at the same Chinese character. Meanwhile, had Huang chosen to use a different IME altogether (as many of his competitors did), his input sequence would have looked entirely different. And if, hypothetically, Huang just happened to be an inventor himself, designing his very own

Figure 0.3 The "Y" and "M" keys on a QWERTY keyboard with Wubi symbol markings

Chinese input method from scratch, he could have fashioned any one of a theoretically infinite number of ways to link a given alphanumeric "primary text" to the "secondary text" of President Hu's speech.

To understand how this can be so—how for any Chinese character, there can exist an infinite number of ways to input it—we need only scrutinize the logic of Wubi input a bit further. It becomes immediately apparent that although Wubi inventor Wang Yongmin chose to "set equal" the Latin letter 'Y' and the Chinese character component 亠, there exists no heaven-ordained law to tell us this must always be so. Likewise, there is no intrinsic property of the universe that tells us, for example, that "M" should equal 冂. All of these pairings of Latin letters and Chinese character components are *arbitrary decisions*. One could imagine entirely different ways of governing the relationship between the "primary transcript" of an input sequence and the "secondary transcript" of Chinese characters.

Contrast this with the computational inputting of English words, where convention dictates that there is one correct and accepted way to achieve a desired output. To input the word "electricity," let's say, we know to enter e-l-e-c-t-r-i-c-i-t-y. If I were I to enter "elctricity," "electrcty," "elec," or otherwise, these would be either typos or abbreviations. By comparison, Chinese input system designers from the 1950s onward have developed over 1,000 ways to input the single Chinese character 电 (*dian* "electricity"). If using "Area-Position Code" input (*Quwei ma*), for example, the correct input sequence would be "2171." Within "Taiji Code" input, the correct input sequence would be "NY." Within "OSCO" input, "D79." And the list goes on (see table 0.1).

Through the story of one engineer in particular—Zhi Bingyi, whose input system emerged out of his experience in a Cultural Revolution–era prison cell—chapter 4 examines how inventors in this period went on to design hundreds of different Chinese IMEs.

Putting aside Huang's QWERTY keyboard for a moment, was there something else about his computer that distinguished it from ones used in the Anglophone world? And what exactly makes a computer "Chinese?" Is a "Chinese computer" any computer manufactured within the political boundaries of China or the Sinophone world? Do Chinese computers operate by some kind of alternate logic that sets them apart from

Table. 0.1 More than two dozen of the thousand ways to input *dian* (electricity) using different Chinese IMEs

Inputting the Character 电 (*dian*) on a Computer	
Using this Input Method Editor (IME) . . .	**You enter . . .**
Chinese Transalphabet	d i a n t m v v
Zheng code (*Zheng ma*) [郑码]	k z v v
Five-Stroke input system (*Wubi shurufa*) [五笔输入法]	j n v
Cangjie encoding (*Cangjie bianma*) [仓颉编码]	l w u
Beginning-and-End Sound-Shape Code (*Shouwei yinxing shurufa*) [首尾音形输入法]	d j f z
Double Stroke Sound-Shape input system (*Shuangbi yinxing shurufa*) [双笔音形输入法]	d j j m
Pinyin [拼音]	d i a n #
Stroke-Shape Code (*Bixing bianma*) [笔形编码]	6 0 1[1]
Four Corner-Add Sound (*Sijiao fuyin*) [四角附音]	5 0 7 1 6 d[2]
Double Pinyin Double Radical Encoding System (*Shuangpin shuangbu bianmafa*) [双拼双部编码法]	d q t k
Yi input system (*Yi shurufa*) [易输入法]	r g d
Double Pinyin (*Shuangpin*) [双拼]	d m #
Area-Position Code (*Quwei ma*) [区位码]	2 1 7 1
GB Code (*guobiao*)[国标]	3 5 6 7
Shape-Meaning Three Letter Code (*Xingyi sanma*) [形意三码]	b 1
Weiwu Code (*Weiwu ma*) [唯物码]	」 4 7
Fifty Character Element input method (*Wushi ziyuan shurufa*) [五十字元输入法]	l j d
Chinese Character Stroke-Shape Look-Up Encoding Method (*Hanzi bixing chazifa bianmafa*) [汉字笔形查字法编码法]	6 0
OSCO (*Jianzi shima*) [见字识码]	D D D D
Qian Code (*Qian ma*) [钱码]	d j m
Wubi zixing [五笔字型]	j n
Natural Code (*Ziran ma*) [自然码]	d m l o /

Table. 0.1 (continued)

Inputting the Character 电 (*dian*) on a Computer	
Using this Input Method Editor (IME) . . .	**You enter . . .**
Shape-Describing Code (*Biaoxing ma*) [表形码]	l k k d
Public Code (*Dazhong ma*) [大众码]	d o w w
Hua Code (*Hua ma*) [华码]	d r z
Taiji Code (*Taiji ma*) [太极码]	n y
3F Code (*3F ma*) [3F 码]	; d
Cangjie [仓颉]	m b w u
Wubi xing [五笔型]	j t w
Graduated Four Corner (*Cengci sijiao*) [层次四角][3]	x e
Internal Code (*Neima*) [内码]	B5E7[4]
Ann's System of Coding Chinese Characters	106071[5]
Basic Stroke input (*Jiben bihua*) [基本笔画]	1 0 4 6[6]
Stroke Order Code (*Bishunma*) [笔顺码]	0 1 6[7]

1. Tianjin City Zhonghuan Electronic Computer Company [天津市中环电子计算机公司], *Chinese Character Encoding Manual (Hanzi bianma shouce)* [汉字编码手册] (Tianjin: Tianjin City Zhonghuan Electronic Computer Company, 1982), 187.
2. Tianjin City Zhonghuan Electronic Computer Company, *Chinese Character Encoding Manual.*
3. Lin Shuzhen, *Household Computer: Chinese Character Encoding Rapid Look-Up (Jiating diannao: Hanzi bianma sucha)* [家庭电脑: 汉字编码速查] (Fuzhou: Fujian kexue jishu chubanshe, 1994), 16.
4. Lin Shuzhen, *Household Computer*, 16.
5. T. K. Ann [安子介], *Chinese Character List A: Ann's System of Coding Chinese Characters* (Hong Kong: Stockflows Co., Inc., 1985), 14.
6. Tianjin City Zhonghuan Electronic Computer Company, *Chinese Character Encoding Manual*, 187.
7. Wang, Songping [王颂平], *Complete Illustrated Guide to Stroke Order Code (Bishunma tujie quanji)* [笔顺码图解全集]. Beijing: Zhongguo funü chubanshe [中国妇女出版社], 1998, 5.

computers built elsewhere? Is a computer "Chinese" only if developed by engineers who claim Chinese heritage? Was Huang's computer uniquely "Chinese" in some way that cannot be discerned by the machine's outward appearance?

In other words, if, on the outside, it looked the same as desktop machines found all over the world, was there something *inside* Huang's machine—its CPU, perhaps—that made it uniquely capable of handling this highly complex form of textual input?

The answer to this question, as we will see in chapter 5, is both *no* and *yes*.

No, the computers in this competition were in no way different than Windows-compatible machines found in homes, offices, and schools elsewhere in the world. There was nothing specifically "Chinese" about them, inside or out.

In a historical sense, however, the answer is *yes*. In the twenty-first century, every mass-manufactured personal computer in the world comes equipped to handle Chinese IMEs of the sort examined here. Chinese IMEs come preloaded, in fact, with others available for download (often at no charge). Circa 2013, every store-bought desktop, laptop, and smartphone was "Chinese," in the sense that they were capable of handling Chinese input and output.

This hasn't always been the case, however. In fact, it is a remarkably recent phenomenon. During the early rise of consumer PCs in the 1980s, no Western-designed CPU, printer, monitor, operating system, or programming language was capable of handling Chinese character input or output—not "out of the box," at least. For the better part of the history of computing, computing technology has been biased in favor of certain alphabetic scripts—none more so than the Latin alphabet. In the mid-twentieth century, for example, Western engineers determined that a 5-by-7 dot matrix grid offered sufficient resolution to render legible Latin alphabetic letters on monitors and dot-matrix printouts. To do the same for Chinese would have required engineers to expand this grid to no less than 16-by-16. In the 1960s, the development team behind ASCII (the American Standard Code for Information Interchange) determined that a 7-bit coding architecture and its 128 addresses offered sufficient space for all of the letters of the Latin alphabet, along with numerals and

key analphabetic symbols and functions. Chinese characters, by comparison, would have demanded no less than 16-bit architecture to handle its
more than 60,000 characters. And of course, long ago Western engineers
piggy-backed on the preexisting typewriter keyboard, using the two-
dimensional "shift" key to toggle between lower and uppercase letters
(Chinese, of course, has neither an alphabet nor uppercase or lowercase).[19]
Whether in terms of character encoding, computer monitors, dot-matrix
printers, programming languages, disk operating systems, input surfaces,
optical character recognition algorithms, or otherwise, the early history
of computing has, in many ways, been the story of one digital "Chinese
exclusion act" after the next.[20]

As we will see in chapter 5, what ensued during the 1980s was a
period of hacking and modification—"modding," to use a common
term of art within engineering circles. Western-built dot-matrix printers
were modded. Western-designed disk operating systems and the Basic
Input-Output Software (BIOS) were modded, too. Element by element,
engineers in China and elsewhere rendered Western-manufactured computing hardware and software compatible with Chinese. It was a messy,
decentralized, and brilliant period of experimentation and innovation,
all of which made the moment in 2013—as well as China's current stature
in the world of computing and new media—possible.[21]

Why would Huang bother with *ymiw2klt4pwyy1wdy6* in the first place?
Even if we accept the ubiquity of QWERTY interfaces within Chinese
computing, and even if we accept the central importance of IMEs, this
still leaves open the question: Why didn't Huang decide to input Hu
Jintao's speech the way Hu's speech *sounded* instead of inputting the
structural features of the Chinese characters printed on the page? When
entering the two-character word "高举" ("hold high"), for example, why
didn't Huang enter "g-a-o-j-u," the way one would expect for a compound pronounced *gaoju*. When inputting "中国特色" ("with Chinese
characteristics"), why not key in "z-h-o-n-g-g-u-o-t-e-s-e"? Why bother
with structure-based IMEs at all, in which sequences like *ymiw2klt4pwyy-
1wdy6* bear no relationship to the way Chinese is pronounced?

The answer, in part, is that most Chinese computer users today *do* use
phonetic input (in mainland China, at least). In his 2013 victory, Huang

Zhenyu's choice of Wubi input—despite its widespread popularity in the 1980s and 1990s—placed him in the minority in his day and age. In the twenty-first century, most mainland Chinese computer users employ IMEs like Sogou pinyin, QQ pinyin, Tencent pinyin, Google pinyin, and other phonetic systems based on pinyin Romanization.

In light of this fact, it's natural to ask: If Huang had used a phonetic input system that day in 2013—inputting a more "naked-eye readable" phonetic string like *"gaoju Zhongguo tese shehui zhuyi . . ."* instead of the mysterious *"ymiw2klt4pwyy . . ."*—would the contrast between Chinese and English computing perhaps seem less stark? After all, the underlying system of pinyin input—the official Romanization standard of the PRC, known as Hanyu pinyin—is premised on "spelling out" the sounds of Chinese characters in full, much like one spells out words in English.

We could take this one step further and ask: Given that pinyin input is based on Hanyu pinyin, and given that Hanyu pinyin is a system of romanized spelling, doesn't this mean that pinyin-based input systems are effectively ways of "spelling" Chinese characters using a QWERTY keyboard? Doesn't the rise of pinyin input bring about a figurative "end of history" for Chinese computing, then, assimilating it into the global and long-dominant mode of human-computer interaction? Isn't pinyin input China's version of what-you-type-is-what-you-get?

The answer is *no*. In chapter 6, we examine the rise of pinyin-based phonetic input, dating from the 1990s to the present. Despite its uncanny resemblance to "spelling" in the conventional sense, pinyin input is in fact *still input*—by which I mean, it is still a core part of the half-century trajectory of Chinese computing history explored in the chapters preceding chapter 6. Not only did phonetic input achieve popularity only quite recently in history—after four decades of structure-based input dominance—but once it did, it was never in fact a technique for simply "sounding out" the pronunciation of Chinese characters by means of the Latin alphabet. Put simply: "Hanyu pinyin" and "pinyin input" are completely different kinds of systems, even if the latter is built atop the former.

Consider the Hanyu pinyin Romanization of China's capital city: "Beijing" (or "Běijīng," with tone markers included). Anything other than this spelling—"beij," "bj," "bjing"—is an error or a typo, pure and simple.

Within pinyin-based *input systems*, by comparison, all of these (and others) are perfectly valid options—valid in the sense that any one of them "achieves its aim" of hailing the user's desired two-character term 北京 from character memory for confirmation in the pop-up menu. Whereas Hanyu pinyin is a system of Romanization, "pinyin input" is an approach to computational input based on all of the principles introduced thus far, and which we will explore in depth: multiple input pathways to the same Chinese character output; what you type is *not* what you get; and more. Even more than structure-based input systems, in fact, pinyin input is enmeshed in a broad repertoire of advanced techniques and technologies that are bringing about a sea change in the history of Chinese writing: autocompletion, predictive text, contextual analysis, and most recently, artificial intelligence.

THE HYPOGRAPHIC AGE

In this book I tell the history of Chinese language in the digital age, but I also look beyond China to explore how a new mode of writing—a new "technology of the intellect"—has taken shape over the past half-century.[22] This mode of writing stands at a profound yet often imperceptible distance from writing as we've long understood it. I call this new mode of writing *hypography—hypo* meaning "below" or "beneath," and *graphy* meaning "writing." Chinese input is a form of *hypography*, writing that operates in service of conventional writing or script but at a register beneath.[23]

Orthography—the term we'll use here for conventional writing—is the form of writing we use when addressing birthday cards, submitting job applications, or posting comments on social media. Orthographs, formed with a particular pattern of letters we recognize as words with a particular meaning, appear on street signs, restaurant menus, computer screens, or the pages of a poem. Orthography has existed since the advent of writing itself: hieroglyphics, oracle bone script, cuneiform, woodblock prints, manuscripts, copperplate engravings, moveable type, and more.[24] Whether through the displacement of clay or wax, the subtraction of stone or wood, or the staining of skin or paper, these are signs that are meant to signify concepts.

Hypographs, by contrast, are a new and special class of writing whose one and only job is to help search for and retrieve orthographs from memory. To revisit our examples above, *diantmvv* is a hypograph. So is *kzvv, jnv, lwu, djfz, djjm,* and *dian.* The objective of these hypographs is not to "stand" for any particular concept, the way the word "cat" signifies or "stands for" a particular kind of four-legged mammal. The sole purpose of hypographic signifiers is to help retrieve *other* signifiers—orthographs—and then to let these orthographic signifiers do the semiotic labor of making meaning. Phrased another way: while the advent of print radically transformed the outward appearance of the written word, the rise of hypography has not (and will not) change the appearance of writing *at all.*

What kind of writing is hypography, then? To answer this question we can begin by examining some of its core features. Every single one of the 900 million computer and internet users in the People's Republic of China (PRC)—and hundreds of millions more in the broader Sinophone and Sinographic computing worlds—experience hypography and its characteristics in the following ways:

The symbols a user is actually manipulating—whether by depressing keys on their keyboard or swiping strokes with a finger or stylus—are *never the same* as the Chinese symbols that appear on the screen;

there exists a theoretically infinite number of ways to map this relationship between input and output symbols;

the act of writing depends upon a recursive process, in which the user is constantly presented—and then presented again and again—with an ever-changing set of "candidates";

the successful completion of any text—whether a single character long, or many more—does not depend upon the complete "spelling out" of its corresponding input sequence;

the symbols which each user is directly manipulating—the "primary transcript"—is destroyed the split-second that the desired, *secondary* transcript is confirmed and added to the composition window.[25]

This final characteristic—the purposeful ephemerality of hypography—is perhaps the most startling feature. Unlike ink on paper, chisel cuts in wood, or graffiti on subway cars, hypographs do not persist for even a split second longer than they are needed to fulfill their purpose. They help retrieve something else, then they fall into oblivion—that is their lot in life. Consider the examples of *diantmvv, kzvv, jnv, lwu, djfz, djjm, dian, 601, 50716d, dqtk, rgd, 7193, dm, 2171, 3567, b1, ⌋ 47*, and all of the other input sequences that lead to the Chinese character 电. Once "电" is added to the composition screen, none of these hypographs need stay around any longer, since it is now the job of "电" to do the work of signification. Hypographs live out their days in the confines of the Input Method Editor pop-up window, existing for only a fraction of a second, promptly dying off as soon as they succeed in helping a user retrieve their desired character. Instead of "standing" for anything, then, hypographs are meant to "fall" for the symbols they are designed to retrieve.

The scale and tempo of hypographic ephemerality is breathtaking. During the competition in 2013, Huang Zhenyu depressed somewhere in the order of 40,000 keys on his QWERTY keyboard. Not one of these letters or numerals appeared in the final text, however. All of them were erased the moment they fulfilled their objective: to help the IME to retrieve his intended Chinese characters from memory.

For the estimated 900 million computer and internet users elsewhere in the PRC, the phenomenon is the same. Each day, billions if not trillions of keystrokes and finger swipes give rise, first, to an immense and ephemeral corpus of hypographs, and only thereafter to Chinese script. Every digital Chinese text is produced like this—the exercises of schoolchildren, Weibo microcasts, the mailing address for one's latest Alibaba purchase, text message missives between lovestruck adolescents, or otherwise. Every year, this is to say, a hypographic corpus of almost unimaginable scale—likely quadrillions of alphanumeric symbols long—is created and erased, again and again, in the blink of an eye.[26]

The intentional ephemerality of hypography distinguishes it from a wide variety of written forms with which it might at first seem identical. Consider acronyms, initialisms, and abbreviations: "NASA," "PRC," "congrats," and the like. All three techniques of writing employ compression; none, however, are hypographic. Were "PRC" or "NASA" hypographs, they would never see the light of the printed page, nor appear in any final composition. Instead, they would be converted into the full orthographic forms they were designed to retrieve from memory: "National

Aeronautics and Space Administration" or "People's Republic of China." Then, as with *diantmvv, kzvv, jnv, lwu, djfz, djjm*, and *dian*, they would be disposed of, never to be seen again.

The ephemerality of hypography distinguishes it from transliteration as well—along with slang, hacker language, emoji, kaomoji, constructed language, assistive script, translanguage, code-switching, diglossia, code-mixing, Chinglish, interlingual punning, tranßcripting, txt spelling, text-speak, netspeak, typographic play, and rebus—even if, as we will soon see, hypography sometimes borrows techniques and tricks from many if not all of them.[27]

Chinese computing is ground zero for this new epoch, but not the origin point. Hypography isn't a Chinese invention, that is to say. English-language stenography and stenotyping are forms of hypography dating back to the nineteenth century. In courtrooms and boardrooms across the United States, for example, stenotypists capture court proceedings and depositions at the speed of sound using, not conventional typewriters, but specialized devices that rely on the exact same strategies of non-identity as Chinese IMEs.[28] The stenograph machine, with its 22 keys, does not feature one key for each letter of the Latin alphabet. The Latin alphabetic letter "N," for instance, does not appear on the machine at all. Rather, in order to produce an "N," the stenotypist must type three *other* letters, making sure to depress them *simultaneously*: "T," "P," and "H" (all of which do appear on the keyboard). Similarly, in order to produce the letters "B," "C," "G," and "J," the stenotypist instead depresses the "letter chords" "P-W," "K-R," "T-K-P-W," and "S-K-W-R," respectively. Thereafter, the stenotypist reviews this cryptic primary transcript, and sets about replacing every instance of "T-P-H" with the letter "N," and so forth. In other words, just like Chinese input, the "primary transcript" of English-language stenotyping is never the end goal of the inscription process. The primary transcript is there only to help achieve the next transcript: the secondary, natural-language transcript. When the secondary transcript is achieved, the primary transcript is disposed of, just as it is in Chinese input (figure 0.4).[29] As complex as this may sound to some readers, English-language stenotypists often had little more than a high school education, achieving their skills during relatively brisk training regimens. Nevertheless, they were (and are) able to achieve inscription

Table 0.2 English letters and their corresponding stenotype letters

English letter		Stenotype Chord											
		S	T	K	P	W	H	R	A	O	*	E	U
Initial consonants	*b*				P	W							
	c (soft)			K				R					
	ch			K			H						
	d		T	K									
	f		T		P								
	g		T	K	P	W							
	h						H						
	j	S		K		W		R					
	k			K									
	l						H	R					
	m				P		H						
	n		T		P		H						
	p				P								
	qu			K		W							
	r							R					
	s	S											
	t		T										
	v	S						R					
	w					W							
	y			K		W		R					
	z	S									*		
Vowels	Short *a*								A				
	Long *a*								A			E	U
	aw								A				U
	Short *e*											E	
	Long *e*								A	O		E	
	Short *i*											E	U

Table. 0.2 (continued)

English letter		Stenotype Chord
Vowels	Long *i*	A O E U
	Short *o*	O
	Long *o*	O E
	oi	O E U
	oo	A O
	ou	O U
	Short *u*	U
	Long *u*	A O U
Final consonants	*b*	B
	ch	F P
	d	D
	dz	D Z
	f	F
	g	G
	j	P B L G
	k	B G
	l	L
	m	P L
	mp	F R P
	n	P B
	ng	P B G
	nj	P B G
	nk	* P B G
	p	P
	r	R
	rch	F R P B
	rf	F R B

Table. 0.2 (continued)

English letter		*	F	R	P	B	L	G	T	S	Z
Final conso-nants	*rv*		F	R		B					
	s									S	
	sh			R		B					
	sm		F		P		L				
	st		F						T		
	t								T		
	v	*	F								
	x					B		G		S	
	z										Z
	shun							G		S	
	kshun	*				B		G		S	

speeds that far outpace the maximum speeds achievable on a conventional typewriter. Put simply: however fast English-language typists may be, they can never be faster than the fastest stenotypist.

English-language hypography has endured in the age of computing and new media, moreover. "Graffiti"—a proprietary input system with the once wildly popular PalmPilot personal digital assistant—was hypographic as well. Every letter of the Palm's Graffiti alphabet disappeared as soon as it was successfully resolved to one or another "real" graph in memory (figure 0.4). So too was (and is) T9 input, which enabled users to type using the ten digits of their mobile phones.

The defining feature of the hypographic age is not novelty, then, but scale and mindset. In the digital age, every single Chinese computer user is a hypographer. Every single person in the world of Sinophone computing is, in effect, far more like a stenotypist, a Palm Piloter, or a T9 inputter than a "typist." To be clear, I do not mean to suggest that the average Chinese computer user operates at anywhere near the speed of Huang Zhenyu, any more than the average typist in the English-speaking world approaches the speeds of Blackburn or Wrona. We can say, however, that every single person who inputs Chinese characters on a computer, a smart

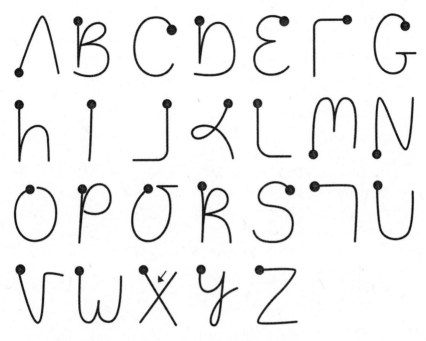

Figure 0.4   Palm Pilot Graffiti alphabet as English hypography

phone, or other digital devices does so by employing fundamentally the same techniques as Huang and his fellow contestants.

When we extend our view to the broader Sinophone world, as well to countries that make use of Chinese character script (Japan and Korea, for example), the scale of hypography grows even further. Then comes the wider world of non-Latin computing and new media, a topic which makes an appearance at the close of this book. When people sit down with their laptops or smartphones in Bangalore, Riyadh, and Phnom Penh, what they do with these devices looks a great deal more like Shanghai, Kyoto, and Seoul, than New York, London, and Paris—what they type is not what they get. What we find, in short, is that the mode of digital writing taking shape in China is quickly and quietly becoming a dominant mode of writing across the globe.

Although hypography itself is not new, the scale at which it now operates is. Hypographic practices are so widespread, in fact, that it practically demands a retroactive naming of the epoch in which human beings have been living throughout recorded history, from cuneiform and Dragon

Bones to copperplate engraving and typewriters—an epoch which, thus far, we have never needed to give a name to, since until now it's all there was. I will refer to this as the epoch of *writing as composing*, or more simply as the age of orthography; to be contrasted with an emerging epoch of *writing as retrieving*, or the age of hypography.

Revisiting the so-called Chinese-character crisis introduced at the outset, then, I will argue in the course of this book that the phenomenon some call "character amnesia" is neither dysgraphia, nor illiteracy, nor amnesia. Rather, writing in China is changing, and fundamentally so. Writing is also changing in many other parts of the world. But the theoretical framework we use to understand what writing is, and what it should be, has yet to change.[30] What you see in Huang's performance in 2013, or in the activities of China's immense and growing class of computer users, is not a question of "illiteracy." We are bearing witness to an entirely new kind of Chinese and, indeed, an entirely new mode of writing altogether—a mode that is still in its emergent phase and remains poorly understood.

This book is a history of Chinese computing. More than that, however, it is a preliminary roadmap to a new era in the history of writing.

Welcome to the hypographic age.

1

WHEN IMEs WERE WOMEN: IBM, LOIS LEW, AND THE DAWN OF ELECTRONIC CHINESE

Thank you for the memories. I am the woman demonstrating the Chinese typewriter.[1]

My pulse double-timed as I read the comment.

In June 2010, I posted on my blog about an old, black-and-white film I'd discovered. The clip featured an IBM electric Chinese typewriter—a fascinating prototype debuted to the world in 1947.[2]

A young woman was seated at the machine, flanked by reporters and a middle-aged Chinese man—Chung-Chin Kao, the machine's inventor.[3] She pulled a sheet of paper from the device, radiating a smile to the camera. Kao was biting his lip, his eyes darting back and forth between the typist and the crowd (figure 1.1). He was nervous and I knew why, having spent years by that point learning everything I could about the man: everything was banking on this young woman's performance, from his professional reputation to his family savings. But who was she? I still knew almost nothing about *her*.

I'd seen her before, that much I was certain of. I began riffling through my files: thousands of historic photographs, technical manuals, patent documents, archival documents, oral histories, antique machines, and more, cobbled together from around the world. The history of Chinese computing does not come with a ready-made archive, and so I had to compile one from scratch. Without intending to, my office, garage, and soon storage unit in San Francisco had accumulated one of the largest collection of materials on modern Chinese information technology in the world.[4]

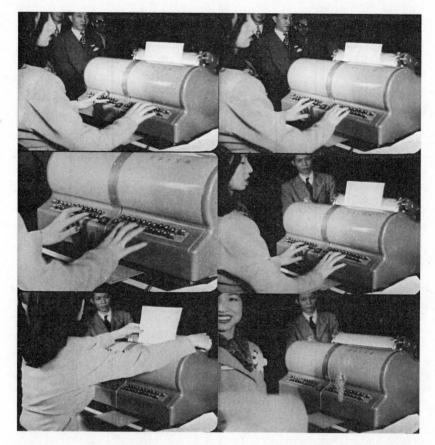

Figure 1.1   IBM promotional film, stills. Reprint courtesy of and © IBM Corporation

Sure enough, I found her in a glossy IBM brochure from 1947. Then again, on the cover of a Chinese magazine. Then again and again. Why did she factor so prominently in the early history of Chinese electro-automation? And how did she master a machine that IBM itself would ultimately abandon as hopelessly complex? (figure 1.2).

Chung-Chin Kao's typewriter was unlike any other in history. On most machines, what we type on the keyboard is what we see on the page. Depress the key marked "L" or "א" or "Д," and we see "L" or "א" or "Д" impressed upon the page. One symbol, one key. Depression equals impression. It goes without saying.

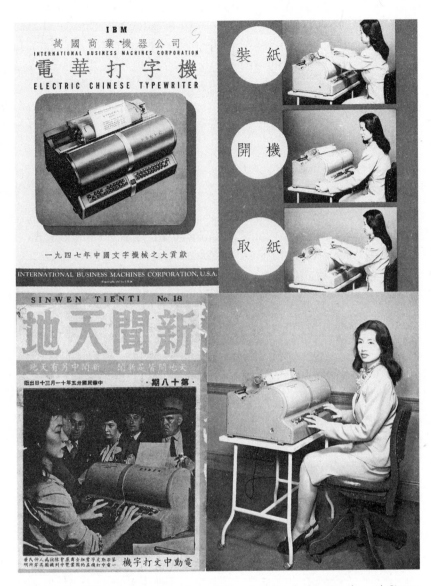

Figure 1.2 IBM promotional brochure, IBM promotional photograph, and *Sinwen tienti* magazine, all featuring Lois Lew. Reprint courtesy of and © IBM Corporation

Not so with the IBM device. Although it outputted Chinese characters on the page, there were no Chinese characters on the keyboard. In
fact, there weren't even alphabetic letters. Oddly enough, there were only
Arabic numerals—36 number keys in all were divided into four banks: 0
through 5; 0 through 9; 0 through 9; and 0 through 9. As if by magic,
these 36 keys were all this young woman needed to produce 5,400 different Chinese characters.

What she typed was *not what she got*.

Nor did this machine "spell" Chinese characters piece by piece—the way
you would type the word "cat," c-a-t. Instead, these cryptic numbers functioned as a form of electronic "address," which the typist entered into
the machine, and which the machine then used to retrieve the desired
Chinese character from memory: a hard drive–like drum that revolved
constantly inside the machine's hulking, gunmetal gray chassis.[5] If you
wanted the character 田 (*tian* "field"), for example, you entered the four-
digit address: 0-2-1-6—depressing the buttons simultaneously, like a pia-
nist playing a chord. To enter 大 (*da* "big"), 果 (*guo* "fruit"), or 聽 (*ting* "to
hear"), meanwhile, you chorded out the numerals 0-1-2-1, 0-4-1-2, or
1-0-8-9. The typewriter then produced full-bodied Chinese characters, a
hammer blow striking the rotating cylinder and transferring the Chinese
character etched on its surface.

The IBM Chinese typewriter was not a "computer," of course—it
didn't run stored program applications, nor did it offer any form of
memory. It was a critical precursor to the era of Chinese computing that
followed, however, embodying the first key question we must confront
in order to understand Chinese in the digital age, and hypography more
broadly:

What is the limit of a human being's ability to operate in code in real-time?
Phrased more directly: *How much code can the average person stand?*

If, in the twenty-first century, hundreds of millions of Chinese computer users can produce Chinese characters via long strings of alphanumeric code—even seemingly nonsensical strings such as Huang Zhenyu's

ymiw2klt4pwyy1wdy6 . . . examined in the introduction—the typist in this 1947 film was doing something remarkably similar, seven decades earlier. She was operating *entirely in code*, and doing it with a smile.[6]

Her job in 1947 was harder than Chinese computer users have it today, in fact. She had to perform these staggering feats of memory blindly, after all, without the benefit of the "pop-up menu" that forms part of all Chinese Input Method Editor systems in the computing age. There was no feedback loop on the IBM machine—no way for users to "check their work" before committing ink to paper. Neither did the machine contain any kind of IME. *She* was the IME, and she either remembered the code correctly or she failed.[7]

Many professions deal in code, of course. Cryptanalysts and security professionals are prime examples. Their codes are not "real-time," however: they are created and cracked over sometimes significant lengths of time. For real-time coding, better examples include telegraph operators, emergency responders, court stenographers, trained musicians, police officers, and even grocery store clerks. These are jobs that require one to memorize codes and wield them on the fly, like sight-reading music without missing a beat.

The complexity of this woman's job far outstripped any of these examples, however. Skilled telegraphers and ham radio operators receive and send Morse code with aplomb, but how well would they fare if Morse contained thousands of dot-dash combinations rather than a few dozen? How long would the grocery check-out line take if employees had to memorize 10,000 product codes? Somewhere there must be a breaking point beyond which the human mind simply can no longer function in code, in real time. But where is that limit? This woman—whoever she was—seemed to be standing at that very precipice.

The question of Kao's four-digit code—that is, the question of whether an average typist would really be able to use it—bedeviled IBM engineers and executives for years, as we will see. They debated it in private communications. They even hired one of the leading minds in Chinese studies to help them answer it. They wanted to break into the Chinese market, desperately, but they just couldn't stifle their doubts: was this a viable mode of human-machine interaction?

This brings me back to the cryptic message I received that August day in 2010.

"Thank you for the memories," it read. "I am the woman demonstrating the Chinese typewriter in the recent restored movie. If you'd like more info please contact me."

Could it really be her? Had a now 90-plus-year-old woman really found my blog? Or was this some peculiar prank concocted by a netizen with too much time on their hands? I had to respond, but with caution:

Dear Ms. Lew,

> My name is Tom Mullaney, and I am writing in response to your recent post on my Chinese typewriter blog. I was extremely excited to receive your note, and just wanted to confirm: you are the person who worked with Kao Chung-Chin (Chung-Chin Kao) to demonstrate the IBM Chinese Typewriter?

> > Thank you very much for making contact, and I eagerly await your response,

> > Tom Mullaney

> > p.s. May I ask what your Chinese name is, in Chinese characters?

My postscript was a shibboleth: a question which, I knew from my years of archival research to that point, no one could answer except Ms. Lew herself or someone who knew her personally.

She responded—accurately.

Doubt dispelled, I responded immediately, overflowing with questions. How did she become involved in the IBM project? What was her background? What was it like to use the machine? How did she manage to memorize all those four-digit codes? I couldn't wait to speak with her in person.

Silence followed. My email to Lew went without a response, for weeks, and then for months. I sent a polite follow-up email. Silence again. Finally, the trail went entirely cold, and I never learned why.

It would be another eight years before I reconnected with Lois Lew, thanks to a friend and former employee of hers. He saw another blog entry of mine and decided to reach out, just as Lew had. We chatted by email, and he agreed to put me back in touch, to see if Lew might be interested in chatting after all. Perhaps because I was vouchsafed, the conversation took place this time.

It was worth the wait.

"You've been looking for me for ten years," she exclaimed as soon as the call began. I could hear her smile through the telephone line, sending my thoughts back to the 1947 film.

She was right. When I first saw a photograph of Lois Lew, I was 29 years old. On the phone with her, I was 40.

She had traveled a greater distance. In the IBM promotional film from 1947, she was 22. On the phone, 95.

I couldn't believe I was finally talking to her.

CODE CONSCIOUSNESS: FROM CHINESE TELEGRAPHY TO THE ELECTRIC CHINESE TYPEWRITER

How did IBM end up building an electric Chinese typewriter based on the idea of cipher-based keyboard input? To understand the remarkable challenge Lois Lew would eventually confront and overcome—memorizing six thousand arbitrary four-digit codes and deploying them in real-time—we need to understand where this coding system came from. For this, we turn first to the machine's inventor, Chung-Chin Kao, and the broader context of mid-twentieth-century Chinese information technology that shaped both Kao's career and the machine's design. The idea of a number-based Chinese typewriter, as we will see, was not without precedent, nor was it as outlandish as it may first appear.

Chung-Chin Kao was born in China in late 1906 and went on to enjoy a long and successful career in Chinese telecommunications.[8] In 1936, he traveled to the United Kingdom to pursue a degree in electrical engineering at Marconi College. In 1938, he continued his studies in Germany, although his plans were cut short by the outbreak of World War II.[9] Kao managed to relocate to the United States in 1942.

Based in the United States, Kao went on to become Chief of the Radio Department of the Central News Agency of China, as well as Research Officer of Radio Communications for China's Commission of Military Affairs. In these roles, the technology of primary concern was not Chinese typewriting, but Chinese telegraphy.

Chinese telegraphy dates back to 1871, when a newly laid telegraphic cable between Shanghai and Hong Kong linked the Qing dynasty (1644–1911) to a rapidly expanding international network dominated by the British Empire.[10] During this period, the revolutionary technology of electric telegraphic communication began to expand far beyond the Latin alphabetic world. Cables were laid from Suez to Aden and Bombay, and from Madras to Penang, Singapore, and Batavia. Morse code came into contact with languages, scripts, alphabets, and syllabaries that it was never intended to handle.

China's entry into global telegraphy, a network dominated by the West, triggered one of the most conceptually vexing of all puzzles in the modern information age: How to send Chinese characters by wire using Morse code? Chinese script is a nonalphabetic writing system, after all, the only major world language that employs neither letters nor syllables in the formation of words. How could one ever hope to "fit" a nonalphabetic writing system with many thousands of characters into an alphabet-centered system that offered only a few dozen code spaces?

Chinese wouldn't fit, the answer turned out. More accurately, the European duo who invented the first Chinese telegraph code book decided to cut the Gordian knot of Chinese Morse with the razor blade of exclusion. The Danish astronomy professor H. C. F. C. Schjellerup and the Shanghai-based French harbormaster V. A. Viguier simply selected a small subset of commonly used Chinese characters, and then assigned each character a unique four-digit code from 0001 to 9999.[11] As for the many tens of thousands of other characters in the Chinese lexicon, those were simply left out of the code book, and thus telegraphy, altogether.

To transmit a character using the four-digit code—*shuo* (說 "to speak"), for example—the telegraph operator first needed to find its corresponding cipher in the code book (5-3-5-6) and then convert these digits into standard Morse dots and dashes, in preparation for transmission. At the other end of the line, an operator received these Morse-encoded numerals, and

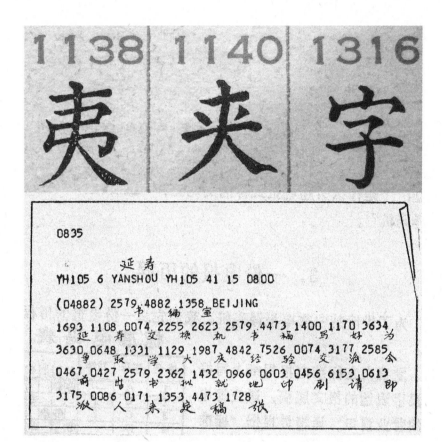

Figure 1.3 Sample of a Chinese telegraph code and telegram, featuring four-digit codes

carried out the process in reverse: from Morse to numerals, then numerals to Chinese characters, once again using the code book to look them up. Chinese was the only major language to rely upon this "in-between" step in the telegraphic transmissions process (figure 1.3).

The 1871 code book posed profound problems for Chinese telegraphers, as we might imagine. Enciphering and deciphering slowed down the transmission process, for one thing, placing Chinese at a disadvantage compared to other languages. Chinese telegrams also cost more to send, insofar as numerals take many more dots and dashes to encode in Morse than letters.[12] Despite such challenges, however, the four-digit telegraph code became an embedded part of China's nascent telecommunication infrastructure, persisting well into the late twentieth century. By the 1930s and 1940s, when Chung-Chin Kao was in the middle of his career, telegraph stations dotted the Chinese landscape, each of them employing operators trained in the four-digit code.[13]

In each of these hundreds of stations—as well as in government offices, newspaper plants, customs offices, banks, business firms, and other sites where telegrams were regularly used—many thousands of Chinese characters were transmitted and received daily, each of them encoded in one or another four-digit cipher. For every single one of these countless millions of Chinese characters sent by wire in the nineteenth and early twentieth centuries, *no operator ever dealt directly with characters*—only with codes. In the decades before Lois Lew sat down in front of the cameras in 1947, that is, generations of Chinese telegraph clerks, postal workers, and others had already learned to live with the Chinese language entirely in code. They were Lois Lew's forerunners.

They were also Kao's inspiration. Beginning in the early 1940s, he began to concoct a scheme as grand as New York City itself: to revolutionize Chinese information technology by using China's four-digit telegraph code, not just for telecommunications but also *inscription*. If the four-digit cipher was already second nature for a sizable labor force in China, and the semiotic backbone of Chinese telegraphy for a half-century, perhaps it could serve as the governing principle for the electrification of Chinese script more broadly.

Such technological convergences were already commonplace in many other parts of the globe, moreover. In the English-speaking world, printing telegraphs, teletypewriters, stock tickers, telegraphically controlled clocks, and other communication devices already bore witness to the blending and remediation of earlier technologies that once existed more or less independently, such as telegraphy, typewriting, radio, and time-keeping.[14] Kao reasoned that electric Chinese telegraphy could serve as a system for controlling a host of interconnected devices and systems as well: electric Chinese typewriters, typesetting systems, typecasting systems, and more. Kao's ultimate ambition, therefore, was not to build an electric Chinese typewriter. His goal was nothing short of the wholesale electrification and informationalization of Chinese.[15]

Splitting his time between a workshop on East 37th Street, and Room 2308 of the *Daily News* building on East 42nd, Kao mapped out a three-phased plan. Phase 1: design an electric Chinese typewriter, to be proto-typed and mass-manufactured by IBM. This device would serve as both a proof-of-concept for his four-digit coding method and a saleable product in and of itself. Phase 2: design the world's first hot metal composition system for Chinese, built and sold by the Brooklyn-based printing giant, Mergenthaler Linotype. Phase 3: conquer the Sinophone world, unveiling electrified systems for textual inscription, reproduction, transmission, and more, all premised upon the four-digit coding system.

Kao's relationship with IBM and Mergenthaler began in the early 1940s, when he secured a meeting with Mergenthaler's president, J. T. Mackey, on December 17, 1943, to discuss ideas for a Chinese ideographic autocasting machine or "Chinese Linotype."[16] Presenting initial sketches of his electric Chinese typewriter to company representatives—again, as a proof of concept—he expressed confidence that the device could secure a price of US$500 on the Chinese market (just under $6000 in today's terms)—making it not only a proving ground for his cipher-based keyboard, but also a potentially lucrative product line for IBM.[17]

Kao's ambitions for an electric autocasting system were even greater. He requested an advanced royalty of $30,000 from Mergenthaler (roughly $450,000 in today's terms), confident that such a machine could fetch a sales price of US$3000 in China (approximately $35,000 in today's terms) as a welcome replacement for Chinese printing houses still dependent on slow, handset type.[18] Even if Kao's machine never exceeded 10 characters per minute, he explained, this would still have been faster than available manual techniques.

As confident as Kao's claims were, and as appealing the idea of unlocking the vast Chinese-language market, Mergenthaler and IBM both had strong reservations. Both companies sought the counsel of Chinese-language experts based in the United States, mobilizing America's emerging Asian Studies brain trust. One scholar in particular held the key to their many questions, executives felt: George Kennedy, then an associate professor of Chinese at Yale University.

Beginning in December 1943, Kennedy served as a consultant to both companies with respect to Kao and his projects—a role he held for years.

In his first letter to Kennedy, Chauncey Griffith—vice president in charge of typographic development at Mergenthaler Linotype—expressed concerns over Kao's code, as well as the limited number of characters that the machine could produce. With Kao's proposed typewriter featuring only 5,400 of the many tens of thousands of characters in Chinese, Griffith wondered, would this be sufficient for a broad market?[19] Then there was the obvious question of Kao's cipher—could such an unconventional, cumbersome system really be used by the average operator? Griffith was relying on Kennedy's expert judgment, either to help allay his fears about the feasibility of Kao's unconventional approach, or to nip an absurd idea in the bud before it took the company down a quixotic and costly path.[20]

Kennedy eased Griffith's mind, at least when it came to the question of vocabulary. A character set of 5,400 was more than sufficient, Kennedy assured. In fact, Kao's estimates about Chinese-character frequency were too conservative, the Yale professor expanded. Kao claimed that the 1,000 most common characters in Chinese account for more than 70 percent of everyday Chinese correspondence—making a set of 5,400 characters more than enough. Kennedy offered Mergenthaler much more aggressive estimates. In research of his own, dating back to the 1930s, Kennedy found that the 1,000 most common Chinese characters accounted for 90 percent or more of everyday correspondence.[21] In that case, 5,400 would be more than enough.

When it came to Kao's coding system, however, Kennedy's support was less than full-throated.[22] "I am skeptical," Kennedy noted.[23] The task of memorizing the 5,400 numerical codes "seems to be gigantic." "Some of us who are interested in the alphabetization of Chinese," he continued, "would feel that a great victory had been won if it could be shown that the Chinese characters cannot serve as an efficient medium of communication for a modern industrial state." Even Kennedy equivocated, however, leaving the door of optimism slightly ajar: "but only a real test can prove anything there."[24]

Griffith could hardly afford to dismiss Kao's idea out of hand. Mergenthaler's presence in the Chinese market was scant—pathetic even: a yawning chasm in an otherwise seamless map of global conquest. The company boasted Linotype machines for Arabic, Hebrew, Cyrillic,

Burmese, Hindi, and more. Their machines could be found across Africa, the Near East, Latin America, and Europe. But for Chinese? Nothing.[25]

The same was true for IBM, the second company in Kao's crosshairs. IBM in the 1940s enjoyed only a wafer-thin presence in the Chinese market—a far cry from its global reputation in the 1980s and beyond. One of its earliest Asian ventures came as late as 1925 and focused on distributing clocks and butcher scales in the Philippines. Even a decade later, circa 1937, "Watson Business Machines" was little more than a skeleton staff of three people working out of an office inside the former Philippine National Bank Building.[26]

Would Kao's invention change all this? Had this talented, driven, and somewhat eccentric engineer hit upon a novel solution to the puzzle of mechanized textual production for Chinese, one that might not be immediately apparent to the uninitiated observer? How long did it take people to realize the genius of Ottmar Mergenthaler, after all, the German American inventor whose name now adorned the halls of Griffith's workplace?[27] Mergenthaler's invention—hot metal composition—once seemed far-fetched as well: A clanking, pipe-organ-shaped, keyboard-controlled contraption in which molten hot liquid metal was injected into moveable matrices, the metal rapidly cooling to form solid "lines of type" which could then serve as the basis for printing (hence the moniker "lin-o-type")? Who would have immediately believed that such a machine would, in the span of a single lifetime, put countless commercial typesetters out of business, make it largely unnecessary for printing houses to purchase and store immense cabinets of metal type, and drive the final nails in the coffin of Gutenberg and the era of handset type? Could Chung-Chin Kao be China's Mergenthaler? Kao certainly seemed to think so.

Kao's communications with IBM and Mergenthaler in early winter 1944 bordered on bravado, exuding a level of confidence born either out of sincere conviction, or out of a desire to overwhelm his counterparts' doubts through sheer charisma. Kao also did everything in his power to keep his audiences separate, a kind of "divide-and-conquer" technique in which Kao sometimes communicated exaggerated (or completely fallacious) claims to one party or another, under the assumption that his counterparts were perhaps not in direct contact—in some cases, even

colleagues in the same company. In a letter to A. P. Paine, associate director in Mergenthaler's department of research and development, Kao
wrote that Mergenthaler was on the verge of commencing formal support of his project. "I am very happy to learn," Kao wrote, "that within a
short time your company will be ready to assist in the exploitation of the
Chinese autocasting machine."[28] Paine immediately relayed Kao's letter
to Mackey, however, expressing concern. Kao seemed to think that his
relationship with Mergenthaler was a "fait accompli," Paine wrote. "I do
not know where Mr. Kao learned this since he has not been in touch with
either Mr. Griffith or myself."[29]

By February 1944, cracks began to show in the relationship between Kao
and Mergenthaler. As Griffith explained to Mackey, Kao had pressed him
on questions of legal control, and on his desire for a payment of $30,000 in
advanced royalties. Such an agreement, Griffith urged to Mackey, should
not be put into place at this time. "I do not recommend any such arrangement, nor a legal contract of any kind until we are reasonably sure of our
grounds respecting the successful development of his scheme."[30]

Exchanges between Griffith and Kao became increasingly tense.
Griffith insisted that Mergenthaler Linotype would need to see Kao's electric typewriter project with IBM brought to a proof-of-concept stage before
Mergenthaler would consider moving on with the far more expensive
typecasting project. Both machines depended upon the same untested
principle, after all: namely, having an operator memorize four-digit codes
to control a set of 5,400 characters. By Griffith's back-of-napkin estimates,
moreover, the typecasting project was "a hazardous undertaking, considering the fact that it could easily involve the expenditure of a quarter
of a million dollars in money and six- or eight-years' time."[31] Caution
was essential.

Kao was outraged. "Mr. Kao objects strenuously to our method of
approach," Griffith reported following communications with the Chinese
engineer, "and has insisted all along that the two interested companies
operate independently." The Chinese engineer "intimated that unless
we were willing to do so," the report continued, "he would approach
Monotype or some other printing machinery concern to undertake it."[32]
Griffith called Kao's bluff. "I told [Kao] very frankly that we would not
be influenced by this line of pressure, and he was perfectly free to follow

his own inclinations. After this his attitude seemed to be a little more reasonable."[33]

In late April 1944, a meeting took place that Kao had been keen to avoid: a direct conversation between the executives of Mergenthaler Linotype and IBM—without Kao in attendance. Executives at Mergenthaler openly critiqued their counterparts at IBM, arguing that Watson and others seemed to be undertaking the electric Chinese typewriter project out of "altruism" (their word), rather than profit.[34] Griffith returned to the keyboarding system, arguing that that Kao's four-digit coding technique "seems to be the only vulnerable detail of the entire system."[35]

Kao was becoming ever more aware of just how deeply IBM and Mergenthaler doubted his coding system. In a lengthy letter dated April 29, 1944, Kao tried again to convince Mergenthaler executives that China was already home to many thousands of people with immediately transferrable skills: namely, people trained in the Chinese telegraph code.[36] "Through years of services," Kao explained, "there are thousands of Chinese telegraph operators who can remember more than 4,000 characters in their numbers." "In addition to this," Kao continued, "there is a special profession in China called 'Telegraph translation clerk' who do such translation work in government offices, banks, and other business organizations." Summarizing his work while at the Central News Agency of China in the early 1930s:

We transmitted and received more than 50,000 words in the form of numbers every evening between the hours of 6 p.m. and 2 a.m. These numbers must be translated into plain Chinese for publications and at the same time translate the Chinese characters into numbers for communication. The clerks in my office translate numerical messages of approximately fifteen words a minute without the necessity of looking them up in the telegraph dictionary. There are, of course, several characters which are not remembered, but it is less than one per cent.[37]

Kao assured Mergenthaler's Chauncey Griffith that "it would not take more than two months to train an operator to remember the numbers of these 1,000 characters."[38] Full proficiency would take no longer than three months, he claimed.

Kao closed on a much more personal note: "I don't know how long I can stay in this country. It will therefore be a great favor to me if you can kindly consider to arrange a preliminary agreement between your company and myself under the conditions which your company may suggest."[39]

The more time Kao spent explaining, however, the more uncertain Mergenthaler executives became. No matter Kao's description of Chinese telegraphy, nor his portrayal of China as home to thousands of Chinese operators who from morning to night blipped out millions of four-digit codes, they just could not be certain. Perhaps Kao's machine was the key to unlocking the Chinese-language market, but perhaps it was absurd, pure and simple. Everything boiled down, not to the mechanical design of the proposed device, but to the questions raised at the outset of this chapter: How much code could a human being really stand? Was real-time coding truly possible when thousands of ciphers were in play? These were questions that couldn't be answered by means of engineering schematics, Chinese history lessons, or reports by Ivy League consultants. It could only be answered by testing the machine out on an average typist—on someone like Lois Lew.

GRACE UNDER PRESSURE: THE FIRST CHINESE INPUTTER

The New Yorker Hotel convention hall was teeming with people, all eyes on a young woman seated at the front of the room. The Chinese Consul General was there. So were the *New York Times* and the *Herald Tribune*. The fate of Kao's entire project rested on this woman's shoulders—his chance to convince a skeptical world that his four-digit coding system could work.[40]

This wasn't Lois Lew, however. It was Grace Tong, Kao's first assistant and the first "electric Chinese typist" in history.

The prototype was something to behold, completed in January 1946 thanks to Kao's perseverance and the ingenuity of IBM's engineering staff.[41] Weighing in at 65 pounds, it operated on a 110-volt alternating current, driving a rotating drum inside the machine's chassis at a continuous speed of one revolution per second. Measuring 7 inches in diameter, and 11 inches in length, the drum's surface was etched with 5,400 Chinese characters, letters of the English alphabet, punctuation marks, numerals, and a handful of other symbols (figure 1.4).

Figure 1.4 Completed prototype of the IBM Electric Chinese Typewriter, Mergenthaler Linotype Company Records, Archives Center, National Museum of American History, Smithsonian Institution (hereafter MLCR)

The numbers on the keyboard controlled the movements of the drum. Within each four-digit code, the first and second digits—0 through 6, and 0 through 9—controlled the lateral, left-to-right movement of the drum inside the chassis. This lateral movement brought a subset of the drum's 5,400 characters into the typing position. The third and fourth digits—0 through 9, and 0 through 9—further specified the typist's desired character. When the correct character passed briefly into the typing position, a single hammer blow—delivered in 1/600th of a second—registered that character on the page.[42]

Kao wrote a letter on the machine, addressed to the president of IBM on February 2, 1946:

Dear Mr. Watson: It is my great privilege to write you the first typewritten Chinese letter with the typewriter which was achieved only through your foresight and generous support.[43]

This was not only the first letter ever written on the IBM electric Chinese typewriter, but the first Chinese text in history to be produced on an electro-automatic device.

It was also one of the last letters that Kao wrote on the machine, at least publicly. Faced with persistent doubts by both IBM and Mergenthaler, Kao knew it would convince no one to see the machine operated by the inventor himself—no one knew the device better than he did, after all. He had to prove to the world that *anyone* could use the machine. As early as March 1946, Kao launched a training course with the goal of creating a female labor pool capable of handling the machines.[44] If the profession of Chinese telegraph clerk was one long dominated by male operators, the world of the electric Chinese typewriter would be populated by "young Chinese girls," as he put it.[45]

Central to Kao's training regimen was his code book, which provided the numeric ciphers for all of the characters, letters, numerals, and symbols in his system.[46] In many respects, this code book—and the broader training regimen—was identical to those found in the domain of Chinese telegraphy for decades by that point. Confronted with thousands of four-digit codes, each one linked to a different character, letter, numeral, or symbol, the job of these "young girls" would be akin to those of countless thousands of Chinese telegraphers-in-training who came before.

In key respects, however, the job of a typist using an electric Chinese typewriter was far more strenuous. Kao's typewriter code was less forgiving than the original Chinese telegraph code, for one thing. In telegraphy, the process of converting a Chinese-language message into a transmittable, cipher-based code was a multistep process involving more than one clerk. No operator was expected to be able to read a Chinese message and, from memory, convert it into a transmittable sequence of digits. First a "translation clerk" (*yiyuan*) converted the message into ciphers. Only then did a transmission clerk convert these numerals into dot-dash pulse patterns and send them out over the wire.

For Kao's typists, however, expectations were "real-time." While typists were permitted to refer to the code book in the case of rarely used Chinese characters, they were expected to convert most characters into codes immediately, and flawlessly.[47] In fact, demonstrations of the prototype—as we'll examine shortly—regularly involved spectators shouting out Chinese characters with an expectation that the "typewriter girl" on stage would be capable of conjuring to mind the corresponding code on the fly, and entering it on the machine flawlessly. To use a present-day metaphor, these typists were expected to function like human Input Method Editors: human "middleware" capable of translating between the hardware of the machine and the commands of software (meaning one's employer)—without referring to a code book along the way.

There was no tolerance for error, moreover. When the typist depressed the four-key chord, the machine immediately registered a Chinese character—either the correct one, or not. There was no feedback loop by which potential mistakes could be fixed in advance.

Much to Kao's joy, he found a person who fit the bill in every imaginable way: educated, connected, and attractive. Her name was Grace Tong.

Born Grace Kwoh in 1920 in China's Shandong province, she arrived in Seattle in the fall of 1937, the ship manifest listing her as a 17-year-old student, bilingual in Chinese and English.[48] In 1943, Kwoh married Yanghu Tong, the young Chinese engineer who Kao was working with on his Chinese Linotype project.[49] This marriage brought Grace into a family of impeccable credentials, Yanghu being the son of the well-known Chinese diplomat and journalist Hollington Tong—himself a regular associate of the Chinese head-of-state Chiang Kai-shek. To hire Grace, then, was not only to hire a talented and educated individual, but also to reinforce connections valuable to Kao's broader networking campaign.[50]

Kao was quick to capitalize on these connections. Eugene B. Mirovitch, a Mergenthaler executive, wrote to the company president that according to Kao, "the construction of the typewriter by International Business Machines Corporation has made an excellent impression on Generalissimo Chiang Kai-shek and the Chinese Government educational authorities, and that he has had extremely encouraging comments from all of them." Nevertheless, "there still remains the bigger and more important problem of mechanizing Chinese language composition; in other words, the designing and building of a Chinese typesetting machine."[51]

The strategy bore fruit. Evidently intrigued by Kao's invocation of the name "Chiang Kai-shek," and Kao's reports of official Chinese reactions to the IBM machine, Mirovitch asked Kao if he could guarantee that the Chinese government would purchase a certain number of autocasters.

Kao confirmed, offering assurance that China could "absorb some 300 typesetting machines to start with, after which there would be a gradual increase of sales from year to year," and that the Chinese government would even be willing to fund part of the development costs.[52] There was even a glimmer of hope, for at least one Mergenthaler executive, that the company was beginning to change its position on the question of real-time coding and its limits. "The human mind," F. C. Frolander wrote, "is generally very much underestimated, as to its ability to instantly detect and remember unlimited detail and their pertinent associations. There is no conscious pattern of reasoning used in flexing the vocal muscles in producing the words and the associated tones in singing. The human mind is the most marvelous instrument of all, it only needs to be put to use, through the bodily senses."[53]

Encouraging as these sentiments may have been for Kao, however, they were still in the minority. "It was rather startling," Frolander continued, bordering close to self-contradiction, "that Dr. Kao had chosen to resolve the great number of characters required in the Chinese graphic mode of communication and expression into combinations using Arabic numerals, 0 to 9 inclusive." Mirovitch continued to share the same worries, according to his colleague. Mirovitch "frankly stated that he was very doubtful of the wisdom of utilizing such a system," being chiefly concerned with the ability of the human mind to cover in memory a range of 5,400 Chinese symbols or characters and their numerical equivalents.[54]

Grace Tong seemed to prove such doubts wrong, however, handling the machine masterfully. "While we were there," Mirovitch couldn't help but note, "the young lady typist seemed to have no trouble in locating any one of the 1,000 most-used characters and operating the machine."[55] Evidence suggests, moreover, that Kao's earlier "Chinese history lessons" to Mergenthaler—his explanations regarding Chinese telegraphy, that is—were having some effect, convincing at least some executives that there may indeed be a deep talent pool in China already familiar with four-digit ciphers. As Mirovitch wrote privately to Mackey, "I had occasion to talk to Mr. J. T. Wilson, Manager of the World Trade Division, IBM, on the subject of C. C. Kao's experimental typewriter. . . . He spoke favorably of Kao's system and saw no objection in the use of numerals

Figure 1.5   Grace Tong demonstrating the IBM Electric Chinese typewriter, MLCR

to identify the 5,278 ideographs of Dr. Kao's typewriter." Mirovitch then continued, "Wilson added that during his many years of railroad and communication experience he found that a certain type of employee was quite able to memorize several thousands of code numerals designating railroad stations and other words."[56]

Grace Tong featured heavily throughout the early stages of Kao's tour. Two photographs in particular capture what these demonstrations were like. One featured Tong at the machine, semicircled by Kao and a small crowd of men. The second found Tong at the machine, this time being observed by the Chinese Ambassador to the United States, Dr. Wellington Koo, during a July 23 presentation (figures 1.5 and 1.6).[57]

As hard as Kao worked to impress figures like Watson, his main target remained Mergenthaler Linotype: they were the gatekeepers whose support, or rejection, would decide the fate of Kao's master plan. Kao launched an all-out charm offensive. Crafting an attractive, bound album, he excerpted highlights from the American tour and sent them along to J. T. Mackey in Brooklyn: press clippings from major media outlets, and

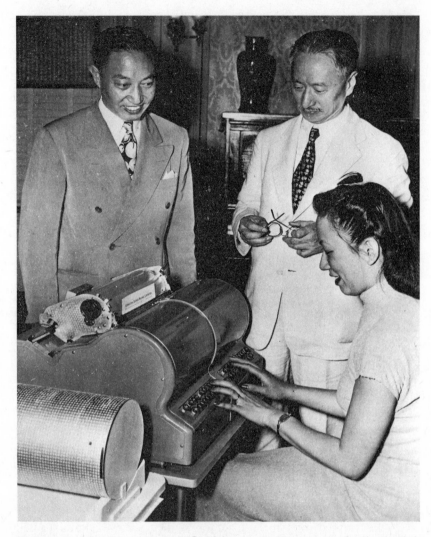

Figure 1.6 Grace Tong demonstration of the electric Chinese typewriter for Dr. Wellington Koo, Chinese ambassador, MLCR

Figure 1.7 Photographs of Chung-Chin Kao during American tour. Reprint courtesy of and © IBM Corporation

crisp photographs of Kao alongside IBM president Thomas J. Watson, Ambassador Koo, and other onlookers (figure 1.7).[58]

Mergenthaler failed to respond—again.

A week later, a nervous Kao wrote to Griffith once more, his tone more desperate than ever. In a rare moment of vulnerability, he reminded Griffith of the nearly three years' worth of back-and-forth communication between Kao and Mergenthaler—three years in which the Brooklyn company expressed interest, offered encouragement, and kept their executives' doubts largely to themselves. "During our conferences in 1943 and 1944," Kao wrote, with overtones almost accusatory, "you mentioned that the first step of the research for the Chinese typecasting machine was the successful accomplishment of the Chinese typewriter—to which I agreed. Now, with the support and cooperation of the International Business Machines Corporation, I have finally completed the first electro-automatic Chinese typewriter in commercial form. The announcement of this machine was made last month and the production and sales programs are now under consideration."[59]

The *New York Times* and the *Herald Tribune* celebrated his machine as a "boon to China" and "the first of its kind," Kao emphasized.[60] His typewriter—*including* the four-digit code upon which it relied—impressed engineers and journalists alike, and received continued encouragement by high officials in the Chinese government who were eager to see Kao extend his work into the development of Chinese typecasting.[61]

In late September 1946, Kao pressed forward, hitting the road to demonstrate his typewriter in Boston at the annual convention of the National Association of Cost Accountants.[62] From September 30 to October 5, Kao and the machine returned to New York for an exhibit at the National Business Show in Grand Central Palace (the same show, incidentally, where IBM debuted its 603 Electronic Multiplier, later celebrated by IBM as "the first electronic calculating machine ever placed in production").[63]

Kao sent four tickets to R. H. Turner at Mergenthaler, asking them to attend.[64]

No one came.

Not all American media coverage was glowing, moreover, which compounded Kao's worries. "The Chinese have a practical reason for believing that one picture is worth a thousand words," a *Time* magazine punned on July 15, 1946, "it takes so long for them to write the words."[65] The article continued:

The machine, which would give a US stenographer the heebie-jeebies, has 5,400 characters (the most commonly used of the 80,000 in the Chinese language), mounted on a drum. . . . It takes two months for an operator to learn to write simple sentences, four months to achieve the machine's top speed—45 words a minute (par for a fast typist in English: 120 words).[66]

Perhaps the only help that such media coverage offered Kao was of a somewhat dubious nature. Most of these articles heaped smug ridicule on the invention, it was true, but they reserved their most acute scorn for existing practices of textual reproduction for the Chinese language. "The machine depends on good memory for efficient operation and probably will be costly," as one *New York Times* article noted, "but Chinese business,

Government and communication agencies are now dependent primarily on tedious hand work to inscribe thousands of complex ideographs."[67] Phrased differently: Kao's typewriter might be absurd, but other Chinese techniques were even worse.

In the midst of all this, Kao's plans were dealt a major blow: Grace Tong withdrew as demonstrator for the machine. In our conversations, Lew recalled hearing that Tong had fallen ill, unable to continue with Kao in the tour itinerary. It is also possible that Tong's professional life was being steadily consumed by familial expectations and responsibilities.[68] Whatever the specific reason, the fact remained: Kao was now in need of a replacement, one as capable as Tong if not more, lest his ambitions to electrify Chinese come to a screeching halt.

LIVING IN CODE: LOIS LEW IN CHINA

When the IBM Chinese typewriter was debuted to the world, Lois Lew was a worker in Department 76 of Plant 3 of the IBM offices in Rochester, New York.[69] Born Lois Eng on December 21, 1924, in Troy, New York, her life trajectory could hardly have been more different than that of Grace Tong. Unconnected, and without a formal education, her early life was marked by struggle, political turmoil, and near constant movement. Shortly after she was born, her family returned to China in the years leading up to the outbreak of war with Japan in 1937.[70] When war erupted, Lew's family fled south, largely on foot, on a perilous trek from north China to Hong Kong. Along the way, Lew recalled to me, there were times when she had to carry a sibling on her back.

In Hong Kong, her mother took notice of a family in the neighborhood that struck her as financially stable. Engaging the help of a matchmaker, she inquired as to whether any of the family's sons were eligible bachelors. She provided a photograph of Lois, and after some time received an answer in the affirmative.

Lois's mother matched three of her daughters this way, one to a man in Chicago, a second to a man in San Francisco, and a third—Lois—to a man in Rochester. All of these men, her mother was assured, were financially well-off and more than capable of taking care of their brides-to-be.[71]

At the age of just 16, Lois ventured out upon a trans-Pacific voyage all by herself, disembarking at San Francisco and boarding a train bound for Chicago. There, her soon-to-be brother-in-law was awaiting her, accompanying her for the remainder of the journey. She could speak and understand barely a few words of English.

Upon arriving in Rochester, Lois learned the truth about her new husband-to-be—a man named Yuen Lew—and his financial situation. It was nothing like she or her mother expected. Instead of a comfortable life, Lois slept in the back room of the young man's laundry shop, along with her new sister-in-law, Gay.[72] Being too young to marry legally in the state of New York, the couple traveled to New Jersey to form their union. Lois Eng became Lois Lew. And while her sister-in-law would go on to attend high school, Lois was told that married women like her didn't do that kind of thing.

Lois and Gay were among only a few Chinese women living in Rochester at the time, which may in fact have contributed to their being hired by nearby IBM. "In those days, you didn't see many Chinese girls," Lew recounted to me in our conversation. "They would just use us for show: Chinese girls using an American typewriter." Lois became a capable typist, as did Gay.

Then the IBM Chinese typewriter was unveiled to the world, setting off a chain of events in motion that led to Grace Tong's employment—as well as Grace Tong's sudden decision to stop. Suddenly, IBM—and, above all, Kao—was in desperate need of Chinese-speaking typists to help demonstrate the prototype both in the United States and in China.

Lois and Gay were summoned to Manhattan to meet with Kao in person. Fate—and illness—struck once again, however. Gay contracted tuberculosis and needed to be hospitalized—and so Lew made this journey alone, just the latest solo trek in her young life.

Renting a room at a local YWCA, Lois taxied to IBM's World Headquarters at 590 Madison Avenue—an imposing, 100,000-square-foot, 20-story building.[73] Kao read over Lew's resume, and it exhibited nothing like Tong's impressive educational background. He seemed displeased, Lew recalled.

"Do you know how to spell the word 'encyclopedia'?"[74]

It was a peculiar question—a kind of aggressive non sequitur. Lois understood immediately what Kao was after: he was testing her, disappointed in her educational background. She did not, she admitted. Overcome and on the verge of tears, Lois wanted to race back down the elevator, back to the Bronx, back her room at the YWCA, and perhaps all the way back to Rochester.

"Do you want me to go home?," she asked. Kao looked at her, then away. The room fell silent. His decision seemed to take a lifetime. What Lew could not have known at that moment, however, was that the stakes for this meeting were far higher for Kao than for her. She was Kao's only chance, perhaps his last, at winning over incredulous New York executives, newspapermen, and the market at large. If the world was to be convinced of his system's viability—especially the four-digit code on which it relied—he would need Lew to convince them. And while Lew's biography may have left something to be desired for Kao, in many ways Lew was an even better ambassador for the machine than Tong had been. Tong's pathway through life—college-educated, bilingual, connected through family relations to the highest rungs of the Chinese elite—was hardly something Kao could hope to recreate consistently. If history served as a judge, the two existing labor pools most comparable to the one that Kao sought to create—Chinese telegrapher clerks, on the one hand, and Chinese typists on the other (for mechanical Chinese typewriters, that is)—were populated by men and women who rarely boasted anything beyond a lower- or middle-school education.[75] Lew was far more representative of what an "average" Chinese typist might look like, that is. If Kao were really going to establish a reliable stable of operators for his machine, these workers were by and large going to resemble Lois Lew to a far greater extent than Grace Tong.

"There's something else about you," Kao replied at last. He paused and sighed. "I have no choice. I have to try you."

"Take this chart," Kao instructed her. "Go back to your hotel, and memorize the four-digit codes for one hundred characters." Over the next few days, Kao's chart became Lew's whole world. On probation at the YWCA, she pored over the book, vying to memorize the four-digit codes for the first set of common usage characters. She succeeded, and so began the real training regimen. In the span of three weeks, as Lew attempted to learn by heart the four-digit ciphers for 1,000 of the most common Chinese characters, she remained glued to the code book during her long commute from the Bronx.[76]

Lew became Kao's main demonstrator, traveling to shows in Boston, New York, and San Francisco. She walked a fine line, wielding this complex mechanical object through her feats of mental acuity, even as she herself was on display, objectified. When the team was not on the road, for example, she was sometimes stationed by the window of the IBM offices, literally put on display so that passers-by could see a young Chinese woman working on the Chinese typewriter. "My fingernails were red. I had nylon stockings. They'd never seen anything like that," Lew remarked to me.

Then came the voyage to China.

The last time she traveled by ship across the Pacific, she was a young girl, fleeing Hong Kong to meet a fiancé she had seen only in photographs. She returned as an adult, flanked by two IBM engineers and a Chinese inventor. Her meals were paid for, along with a new wardrobe. She was living like a movie star, as Lew put it.

For IBM, the visit to China served a dual purpose: to generate interest in the machine by way of (hopefully) successful demonstrations, and more importantly, to determine the market potential of the machine.[77] For Kao, the trip meant so much more. To say that his hopes were high would be a vast understatement—stratospheric, would be more fitting. As Kao's son recounted to me, the inventor converted an immense sum of money into Chinese yuan, the country's currency at the time. While the total amount is unknown, the cash filled up an entire trunk, measuring 4 feet wide, 2 feet tall, 2 feet deep, and hundreds of pounds in weight. It was Kao's family fortune.[78]

With this immense war chest, Kao would try to secure contracts, establish manufacturing arrangements, and perhaps grease the skids a bit in a country, as was well known to many, racked by systemic corruption and graft. This was Kao's chance—likely his last—to translate these long and grueling years into something that could make him as famous as Ottmar Mergenthaler, Thomas Watson Jr., or perhaps even Johannes Gutenberg (in China, at least). His name, his reputation, and a small fortune were all riding on the trip—and on Lois Lew's performance.

In China, the reception was thrilling. In Shanghai, the mayor was waiting for them at the docks, photographers at the ready. Lew and the team were treated to sumptuous meals, and they stayed in one of the nicest hotels in the city. The first of a series of demonstrations took place at IBM China headquarters, located at 218 Sichuan Road. The following day, on October 20, 1947, Kao and Lew demonstrated the machine at the Park Hotel, then the tallest building in all of Asia.[79] In attendance were scientists, local government officials, and newsmen.[80]

In Nanjing, the reception was even more stirring. The team was met by high-level government officials, and their visit was covered extensively in the Chinese press. Even before the team's arrival in China, in fact, the local media had been watching their American demonstrations from afar.[81] For months, Chinese outlets circulated news of Kao, reporting on his demonstrations across the United States, including those in San Francisco, Harvard University, and elsewhere. An audience of 3,000 people gathered to watch Lew's performance.[82]

Lew was unfazed by the pressure. In front of these thousands of onlookers, and one intensely nervous Chung-Chin Kao, Lew was handed one newspaper article and business memo after the next as she proceeded to transcribe each of them on the Chinese typewriter, on the spot.

Coverage of the IBM demonstrations was overwhelmingly positive. Stories appeared in *Science, Signs of the Times, Municipal Affairs Weekly, Science Pictorial, Science Monthly,* and many other outlets (figure 1.8).[83] In addition to praising the machine, publishers were clearly enamored of Lew's beauty. Her face soon appeared in *Chinese-American Weekly* (*Zhong-Mei Huabao*), IBM's promotional brochures, and its 1947 film, among other venues.[84] "I know how to dress," Lew told me on the phone. "Very sexy looking. I was beautiful."

0275: YOU / 0178: HE / 0314: ME

Winston Kao and his sister sat on the floor of their family's Taiwan apartment, surrounded by hundreds of thousands of Chinese dollars. The children were playing with it like Monopoly money, piling and scattering the bills, sorting and stacking them. Chung-Chin Kao caught sight and scolded them—not because they had value, but because of the bitter

Figure 1.8 Chinese photojournalistic coverage of the IBM Electric Chinese Typewriter

memories they summoned to mind. The bills were worthless, their value cratering after the 1949 Communist victory in the Chinese Civil War. Not even paper recyclers were interested in taking them off Kao's hands. This massive chest of cash, once part of Kao's grand plan to electrify Chinese script, was now being strewn about his living room floor like so many scraps of paper.

The 1949 Communist Revolution had also pushed Mergenthaler and IBM's anxieties to the breaking point. Unconvinced by Kao's four-digit coding system after all these years, shifting geopolitical sands added even more uncertainty. After the failed efforts of a US Army general, George

C. Marshall, to broker peace between the Communists and the Nationalists, the turning tide of the war and a spate of Communist victories, and finally the CCP's "liberation" of Nanjing, Shanghai, and Beijing—the deal was dead. "The Communist takeover in China was well underway at the time," a retrospective article in *IBM News* explained, "and was completed before the typewriter had a chance to achieve significant sales in an understandably nervous Chinese market."[85]

The year 1949 threw Kao's national identity into turmoil as well. He became a man without a country, in effect, being issued a special diplomatic "Red" visa by the United States. His family responsibilities began to grow, moreover, shortening Kao's runway and making "take-off" a virtual impossibility. Winston, his son, was born in August 1952, and there simply wasn't time to wait for IBM and Mergenthaler's potential change of heart.[86]

Something of a respite came by way of the OKI corporation in Japan, which had earlier taken notice of Kao's work on telecommunications as a way, perhaps, to further solidify its place as Japan's foremost manufacturer of teletype equipment.[87] At OKI's invitation, Kao traveled to Japan in 1953 to work on a Chinese-Japanese teletypewriter, which OKI went on to manufacture. The machine was based upon his electric Chinese typewriter, with certain key changes.[88] Two such machines were purchased by the Ministry of Transportation and Communications which, on New Year's Day 1958, had an inaugural transmission between Taipei and Tainan. Kao called this moment the "beginning of a new era in Chinese telecommunications."[89] Success was limited, however, never making inroads into the mainland.

Kao and his family had moved to Taiwan by this point and, as his son Winston recalled, lived an itinerant lifestyle, moving frequently—sometimes once every six months or so. Kao's fiery personality, and his love of the inventor's life, never waned. When inspiration struck, his father would wake at two in the morning, brain dumping on large, 4-by-6-foot pieces of paper, spread out across the room. By the time the family woke, around seven or eight, Kao would have refined whatever it was he was working on, boiling it down to a single sheet of paper, ready to be diagrammed.[90] In 1986, Kao passed away in Taipei.

Whatever his later successes, they fell far short of his once grand vision. This raises the question: Why had his efforts with IBM and Mergenthaler failed? Why did Kao fail to convince engineers and executives that his four-digit coding system worked? Putting aside whatever gut reactions we as twenty-first-century readers might have regarding Kao's cipher system, the empirics of the case suggest—perhaps to our surprise—that it might indeed have been viable, at least to a point. Kao was telling the truth, after all: there truly were thousands of Chinese telegraph operators, deeply familiar with this method of handling Chinese character information processing. There was, as Kao tried and failed to convince IBM and Mergenthaler, a training regimen already in place—one that could in theory have been transposed from the arena of telegraphy to that of type-writing. And as evidenced by the performance of Grace Tong—and, above all, Lois Lew—it seemed clear that typists could not only memorize Kao's code, but wield it in real-time under conditions of unimaginable stress (most typists are not asked to work under the watchful gaze of thousands of spectators, after all).

Returning to the central question of this chapter—*How much code can a person stand?*—the answer appears to be twofold: *Far more than one might expect*; but also *far more than the gatekeepers of modern information technology were willing or perhaps able to accept*. This is not a straightforward matter of blame and blind spots, however. Figures like Watson Jr., Griffith, Kennedy, and others were being asked to contemplate a model of human-machine interaction that had no precedent, nor parallel, in the Western world at that point. They were asked to imagine, as it were, a world in which one signed off, not with a *Yours sincerely* or a *Best regards*, but a cryptic string of Arabic numerals. No matter how strong Kao's case—a case built on the history of Chinese telegraphy, and on the performance of real-life typists like Tong and Lew—engineers and executives operating in the arena of "alphabetic order" just couldn't stifle their doubts.[91]

For Lois Lew, life after IBM took her in a completely different direction. She and her husband started a laundromat of their own, reinvesting their earnings—along with savings from IBM—in the launch of a new Chinese restaurant in Rochester in 1968. In the Adam Brown Block building on 488 East Main Street, they opened Cathay Pagoda.[92] Located two minutes by foot from the famed 2,400-seat Eastman Theater, the restaurant attracted a steady stream of students and young people, and even the occasional star (Katherine Hepburn among them, and possibly Ozzy Osbourne). The restaurant kept its doors open for decades, becoming a mainstay of the city before its nearly forty-year run ended in 2007.

Well into her 90s, Lois Lew swam at the YMCA three hours a week. She remained close friends with former employees from her restaurant (none more so than a man named Steve, whom she described to me as her "best friend"). Looking back on her time at IBM, she told me she had but one regret: "I could have bought IBM stock. Instead, I bought war bonds. Stupid!"

"I still remember the numbers," she added in passing, referring to the four-digit codes from the IBM typewriter. She began to rattle them off to me on the telephone. "'You,' 0-2-7-5. 'He,' 0-1-7-8. 'Me,' 0-3-1-4."

I couldn't help but smile and laugh. Her vivacity and energy were infectious. After almost two hours of speaking, the time had come to say our goodbyes. I expressed thanks, and we discussed meeting in person in New York City, at a museum exhibit I curated on the history of Chinese information technology—an exhibit in which the 1947 film featuring Lois Lew was to be projected on the wall, larger than life.[93]

I hung up the phone.

As I scrambled to write down all of the ideas coursing through my brain, my thoughts turned back to the numbers she had recited to me. Could it be true? Did she really remember Kao's codes seven decades later? Lew had spit them out so quickly during the conversation, without pause, that I felt certain that she was speaking extemporaneously. I dug back through my archival records to track down my own copy of Kao's code manual.

Looking up the three numbers, I could hardly believe my eyes.

0275: 你 ("You")
0178: 他 ("He")
0314: 我 ("Me")

2

BREAKING THE SPELL: SINOTYPE AND THE INVENTION OF AUTOCOMPLETION

It was summer 1959, and the United States desperately needed a Cold War win. Ten years had passed since Chairman Mao stood before crowds in Beijing to hail the formation of the People's Republic—the "Loss of China" as it was known to many in Washington, DC. In the intervening decade, the Communist Bloc threaded together an impressive string of victories. The USSR joined the nuclear club, then leapfrogged ahead in the space race with Sputnik 1. Then came the Cuban revolution, and the founding of a communist outpost less than a hundred miles offshore. If ever there were a time to demonstrate that capitalism was still at the helm of world affairs, it was now.

In Room 3B 747 of the Pentagon, a plan began to take shape: the United States would unveil the world's first Chinese computer, with news of the breakthrough to be delivered by President Dwight D. Eisenhower himself. Capitalism would bestow upon China the gift of computing—a technology that was already transforming the social, political, military, cultural, and economic fabric of everyday life in the United States, the UK, and parts of Europe—thus scoring a "Free World" technological and cultural victory, not to mention a new infrastructure for the global dissemination and translation of Chinese-language material.[1] Whoever possessed such a device could flood the world with Chinese texts at rates never seen—a dream come true for Cold War propagandists and psychological operations specialists alike.[2]

At the center of this geopolitical intrigue was the "Sinotype," a computational Chinese phototypesetting system devised by the MIT professor Samuel Hawks Caldwell and prototyped by the Graphic Arts Research Foundation (GARF). Located in Cambridge, Massachusetts, one of the epicenters of American computing, GARF was founded as a nonprofit research group focused on advancing the novel technology of photo-typesetting, heralded by contemporaries as ushering in the "atomic age of printing." Pioneered in the late 1940s by the French engineers René Higonnet and Louis Moyroud, it melded high-speed photography, jus-tifying typewriters (which keep the right margins of the text in a clean, not ragged, line), telephone dialing circuits, and computing into a uni-fied device capable of printing text by means of photography rather than metal type.[3]

Samuel Caldwell was a man of many talents. Born in 1904, he studied at MIT under Vannevar Bush before becoming a pioneer in his own right in the field of logic circuits.[4] Caldwell joined the MIT faculty as a profes-sor of electrical engineering, advising such luminaries as the information theorist David Huffman, working with Bush on the design and construc-tion of the earliest large-scale analog computer in history, the differen-tial analyzer. He also helped catalyze the formation of the Association for Computing Machinery (ACM), one of the earliest and most influen-tial computing learned societies in the world.[5] When he wasn't advising his students, he enjoyed playing the organ, even making an occasional guest appearance with the Boston Pops, as his granddaughter-in-law Ann Welch recalled to me in our conversation.[6]

Chinese was one talent Caldwell could not claim, however. His intro-duction to the language came by way of informal dinnertime chats with his overseas Chinese students. Betty Caldwell, Samuel's second wife, sometimes invited students over to the house, venturing to make them feel more at home by preparing dishes she learned from a close fam-ily friend: Joyce Chen, the doyen of Chinese-American cuisine.[7] ("Mrs. Samuel Caldwell" was one of the few people to be thanked by name in the best-selling *Joyce Chen Cook Book*.[8])

In between bites of stir fry, Caldwell and his students got to talking about Chinese characters. One student in particular—Francis Fan Lee (Li Fan 李凡 1927–)—made a vivid impression on him, Caldwell later credit-ing him with first exploring the basics of how Chinese writing works. There was a standard, economical set of "basic strokes" used in Chinese handwriting, Lee explained, and that when different people composed the same Chinese character by hand, the basic strokes they used were always the same—with only rare exceptions.[9] "If the strokes are regarded as the 'letters' of an 'alphabet,'" Caldwell later summarized, "the Chinese always 'spell' a word the same way each time it is written."[10]

"Chinese has a 'spelling'"![11] The notion of consistent Chinese "spellings" struck Caldwell, a leading expert on logic circuit design, with the force of an epiphany.

If Chinese character inscription was governed by an unchanging "logic," it stood to reason that Caldwell could build a logical *circuit* to control the process. And if he succeeded in this, Caldwell would have solved a puzzle that dated back to the late nineteenth century: the puzzle of how to "fit" Chinese characters on a standard QWERTY-like keyboard. He would become the first, as well, to computerize Chinese.

Practically overnight, Caldwell's writings began to fill up with an entirely new vocabulary: words like "stroke," "radical," and "character" began to appear, a far cry from his technical work on logic circuits. He decided to take the plunge, stepping into a role that he would occupy until his untimely death: director of research at the Graphic Arts Research Foundation.

Thanks to one fateful dinner, a man who knew not a single Chinese character had now set out to build the world's first Chinese computer.

HOW DO YOU SPELL IN CHINESE?

If Chinese had a spelling, what was the Chinese alphabet, exactly? How many "basic strokes" were there, and what did they look like? How many strokes did it take to "spell" a character on average, moreover? And were some "stroke-letters" more frequent than others? If Caldwell was going to build a Chinese computer based upon this approach, these were some of the many questions he would need to answer.

But Caldwell soon learned enough about Chinese writing to realize that he couldn't push the metaphor of "spelling" too far. The "structures and modes of operation of conventional typesetting machines used for Western European languages cannot be efficiently adapted by a mere change of scale to the composition of Chinese or similar languages," as

Caldwell later phrased it. Not only was Chinese script non-alphabetic; just as importantly, the strokes to which Francis Fan Lee referred didn't behave like Latin letters at all. The strokes that make up Chinese characters change shape, size, and location in ways that printed Latin letters simply don't.

Instead of building a machine that could "type" words as in English, then, his goal would be something else entirely. He would build a machine "adapted for selection of a complete typographical character by multiple-key operation, as distinguished from selection of alphabetic components by single-key operations as in conventional type composition."[12] For a Chinese machine, that is, "spelling" was going to have to behave differently.

The primary objective on the Sinotype, Caldwell summarized, would be "to furnish the input and output data required for the switching circuit, which converts a character's spelling to the location coordinates of that character in the photographic storage matrix."[13] This was Caldwell's technical way of saying: to "spell" any given Chinese character would be, in actuality, to input the "address" of that character in Sinotype's memory, so that the machine could retrieve the character from memory. Sinotype would be a *retrieval* machine, first and foremost. Inscription took place only after retrieval was successfully completed (figure 2.1).[14]

Like IBM and Mergenthaler Linotype before them, Caldwell and his GARF colleagues set out to build their own Sinological brain trust. In addition to Francis Fan Lee, Caldwell sought the assistance of Lien-Sheng Yang, an esteemed professor and first full-time historian of China in the department of Far Eastern Languages at Harvard University.

Yang helped Caldwell determine four key characteristics of Chinese orthography: the number and types of "basic strokes"; the relative frequency of each stroke within the Chinese writing system; which strokes tended to show up at the beginning versus the end of characters; and finally, which strokes tended to be written immediately before and after one another. Regarding the first question—the number of basic strokes— Yang and Lee relied on their firsthand knowledge of Chinese writing rather than formal experiments. A wide variety of calligraphic theories have been put forth in the history of Chinese writing to explain the "fundamental" structure of Chinese characters. Beginning in the Han

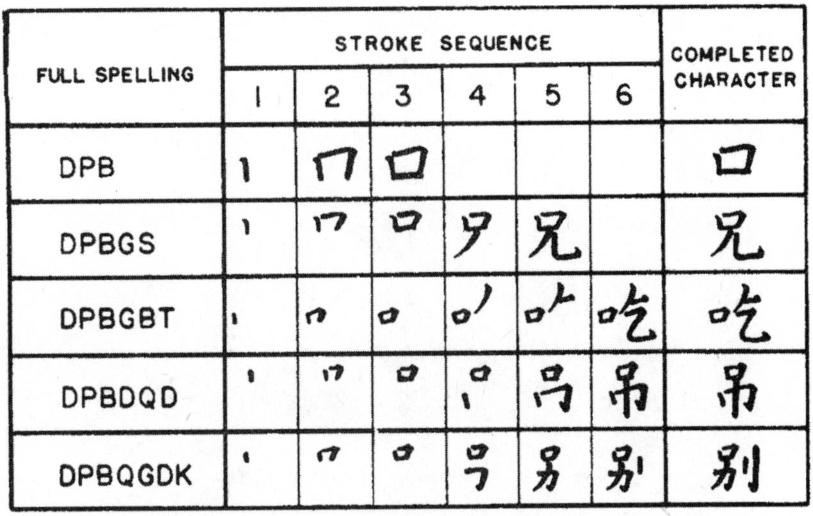

FULL SPELLING	STROKE SEQUENCE						COMPLETED CHARACTER
	1	2	3	4	5	6	
DPB	丨	冂	口				口
DPBGS	丨	冂	口	尸	兄		兄
DPBGBT	丨	冂	口	吖	吖	吃	吃
DPBDQD	丨	冂	口	弔	吊	吊	吊
DPBQGDK	丨	冂	口	弓	另	别	别

Figure 2.1 "Spelling" Chinese characters on Sinotype

dynasty, China saw a proliferation of treatises in which the orthographic features of Chinese characters were categorized.[15] The Jin dynasty calligrapher Wang Xizhi (303–361), renowned as the "sage of calligraphy," put forth perhaps the most widely known of these theories, referred to as the "eight fundamental strokes of the character 'yong'" (*yong zi bafa*). According to this theory, all Chinese characters are composed of eight types of brushstrokes (figure 2.2).[16]

Eight strokes were far too few for the Sinotype project, however. Caldwell was planning on using a standard QWERTY-style keyboard as the interface for Sinotype, after all, so to reduce the number of "basic strokes" to such a small number would vastly under-exploit the interface's available real estate. Yang, Lee, and Caldwell further plumbed the depths of Chinese writing history to uncover other theories, perhaps better suited to the project. Wang Xizhi's eight-stroke categorization was later expanded upon by Li Fuguang, for example, who extended it to include thirty-two strokes in all.[17] Still another Chinese theorist of calligraphy, Lady Wei of the Eastern Jin (265–420), put forth a taxonomy outlining seventy-two fundamental strokes.[18] Yang, Lee, and Caldwell enjoyed flexibility, this is to say, when deciding on a final set of "basic strokes."

Figure 2.2 The eight fundamental strokes of the character *yong* (永 "perpetuity")

The Sinotype team settled upon 24 fundamental strokes. As they con-tinued their initial phase, however, they decided to reduce this number to 21 (by merging 3 of the strokes together with others on the keyboard)."[19] The reduction of strokes from 24 to 21 was related to a second core area of research, namely the relative frequency of "basic strokes." Across all Chinese characters, they asked, how often did the long, vertical brush stroke *shu* appear? What about the horizontal stroke *heng*? What about the *dian* (dot), the *gou* (hook), and so forth? Since Caldwell intended to assign each of these Chinese brushstrokes a unique binary digit code,

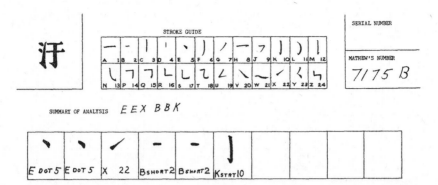

Figure 2.3 Data collection card for Caldwell-Yang Chinese stroke analysis

he needed to know precisely how often each of these "Chinese letters" appeared so that he could assign the shortest binary values to the most frequent strokes (much like Morse code assigns the shortest dot-dash patterns to the most common letters).

Caldwell relied on Yang to conduct a thorough analysis of the structural makeup of the roughly 2,300 Chinese characters he intended to store in the machine's photographic database. Creating an index card with each character at the top left, Yang and his assistants wrote out the character stroke order along the bottom (figure 2.3).[20]

Although Caldwell's analysis of Chinese strokes was unlikely startling to anyone familiar with Chinese script, it produced results that had never been documented before in such precise fashion or by such means. For instance, the frequency of Chinese strokes varied dramatically, a fact that would have "profound significance in the design of the keyboard and in the design of the binary codes to represent strokes." The horizontal stroke (*heng*), Caldwell calculated, accounted for an astounding 33 percent of all Chinese "letters" in his sample, while the vertical stroke accounted for an additional 18 percent—already more than half of all orthographic components. At the other end of the spectrum, by comparison, frequencies of other stroke-types decreased "so rapidly that 90 percent of all Chinese writing is accounted for by only nine of the twenty-one basic strokes" (figure 2.4).[21]

The frequency of basic strokes, while essential to the efficient design of Caldwell's logical circuitry, was only one of the most basic findings. Since

SYMBOL	STROKE	FREQUENCY	SYMBOL	STROKE	FREQUENCY
B	一 一	·329	M	∫	·0073
D	∣ ∣	·183	S	Ｌ∪	·0071
G	ノ ノ	·141	H	一	·0048
E	ヽ ヽ ∕	·101	L	ｊ	·0047
P	フ フ	·073	N	∪	·0032
A	末	·041	I	一	·0027
V	＼	·024	Z	㇇	·0023
Q	フ	·023	R	L	·0014
J	フ フ	·0124	2	二	·0011
K	∫	·0123	W	～	·0008
Y	く	·0085	T	Z	·0003
X	ノ	·0083	3	三	·0003
U	㇄	·0077			

Figure 2.4 Chinese stroke frequency analysis by Samuel Caldwell and Lien-Sheng Yang

operators would be inputting Chinese strokes sequentially, Caldwell also needed to conceptualize "Chinese spelling" according to its temporal dimensions—as a sequence of first, second, and third keystrokes. To this end, Caldwell, Yang, and Lee undertook a second-order analysis of stroke frequency, focusing this time on which "basic strokes" tended to appear towards the beginning of the "spelling" of Chinese characters, versus the end.

The results were startling. Only fifteen of the basic strokes ever appeared as the first stroke in a Chinese character, Caldwell and his team determined—a feature that would enable them to simplify the architecture of the machine's logical circuitry by ruling out certain orthographic possibilities that never occurred in real life. Among these fifteen strokes, moreover, some appeared quite frequently at the beginning of characters, and others far less. Over half of the 2,121 characters they tested, for example, began with one of two strokes: stroke "B" (*heng*) or stroke "G" (the letter Caldwell assigned to *pie* [ˊ]).[22] At the other end of the spectrum, though, strokes R (㇄), T (㇄), H (*heng gou* ㇀), and M (*shu ti* ㇄), appeared at the beginning of a mere 3, 2, 1, and 1 characters out of the entire set of 2,100-plus characters, respectively.[23]

When it came to logical circuit design, insights such as these were invaluable. By understanding the orthographic "unevenness" of Chinese—some strokes appearing far more frequently than others—Caldwell's could fine-tune Sinotype to make it as efficient a retrieval system as possible.

WAGING PEACE: THE WEAPONIZATION OF SINOTYPE

A Chinese computer was possible, Caldwell and GARF's preliminary findings seemed to show. Now he and his colleagues needed money. In 1953, GARF began to apply for support, setting their sights first on the Carnegie Institute and the Ford Foundation. They requested $30,000—roughly $280,000 in today's terms.[24]

The scale of their proposal was ambitious—global, even. Titled "Proposal for Studies Leading to Specifications for Equipment for the Economical Composition of Chinese and Devanagari," it revealed that Caldwell's hopes for Sinotype extended beyond Chinese script itself to encompass practically the entire non-Western world. Sinotype held the key, Caldwell and his colleagues seemed to believe, not only to the computerization of Chinese, but also many non-Latin scripts that had been systematically left behind by Western-designed information technologies from the nineteenth century onward.[25]

"Nearly all of the advancements of mechanization in the type-composing art," Caldwell wrote in his patent application for Sinotype, "have been made to facilitate composition in languages having relatively few distinct typographical characters as compared with the numbers of comparable characters in ideographic languages, such as Chinese."[26] "It has also been recognized," Caldwell continued, "that the structures and modes of operation of conventional typesetting machines used for Western European languages cannot be efficiently adapted by a mere change of scale to the composition of Chinese or similar languages. For example, it is obviously impractical to provide a keyboard having a key for each typographical character."[27]

Sinotype held the solution, Caldwell felt, to overcoming the challenges faced by *all* disadvantaged non-Latin orthographies, both because of its "ability to store a very large number of characters and to select any desired character from that storage," and because the technology of phototypesetting enabled something that was prohibitively complex for conventional typesetting: "the ability to form characters or character groupings by multiple exposure." Here Caldwell was referring to the ability within phototypesetting, in theory, to "build up" the various kinds of complex conjuncts found in multiple South and Southeast Asian scripts thanks to the technology's ability to superimpose multiple shapes atop one another—something unfeasible when dealing with metal moveable type. "We believe," GARF's proposal proclaimed, that "the basic specifications of the economical composition of all language forms can be obtained by selecting the two typical types: Chinese and Devanagari of India. It appears that the principles involved in the solution of Chinese composition would apply equally to all ideograph forms, varying only in detail, and for Devanagari would apply broadly to all Sanskrit-root and Hebrew-Arabic."[28]

Caldwell's hemispheric ambitions hit a brick wall in 1954. Carnegie rejected GARF's application for funding, as did Ford. In both cases, it seemed, review committees balked at GARF's overtly international, philanthropic focus. "They cannot spend the money for benefits out of the United States," Vannevar Bush explained.[29] Serving as Carnegie president, Bush noted in a letter to Garth Jr. that the institute's charter prevented it from supporting work on foreign countries and subjects such as Chinese and Devanagari. "Apparently," Bush surmised, about both the Carnegie and Ford rejections, "they both looked at the subject as though we were developing machines which would then be utilized, and possibly manufactured, in remote areas."[30]

Bush offered a suggestion: Why not reorient the proposal toward Cold War–era concerns? Specifically, why not reapply to both Carnegie and Ford, only this time placing emphasis on the project's potential benefits for the US military? Garth Jr. already had a preliminary contact in the CIA, after all, which in December 1953 dispatched a representative to visit the offices of GARF. While the meeting was only introductory, Garth Jr. had made mention of GARF's work on so-called complex languages, underscoring the geostrategic importance of being able to economically produce texts in Chinese, Sanskrit, Hebrew, Arabic, and more. "Is it not possible," Garth Jr. wondered in a 1953 letter to Bush, that news of a machine for complex languages might have great impact "if this were identified with the United States Government?" "It seems to me that in this program there exists a means for making a great many friends for the United States in those countries where we need friends."[31]

The placement of such equipment in friendly hands should swell the volume of friendly information and propaganda in these countries whereas placing equipment in the wrong hands could do just the opposite.[32]

Caldwell, too, already had ample experience with the military. During World War II he served on the National Defense Research Committee, dedicating his time to the study of switching circuits and logical design— exemplified in the famed differential analyzer project. Throughout the period, indeed, he was in regular contact with—and often in direct service of—various branches of the US military.[33] Perhaps it made sense to tap back into the ample resources of military and intelligence circles.[34]

Caldwell and Garth Jr. took Bush's advice, their new applications underscoring the fact that the United States Air Force, the Psychological Warfare Division, and the CIA had all expressed interest in Sinotype for the purposes of psychological operations. Framed in such a way, Sinotype would be read less as a project meant to benefit foreign countries than one meant to augment the reputation of the United States on the global stage.

The strategy paid off immediately. Carnegie awarded GARF an appropriation of $30,000—the entire amount originally requested. Then it paid off again, in a combined $150,000—over 1.5 million dollars in today's terms—from the US Army and Air Force.[35] The Sinotype project was now fully funded for at least five years.

Overjoyed though they were, Garth Jr. and Caldwell were also wary about how military funding sources might be perceived by the wider world. "Many will wonder why this work was ever done or why our military establishment devoted substantial funds and attention to the project," Caldwell wrote in a report on the Sinotype project for the *Journal of the Franklin Institute*, anticipating questions he expected might arise. "The answer to this question seems simple and clear," he assured his readers. "In selling the idea to the military authorities, the writer had only one real argument. To be sure, it was a fascinating project, but mere fascination was not a sufficient reason for supporting it. The argument that counted was to the effect that a machine for composing Chinese would improve communication among men, and that no improvement of communication ever harmed the cause of peace among men. . . . The writer is burstingly proud of the way the military establishment of the United States of America has supported, both in funds and in enthusiasm, this project to wage peace."[36]

The United States military was concerned about optics as well. "The fact that the machine was developed under contract to the Army and Air Force to meet requirements of psychological warfare," a confidential memo noted, "should be superseded in public exploitation by its potential value in the cultural, educational, and economic fields."[37] Sinotype may have been a weapon of psychological warfare—from their standpoint—but this wasn't something they necessarily wanted to advertise to the world.

Over the course of the mid- and late 1950s, one Pentagon meeting after another took up discussion of Sinotype—known also as the Chinese Ideographic Composing Machine. On May 29, 1958, representatives of the Office of Special Warfare and the US Information Agency met in Cambridge to review progress on the machine.[38] And during early 1959—in the lead up to a potential public announcement by Eisenhower—representatives of the Special Warfare Division, the Department of Defense, the State Department, the United States Information Agency (USIA), and the Operations Coordinating Board (OCB) held regular discussions in preparation for an upcoming presidential address.[39]

The "significance of the breakthrough," a high-level official wrote about the device in 1959, "and the present state of development of the machine are considered appropriate to warrant presidential announcement in order that the psychological value of this achievement may be exploited to the best interests of the United States." Eisenhower's address would be attended by the ambassadors of China and Japan, they imagined, as well as representatives of the United Nations, with the presidential announcement being followed the next day by a more thorough and technical presentation at the Pentagon to the media and the engineering community. "This initial exploitation," one memo proclaimed, "is a 'one-shot' action which by its very nature should be accomplished as early as possible to avoid possible compromise of the psychological impact."[40] There was, the committee explained, a "risk of being 'scooped' by the communists from whom knowledge of the invention cannot be withheld."[41]

Thanks to this new and ample funding, Caldwell and his team could now afford to dedicate themselves full-time to the Sinotype project, building upon their preliminary findings, and working to refine the logic circuit design of the machine. Caldwell had the green light to perfect his Chinese "spelling."

SPLITTING THE SCREEN: FEEDBACK LOOPS, DISAMBIGUATION, AND THE ORIGINS OF THE "POP-UP MENU"

Unexpected problems emerge when referring to the term "spelling" hypographically: to retrieve text from memory, rather than to compose it, that is. What should a Sinotype operator do, for instance, when two or more Chinese characters had the same "spelling?" The characters *niu* (牛 "ox") and *wu* (午 "meridian"), used in the common word 下午, meaning "afternoon"), were both spelled "GBBD" according to Caldwell's system, for example. "BDB," meanwhile, described not just *gong* (工 "work/labor"), but also *shi* (士 "gentry") and *tu* (土 "earth/land"). How would the machine discriminate between these and other "homographs"—or "spell-alike-look-different" characters, as Caldwell called them?

And what about Chinese characters whose spellings contained the codes for other Chinese characters? Take *chi* 吃 (to eat), for example. One of the graphical components of this character is itself a stand-alone Chinese character: *kou* 口 which can appear either by itself, in which case it means "mouth," or as part of other Chinese characters (such as *ku* 哭, "to cry," and *chang* 唱, "to sing," among many others). Chinese abounds with reoccurring structures like this, commonly referred to in English as "radicals." Upon receiving the keystroke instructions "D-P-B," then, how would Sinotype know whether to print "mouth" (*kou* 口), or to await further keystroke instructions in order to retrieve characters like "to eat" (*chi* 吃), "to blow" (*chui* 吹), "to drink" (*he* 喝), and so on?

In conventional typewriting, problems of this nature don't exist, of course. Composing the homographs "bass" and "bass" is not something that might "confuse" a Western-style typewriter, even if the former refers to a musical instrument and the latter to a fish. Nor does it matter that "bass" is both a stand-alone word, and part of longer words like "bassinet." For Sinotype, however, these were challenges Caldwell would need to address.

To help tackle the problem of Chinese homographs, Caldwell decided to dedicate some of the keyboard's precious real estate to keys used purely for disambiguation. Three keys in particular were added—1 (*yi*), 2 (*er*), and 3 (*san*)—with which a user could differentiate between characters with identical spellings. The full spelling for "gong" (工) would therefore be "BDB1," so as to distinguish it from "BDB2" (土) and "BDB3" (士).

To resolve the issue of "nested" spellings—such as "mouth" and "to eat"—Caldwell outfitted his keyboard with a fourth disambiguation key labelled "末" (*mo*)—a Chinese character meaning "ending." If one did indeed want to enter the character "mouth," full stop, they could enter the spelling "DPB" followed by the "ending" key. Sinotype would thereby know to produce the standalone character 口 (*kou*), rather than awaiting further instructions.

Even with these improvements, however, a much larger problem remained. How would operators be able to use Sinotype if they were required to memorize—and then input blindly—lengthy input sequences? In this sense, Caldwell needed to confront the exact same challenge we saw with Chung-Chin Kao, IBM, and the electric Chinese typewriter in chapter 1: the question of whether an average operator would be able to shoulder the burden of deploying real-time codes in an "all or nothing," "sink or swim" way, where each input sequence either succeeded in producing the desired Chinese character or failed.

In addition, Caldwell's coding system made the problem somewhat more complex than Kao's. Unlike Kao's fixed-length code, where every character corresponded to a four-digit cipher, Sinotype's input codes were variable in length, introducing more opportunities for user error (especially given the issues of homographs and nested sequences). There were also more code elements on Sinotype: twenty-one potential keystrokes (plus disambiguation keys) as compared to Kao's ten numerals. While Caldwell's system piggy-backed on a much longer-standing practice of Chinese composition than Kao's—the notion of strokes and stroke order, which arguably made it more intuitive or "natural" for any literate Chinese operator—nevertheless the added challenges baked into Sinotype's technique exacerbated the limits of blind, "all-or-nothing" input.

Caldwell hit upon a solution: a built-in screen, or "flashback window" as he called it, designed to allow the user to view Chinese character "candidates" on the display screen and check the accuracy of an input sequence before committing the character to print. As seen in figure 2.5, a beam of light was passed through a condensing lens, and then through the Chinese-character photographic matrix. Thereafter, however, the beam was split into two: one copy of the Chinese character refracted via the "flashback prism matrix" toward the "flashback window," and another copy passed through the shutter on its way through a series of lenses and prisms en route to the film magazine. Splitting the beam and copying the Chinese character in two places made it possible for the operator to preview the selected character prior to committing it to photographic registration.

In other words, Sinotype—the first Chinese computer—was already outfitted with a "pop-up menu" of sorts, a mechanical predecessor to the kind that appears as part of all Chinese Input Method Editors (here as a physical component of the machine, rather than a middleware application).[42] The pop-up menu has been part of the story of Chinese computing since the very beginning.

CHINESE IS REDUNDANT: "MINIMUM SPELLING" AND THE INVENTION OF AUTOCOMPLETION

The longer Caldwell explored this new and peculiar world of retrieval-writing—hypography, that is—the more it dawned on him: although this approach was vexed by a host of complications not found in conventional

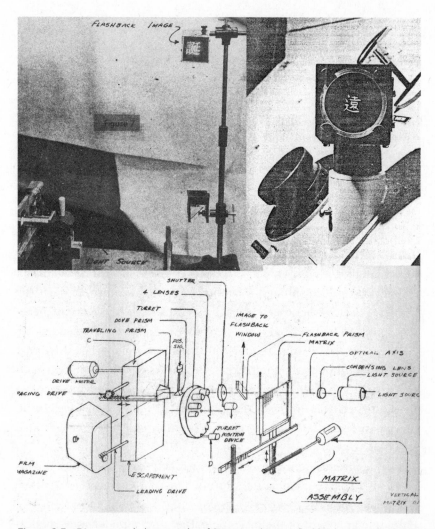

Figure 2.5 Diagram and photographs of Sinotype showing flashback prism matrix and "flashback window"

composition, there were aspects of hypography that afforded certain unique advantages. Thinking back to the problem of nested characters—the example above of "to eat"—the regularity of certain Chinese components offered up the possibility of additional economization. If shapes like *kou* and others appeared so frequently, perhaps this was a resource to be exploited rather than an obstacle to be resented.

Reviewing his spelling data, Caldwell took note of other common three-, four-, and even five-letter sequences. In addition to "mouth" (*kou*, or "DPB" in Sinotype's input code), there was also "sun" (*ri* / "DPBB"), "grain" (*he* / "GBDGE"), "silk" (*si* / "VUE"), and "language" (*yan* / "BBBBDPB"), among others. These "multiple-stroke combinations," Caldwell explained, were "somewhat similar to English syllables such as -ing, -tion, or -ous." He began to refer to them simply as "entities"—and studied how they might be exploited to make Sinotype run faster.

Caldwell experimented with the creation of special "entity" keys: keys which, when depressed, fed Sinotype's logical circuit with more than one "stroke-letter" at a time, thereby reducing the overall number of keystrokes needed to retrieve certain characters from memory.

The experiment met with success. "Entities" proved so useful, in fact, that Caldwell and his team quickly expanded the number from six to twenty, as seen in the top row of the final design of his keyboard (figures 2.6 and 2.7). Moving forward, a total of ten keys on the Sinotype would be dedicated to "entities," ten entities assigned to the "lowercase" level, and ten more to the "uppercase" or "shift" level.

After the entity keys were installed, the analysis by Caldwell and his team revealed that the average Chinese character comprised 4.7 keystrokes. This number carried with it special significance, Caldwell emphasized, as it "compares remarkably with generally accepted figures for the average length of English words."[43]

Further tests proved even more encouraging. After 20 hours spent practicing on Caldwell's keyboard, student subjects were achieving speeds of 4.3 characters-per-second on average.[44] Caldwell had discovered, in other words, that it was theoretically possible for Chinese input to approach speeds comparable to English typing.

But Caldwell's most significant breakthrough was still to come.

Figure 2.6 "Entities" developed by Caldwell

Figure 2.7   Keyboard of the Sinotype

In the course of researching Chinese homographs, entities, flashback windows, and more, Caldwell made a discovery as profound as it was simple: "the spelling of Chinese characters is highly redundant."[45] It was almost never necessary, Caldwell found, for a user to enter the full spelling of a Chinese character in order for Sinotype to achieve a successful match. "Far fewer strokes are required to select a particular Chinese character than are required to write it," Caldwell summarized it.

An English-language analog here might be the words "crocodile" and "xylophone." If the goal were to compose these words in full, then there would be only one way to do so correctly: c-r-o-c-o-d-i-l-e and x-y-l-o-p-h-o-n-e. Anything else would be a typo or an abbreviation. To retrieve these words is a different matter, however. Since few words in English start with "croco" or "xylop," it becomes unnecessary for a user to continue the selection process beyond the establishment of an unambiguous match. What takes nine letters to "spell" in the conventional alphabetic sense might therefore take only five letters to "retrieve."

Caldwell and his colleagues introduced a new set of terms to help synthesize their findings:

"minimum spelling," or "minimum length," defined as the smallest number of keystrokes an operator needed to enter before the machine produced an unambiguous match in the character database; and

"maximum spelling," also known as "maximum length" and "full spelling," being the total number of strokes in a given character.

In Chinese, Caldwell found, the difference between "spelling in full" and
"minimum spelling" was dramatic. For a character containing fifteen
strokes, for example, an operator could enter it with as few as five or six
strokes. In another case, a character with twenty "letters" could be unambiguously matched with only four.[46] Within retrieval, the "redundancy"
of writing became a vast resource to be exploited—a resource that doesn't
exist within the technolinguistic framework of conventional typing.

Caldwell pushed these observations further. Plotting the total versus
minimum spelling of more than two thousand Chinese characters, he
determined that the median "minimum spelling" of Chinese characters fell between five and six strokes; whereas the median total spelling
equaled ten strokes. At the far end of the spectrum moreover, no Chinese
character exhibited a minimum spelling of more than nineteen strokes,
despite the fact that many Chinese characters contain twenty or more
strokes to compose.

Thus in addition to developing the first Chinese computer in history,
Caldwell and his colleagues also unwittingly invented what we now
know as "autocompletion" (table 2.1).[47]

Happening upon this novel feature of retrieval-based composition,
Caldwell and his team decided to exploit such affordances further. It
became the very foundation of Sinotype, in fact. As noted in a 1958 progress report submitted to the Quartermaster Research and Development
Section, in Natick, Massachusetts, the decision had clearly been made
to exclude the usage of "maximum spellings" whenever a "minimum
spelling" was available. "In actual operation," wrote Robert G. Crockett,
one of the team members working at GARF, "the full-length spellings
would never occur because the keyboard automatically locks to prevent
further typing whenever the minimum required spelling is reached."[48]
In other words, it was decided that there was little reason to permit users
to overindulge in the fiction that they were "spelling" Chinese at all. As
soon as the machine produced an unambiguous match for "crocodile"
or "xylophone," to return to our earlier analogy, there was no reason to
permit the operator to continue depressing the keys marked "d-i-l-e" or
"h-o-n-e."

Table 2.1 Comparison of full and "minimum" spellings for sample characters

Mathews' Number	Full Spelling	Minimal spelling
851	BDG EGV BDP BDP BGE GE	BDG EGV B
680	BDG EGV EPB BGD PB	BDG EGV E
4899A*	BDG EGV GVB DGE BGV GBB K	BDG EGV G
5552	BDG EGV UE	BDG EGV U
3201	BDG EPB BME BRG S	BDG EPB BME
3328	BDG EPB BMG V	BDG EPB BMG
87	BDG ES	BDG ES
411A*	BDG EYU EYU EBG ENG E	BDG EY
5890	BDG SEE EPW	BDG S
4593	BDG VA	BDG VA

THE DEATH OF SINOTYPE

Summer 1959 came and went, yet Eisenhower remained silent on the subject of Chinese computing. No press conference was held, no announcement was made, and no Cold War victory lap was enjoyed. Despite the initial momentum of the Sinotype project, the enthusiasm of military planners steadily succumbed to doubt. Was it truly "the greatest step forward in printing of ideographic languages since the Chinese invented movable type in the eleventh century," as early reports glowed? Was it truly going to be "as important to the Chinese language as the inventing of the linotype machine was to Romance language printers"?[49] Like IBM and Mergenthaler executives before them, who worried themselves to a standstill over Kao's four-digit coding system, military representatives and presidential advisers simply couldn't stifle their doubts about Sinotype. Would it prove viable for the average Chinese user? Was it potentially field-changing, as the designers had come to believe? For, while the unveiling of a functioning Chinese computer would bestow incalculable prestige upon US engineering and the capitalist camp, a public failure would bring humiliation and a public relations coup for the Communist Bloc. There were just too many question marks.

May 18, 1959, was a turning point. Representatives of the Department of Defense, State Department, USIA, OCB, and the National Science Foundation gathered once again at the Pentagon to discuss Sinotype, but their tone had changed markedly.[50] The review committee began to hesitate over the "operational capability of the existing machine and as to the degree of 'breakthrough' which it represents at this time." There were concerns that a fully "improved machine" might require as many as two to three years, and as much as $300,000 in additional research support. "No government agency has yet been found," the ad hoc committee noted, "which is prepared to sponsor this continuation of development to obtain a more useful machine."[51]

The risk of premature announcement was simply too great, it was decided. On May 20, the committee issued a memorandum titled "Chinese Ideographic Composing Machine—Briefing Memo on Deferral of Board Consideration."[52] With regard to the Chinese Ideographic Composing Machine, a follow-up summary phrased it bluntly: "a report to the Board is now indefinitely deferred."[53]

In early 1960, the project was dealt its heaviest blow. Samuel Caldwell, whose health had begun to decline over the preceding year, passed away at just 56 years of age. "Not only has the MIT family lost a distinguished faculty member, wise and honored teacher, and outstanding engineer," his associate Louis Rosenblum wrote in a letter to the editor of *MIT Technology Review*, "but also the world of education has lost a great pioneer at the prime of his powers and career."[54] Without the project's foundational figure, and with no obvious successor waiting in the wings, Sinotype was mothballed.[55]

Even with its discontinuation, however, the Sinotype project continued to exert influence over the trajectory of Chinese computing from that point on—both materially and conceptually. Materially, the machine itself lived on, taken up by other organizations and re-christened along the way: first as the Sinowriter, then the Chicoder, then the Ideographic Encoder, and finally as Sinotype II and Sinotype III (machines we will explore in later chapters). The conceptual legacies of Sinotype have been even more consequential. Whether in terms of "minimum spelling," or Sinotype's reliance on a recursive, two-screen approach to human-computer interaction, Caldwell's device laid the groundwork for Chinese computer design well into the present day.

Most importantly, Sinotype served as evidence of hypography's most unexpected feature, without which we cannot understand how Huang Zhenyu could have possibly achieved such remarkable speeds in the twenty-first century: within hypography, it is possible to *subtract time by adding steps*. In chapter 1, we watched as Chung-Chin Kao's system of mediation added time and distance to the process of Chinese textual inscription, requiring typists like Lois Lew to move through multistep processes of translation between Chinese characters and four-digit ciphers. Given all these extra steps, there was simply no way that a typist—even one as talented as Lois Lew—could ever hope to compete with the kinds of speeds clocked within Anglophone typing. Based strictly on the example of the IBM electric typewriter, then, it might appear that hypographic "mediation" is, by definition, something that always increases the distance and time it takes to get from input to output.

But the design of Sinotype upended this assumption by throwing into question some of the deepest assumptions surrounding "mediation" as a concept. Derived from the Latin to describe something that comes "in between" other things, "media" would seem to describe objects such as window panes, fences, gates, and insulation—layers or barriers that, when added to any configuration, increases distance. If true, this would imply that each additional step in the realm of textual input should come with an increased penalty.

Caldwell's Sinotype painted a different picture, however. Even with all of its elaborate retrieval protocols, disambiguation keys, and pop-up menus, this hypermediated system of retrieval-writing revealed a vast, untapped resource unavailable to engineers in the domain of conventional typewriting: the technique of "minimum spelling" by which human operators could exploit the intrinsic "redundancy" of script and writing.[56]

As common as it might be to idealize *what-you-type-is-what-you-get* human-computer interaction, then, Sinotype proved that a hypographic system relying on many more layers of mediation could result in levels of speed and efficiency unseen within ostensibly "immediate" operation. Although Sinotype itself may have failed, Caldwell and his GARF colleagues had "broken the spell" and "split the screen."

3

FAREWELL, QWERTY: THE QUEST FOR A CHINESE KEYBOARD

"This will destroy China forever," a young Taiwanese cadet thought as he sat in rapt attention.[1] The renowned historian Arnold Toynbee was on stage, delivering the final lecture of his 1958 residency at Washington and Lee University. "A Changing World in Light of History" was the theme, ploughing the professor's favorite field of inquiry: the genesis, growth, death, and disintegration of human civilizations, immortalized in his magnum opus *A Study of History* (1934–1961).[2] Tonight's talk threw the spotlight on China.

China was Toynbee's outlier: ancient as Egypt, it was a civilization that had survived the ravages of time. What was the secret to China's continuity? Character-based Chinese script was a key to the mystery, Toynbee argued. Character-based script served as a unifying medium, the argument went, placing guard rails against centrifugal forces that might otherwise have ripped this grand and diverse civilization apart. In a land where Cantonese, Fujianese, and other so-called dialects are more like mutually unintelligible languages, China's character-based writing system—a system that bears no direct relationship to the spoken Chinese word—served as a kind of universal glue holding the peoples of China together in the face of profound internal diversity, and bringing them back together after periods of invasion, rebellion, and fragmentation.[3]

This millennial integrity was under threat, Toynbee argued, by a new specter haunting China: communism. Mao Zedong was calling for the abolition of Chinese characters, echoing earlier iconoclasts like the renowned author Lu Xun who had famously proclaimed, "If Chinese characters are not exterminated, there can be no doubt that China will perish."[4] Indeed, as Toynbee spoke, the new government in Beijing

was busily deploying Hanyu pinyin, a Latin alphabet–based Romanization system; although its primary purpose was to assist Chinese readers
in learning the "standard" pronunciation of Chinese characters, some
regarded it as an eventual replacement for Chinese characters altogether.[5]
If such a radical break with the past ever happened, Toynbee emphasized,
the glue holding Chinese civilization together could crack and crumble.

The cadet was Chan-hui Yeh, a student of electrical engineering at the
nearby Virginia Military Institute (VMI), and that evening with Arnold
Toynbee forever altered the trajectory of his life. It changed the trajectory
of Chinese computing as well, triggering a cascade of events that decades
later led to the formation of arguably the first successful Chinese IT company in history: Ideographix, Inc., founded by Yeh just over a decade
after Toynbee stepped offstage.

During the late 1960s and early 1970s, Chinese computing underwent
three sea changes, one of scale, two of design. No longer limited to small-
scale laboratories and solo inventors, the challenge of Chinese computing was taken up by engineers, linguists, and entrepreneurs across Asia,
the United States, and Europe—including Yeh's adoptive home of Silicon
Valley, still in its infancy.[6] This was the era when Chinese computing
exploded, with multiple sites engaged sometimes in direct competition
with one another.

The design of Chinese computers also changed dramatically. After
meeting Chung-Chin Kao and Samuel Caldwell in chapters 1 and 2, it's
tempting to assume that in this chapter we'll follow the next, incremen-
tal step in an ongoing perfection of QWERTY-based Chinese input—
perhaps a refinement of Caldwell's "minimum spelling" technique, or
an improvement upon Kao's cipher-based input system. To the contrary,
the experiments in Chinese computing we'll encounter during the 1960s
and 1970s marked almost complete departures from anything witnessed
thus far. Nowhere in this chapter will we find a QWERTY keyboard, for
example, or anything even resembling one in terms of shape or size.
Instead, one of the most successful and celebrated systems from this era—
the IPX system, designed by Yeh—featured an interface with *120 levels* of
"shift," packing nearly 20,000 Chinese characters and other symbols into
a space only slightly larger than a QWERTY interface. Other systems from
the era, built in mainland China, Hong Kong, and elsewhere, featured

keyboards with anywhere from 256 to 2,000 keys. Still others dispensed with keyboards altogether, employing a stylus and touch-sensitive tablet, or a grid of Chinese characters wrapped around a rotating cylindrical interface. It's as if every kind of interface imaginable was being explored *except* QWERTY-style keyboards.

The third sea change pertains to the sources of inspiration where so many inventors of the era—indeed, practically all—got their ideas for new designs. In a word, engineers abandoned QWERTY and turned their attention to the deeper archives of modern Chinese information technology—nowhere more so than the world of mechanical Chinese typewriting. As we will see, as bizarre and as unprecedented as 1960s and 1970s Chinese computing interfaces might seem to us at first glance, many of them were in fact spitting images of mechanical Chinese typewriters dating as far back as the 1910s. Whether knowingly or not, engineers from this era set out to computerize Chinese by, in effect, computerizing the Chinese typewriter.

Why this shift? Given the global ubiquity of QWERTY and QWERTY-style keyboards, why would engineers suddenly abandon this most universal of all interfaces? Why design bespoke devices unseen anywhere else on earth? Given the tantalizing potential of input, moreover—a potential that Caldwell's work on Sinotype was already starting to reveal—why would engineers attempt to bypass hypography and build entirely new species of interfaces?

To answer these questions returns us to the enduring prestige and allure of "immediate," Anglophone-style human-computer interaction, as well as input's marked status as an intrinsically "compensatory" approach to computing—something whose sole purpose was to make the best of a bad situation. Chung-Chin Kao believed wholeheartedly in his four-digit code, for example, as did Samuel Caldwell in his Sinotype system, but never once did they suggest that these systems might be superior to conventional modes of textual input. Were opportunities to arise that might render such hypographic "compensations" unnecessary, after all, why would anyone want to continue advancing an inherently "assistive" technology?

The late 1960s and early 1970s presented just such an opportunity. Inspired by the minicomputing revolution—advances in processing speed, memory capacity, graphics, and more—engineers around the world arrived at practically the same idea independently: to harness the toolkit of minicomputing in an attempt to bypass QWERTY-based hypographic input.[7] At long last, they would finally be able to fit Chinese and its tens of thousands of characters—a great many of them, at least—on a user-friendly, tabletop device. Ironically, then, in their effort to make Chinese human-computer interaction "the same" as it was in the Anglophone world, they decided to make Chinese interfaces as different from QWERTY as possible, severing ties with global IT trends and instead pursuing what might be termed "interface autarky."

CHAN-HUI YEH, IPX, AND THE 120-DIMENSIONAL HYPERSHIFT KEYBOARD

The Taiwan-born Chan-hui Yeh's tortuous path to the United States wound through the Korean War when he served as a translator for Allied forces. He moved to the United States with dreams of becoming a pilot—a dream he realized—before applying to VMI at the urging of his father. Aeronautics and flight were his first loves, but history was a close second (hence his attendance at Toynbee's lecture).[8]

Yeh graduated from the VMI in 1960 with a BS in electrical engineering, with a focus in military science. He went on to pursue graduate degrees at Cornell University, receiving his MS in nuclear engineering in 1963, and then his PhD in electrical engineering in 1965.[9] Yeh joined IBM, not for the purposes of developing Chinese text technologies like Chung-Chin Kao before him, but to draw upon his background in automatic control to help develop computational systems for paper mills, petrochemical refineries, steel mills, and sugar mills. He was stationed in IBM's still relatively new offices in San Jose to work on developing simulations of large-scale manufacturing plants.[10]

Toynbee's lecture stuck with him, though, as Yeh explained to me in our spring 2010 conversations. "I thought that, with all the knowledge I had about technology—mechanical, electrical, electronic—and a Chinese person's understanding of Chinese characters themselves, I must do something to preserve this culture." While working with IBM, he spent his spare time exploring the electronic processing of Chinese characters. He felt convinced that the digitization of Chinese must be possible; Chinese writing could be brought into the computational age. Doing so, he felt, would safeguard Chinese script against those like Mao, who seemed to equate Chinese modernization with the Romanization of Chinese script. The belief was so powerful that Yeh eventually quit his good-paying job at IBM to try and save Chinese through the power of computing.

Yeh started with the most complex parts of the Chinese lexicon, working back from there. He fixated on one character in particular: *ying* 鷹 ("eagle"), an elaborate graph that requires 24 brushstrokes to compose. If he could determine an appropriate data structure for such a complex character, he reasoned—one that struck the right balance between economy and aesthetics—he would be well on his way. Through careful analysis, he determined that a bitmap comprising 24 vertical dots and 20 horizontal dots would do the trick: 60 bytes of memory, excluding metadata (figure 3.1). By 1968, Yeh felt confident enough to take the next big step: to patent his project, nicknamed "Iron Eagle," and to form a corporation.[11]

Yeh's ultimate objective was not printing, however—this was just the first of a multiphase plan, reminiscent of Chung-Chin Kao's ambitions before him.[12] "The original intent of my system," he explained to me, "was to automate telegraph operation." "Once you can transmit, he continued, "you can communicate with computers. Then you can do data processing. And then everything becomes possible.[13] Like Kao, Yeh's goal was nothing short of the full-scale informationalization of Chinese.

Like Sinotype, moreover, this "Iron Eagle" project of Yeh's quickly garnered the interest of the military—the Taiwanese military, in this case—which approached Yeh circa 1968.[14] "When I applied for the patent, they immediately notified the military," Yeh explained, "because this was of significant consequence for military operation."[15] In 1972, with the promise of government funding, Yeh founded Ideographix, Inc., opening its doors in Sunnyvale, California at a time when Silicon Valley was only beginning to gain the reputation and stature it now enjoys.[16] Yeh would serve as president and chief engineer, and his brother Chan Jong ("CJ") as vice president.

The flagship product of the Ideographix corporation was the IPX, a computational typesetting and transmission system for Chinese built upon the complex orchestration of multiple "subsystems." The Scanner

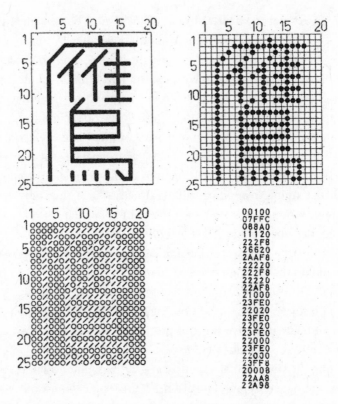

Figure 3.1 Bitmap diagram of the character *ying* (鷹 "eagle")

Subsystem, referred to in IPX literature as the "inexhaustible character generator," enabled an operator to digitize hand-drawn characters and convert them into bitmaps. The "composing subsystem" enabled an operator to compose and edit Chinese-language texts, featuring common word processing functions like deletion, insertion, backspacing, page breaking, and more. The phototypesetting subsystem outputted Chinese character texts onto photographic paper, which could then be converted into printing plates.

To develop the IPX system, Chan Yeh sought the aid of Systems Concepts, the San Francisco–based firm founded by Stewart Nelson and Mike Levitt and best known for its pioneering work on the "Mars" family of PDP-10-compatible computers. Along with Peter Samson, who joined the company in 1970, Nelson worked with Ideographix to integrate Yeh's keyboard with a Data General Nova-series minicomputer and an ABDick printer. In spring 1973, Yeh, Nelson, Samson, and a small coterie of technicians traveled to Taipei by way of Tokyo, to demonstrate the IPX prototype to top Taiwanese military and political authorities, including Taiwanese premier Chiang Ching-kuo. The demonstration took place in a Taiwanese military office, Samson recalled, where a military officer had set up folding chairs.[17]

The marvel of the IPX system was undoubtedly the keyboard subsystem, however, enabling operators to enter a theoretical maximum of 19,200 Chinese characters despite its modest, desktop size: 23 inches wide, 14.5 inches deep, and 4.5 inches tall.[18] To achieve this remarkable feat, Yeh and his colleagues decided to treat this keyboard not as merely an electronic peripheral, but as a full-fledged computer unto itself: a microprocessor-controlled "intelligent terminal" completely unlike conventional QWERTY-style devices.[19]

Seated in front of the IPX interface, the operator looked down on 160 keys arranged in a 16-by-10 grid. Each key contained, not a single Chinese character, but a cluster of 15 characters arranged in a miniature 3-by-5 array. With 160 buttons in all, and 15 characters on each, this brings us to a running total of 2,400 Chinese characters.

Chinese characters were not printed directly onto the surface of the keys themselves, the way that letters and numbered are emblazoned on the keys of standard QWERTY and QWERTY-style devices. Instead, these 2,400 Chinese characters were printed on pieces of laminated paper, bound together as a multipage, spiral-bound booklet that the operator laid down flat atop the surface of the IPX interface. When the spiral-bound booklet was removed, the 160 keys themselves were blank (figure 3.2). The keys on the IPX weren't buttons, as on QWERTY devices, but pressure-sensitive pads. With 3 ounces of pressure, and a displacement distance of just 0.007 inches, an operator could push down on the spiral-bound booklet, and thereby depress whichever key membrane was directly underneath.

To reach characters 2,401 through 19,200, the operator simply turned this spiral-bound booklet to whichever page contained their desired character. With booklets containing anywhere from four to eight pages in all—each page containing 2,400 characters—the total number of potential symbols reached just shy of 20,000.[20]

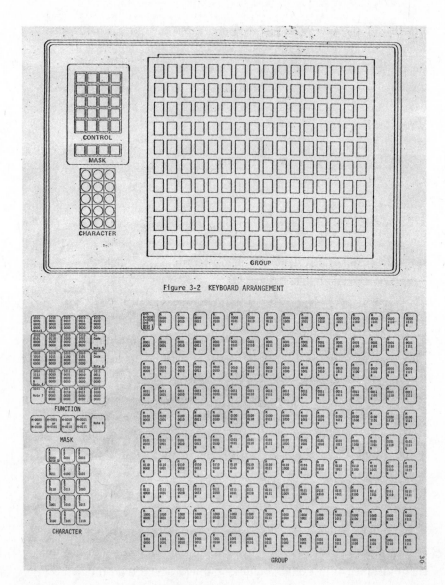

Figure 3.2   Diagram of the IPX keyboard

Figure 3.3 Close-up of IPX keypad containing the character *zhong* (中 "central")

How did the operator type? Consider the character 中 (meaning "central," as well as the first of the characters 中國 (meaning "China"). As shown in figure 3.3, this character was located on the top row of the key's 3-by-5 grid, in the middle column. To enter this character, the operator first pushed one of fifteen number keys found in a bank on the lower-left flank of the keyboard—the number "2," in this case, indicating to the IPX that the operator wanted to enter whichever character was located in "position 2" of the key they were about to depress. Then the operator depressed the key itself, completing the entry process.[21]

To reach any of the other 16,800 characters—characters located on other pages of the spiral-bound keyboard booklet, that is—the operator followed the same two-key process, but this time preceded it by one or two additional keystrokes to alert the IPX that the keyboard booklet had changed pages: first by turning the booklet to the page containing the desired character, and then by depressing an additional number key—key 1 through 4, located in another bank of keys; this instructed IPX as to which page of the booklet the operator was using, and thus which set of 2,400 Chinese characters it should load into memory and prepare to produce.

Figure 3.4 IPX promotional film, stills

However unprecedented, even outlandish, the IPX interface might seem when viewed from the perspective of QWERTY-based computer input, a host of "family resemblances" become apparent as soon as we examine it within the broader context of modern Chinese information technology. In particular, Yeh drew inspiration from the archives of modern Chinese technolinguistics, nowhere more than for the shape and design of mechanical Chinese typewriters. The layout of Yeh's keyboard was one such inheritance. Were we to focus on just the number and distribution of pressure-sensitive keys (that is, the 16-by-10 grid of buttons), the IPX keyboard appears to be a dramatic departure from the tray bed of the mechanical Chinese typewriter and its array of 2,450 character slugs. But when we remember that each of the IPX keys is outfitted with 15 characters, each in a 3-by-5 grid, we are confronted with a

Figure 3.5 Comparison of tray bed of mechanical Chinese typewriter (Double Pigeon brand) and keyboard of the IPX by Ideographix

layout practically identical to mechanical Chinese typewriters from the mid- through the late twentieth century. On the Double Pigeon Chinese typewriter, the typewriter tray bed was 35 characters high and 70 characters wide, for a total of 2,450 characters. On the IPX keyboard, the grid of characters was 50 characters high by 48 characters wide, for a total of 2,400 characters—albeit arranged in a square rather than rectangular format. Just 50 characters, in other words, separated the IPX keyboard from one of the leading mechanical Chinese typewriters of the era (figure 3.5).

The resemblance between Yeh's IPX keyboard and the tray beds of mechanical Chinese typewriters was more than mere coincidence. Chinese typewriters formed an active part of Yeh's early thinking and development process. In a 1976 patent application, for example, Yeh proposed a revised mechanical Chinese typewriter in which each of the machine's metal slugs was outfitted with a machine-readable binary code that would accompany the normal, human-readable Chinese character (figure 3.6). As Yeh imagined it, 14 bits of information could be encoded in a special 7-line code.

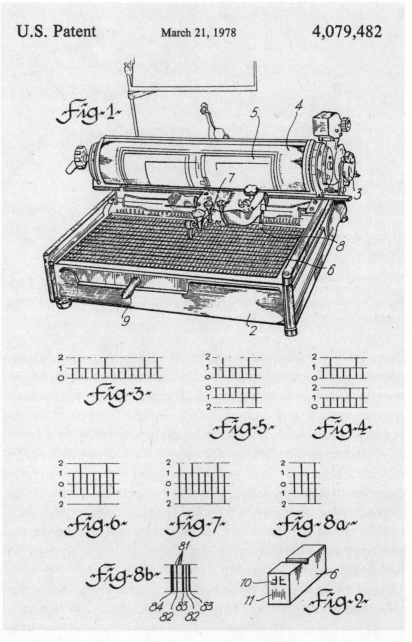

Figure 3.6 Patent featuring machine-readable codes printed with mechanical Chinese typewriter

While Yeh never pursued the idea of a machine-readable Chinese typewriter—there is no mention of it in the relatively abundant materials on the Ideographix corporation, at least—the patent itself illustrates the degree to which Yeh was drawing upon the deep repository of precomputing Chinese information technology, even as he helped pioneer never-before-seen advances in the domain of Chinese minicomputing.[22]

For the first seven years of its existence, "Iron Eagle"—now renamed IPX—was limited to use by the Taiwanese military. As years passed, and as exclusivity was relaxed, Yeh began to seek out customers in both the private and public sectors.[23] Yeh's first major nonmilitary clients included Taiwan's Telecommunication Administration and the National Tax Administration of Taipei. For the former, Ideographix helped process and transmit millions of phone bills, and dramatically reduce the time needed to produce telephone directories. For the latter, IPX enabled the production of tax return documents at an unprecedented speed and scale.[24]

Success followed success, and the reputations of both Yeh and Ideographix began to balloon. "It was news just like Apple PC or the IBM computer in this country," Yeh reminisced to me. "A sensation," he exclaimed. "It was major news for a year."[25] Soon, media outlets started to approach Yeh, including one of Taiwan's largest daily newspapers, *United Daily News* (*Lianhebao*). At this newspaper, an immense labor force, perhaps as many as four hundred typesetters, spent all evening setting type for the daily edition. The introduction of the IPX system reduced this labor force to a mere fifty as well as the amount of time spent on the job.[26]

Exploiting this added speed, *United Daily News* was also able to wait longer to go to press, which meant closing the day's news cycle at two in the morning, as compared to their competitors, who closed their cycles in the late evening. This allowed the paper to "scoop" its competitors and include in its morning edition the breaking news that came in too late for other papers to cover. The IPX system also enabled the outlet to expand upon its print production, both in terms of additional sections and even entirely new newspapers.[27] Indeed, so momentous was the impact of this new system that *United Daily News* dedicated the entire front page of its September 16, 1982, issue—the thirty-first anniversary of the paper itself—to a celebration of IPX and its implementation at the press.[28] In effect, that day's news cycle was dominated by self-referential "metacoverage" focused on *how* that day's paper was printed.

ONE KEY, MANY USES: THE "MEDIUM-SIZED" KEYBOARD IN MAINLAND CHINA AND BEYOND

By the mid-1970s, the People's Republic of China was far more advanced in the arena of mainframe computing than most outsiders realized. The country's achievements in mathematical and scientific computing shocked a delegation of American computer scientists, in fact, who visited the PRC in 1972 just months after the famed tour by US president Richard Nixon. A veritable blue-ribbon committee arrived in July, comprising the Nobel laureate Herbert Simon, the Turing Award–winner Alan Perlis, and a half-dozen other luminaries. Over the course of three weeks, the delegation visited China's main centers of computer science at the time—including the Shanghai Computing Research Institute and the Institute of Computing Technology in Beijing. Upon learning what their counterparts had been up to during the many years of Sino-American estrangement, the delegation was stunned.[29]

The year 1956 marked the formal beginning of mainland China's computing program, the delegation learned, with the launch of the Twelve-Year Science Plan and the formation of the Beijing Institute of Computing Technology (established under the aegis of the Chinese Academy of Sciences). By 1958, engineers completed the country's first vacuum-tube computer, featuring 32-bit architecture, 4K core memory, and a speed of 180 operations per second. The same year, engineers based at the Department of Automatic Control at Tsinghua University began production of nonlinear analog electronic computers, basing their work on two years of close examination of more advanced Soviet systems.[30]

The USSR served as China's technological lifeline throughout these early years. This lifeline was severed during the Sino-Soviet Split of 1960, however, a dramatic collapse in Sino-Soviet bilateral relations that culminated in Moscow's rapid withdraw of its advisers and technical experts, and later even talk of potential military conflict.[31] Research endured, however, as seen in Tsinghua's continued work on the "DJS" mainframe—geared primarily toward the needs of the Chinese Statistics

Bureau and large-scale statistical analysis more generally.[32] Two years later, China boasted its first computer based on vacuum tube circuitry, also an extrapolation of Soviet models.[33] The post-Split period was a time of continued institution-building, as well, notably with the formation of the Beijing Institute of Electronics in 1963.[34] In 1964, the Chinese Academy of Sciences debuted the country's first self-developed large-scale digital computer—the 119—with computing speeds of 50,000 operations per second. This machine played a core part in China's first successful test of a nuclear weapon that same year.

Even during the Cultural Revolution (1966–1976), a period marked by the rupture of higher education and the political persecution of experts, computer engineering was deemed too militarily essential to be disturbed, and thus somewhat shielded from the radical politics of the era. The year the Cultural Revolution was launched, in fact, China made the transition from vacuum tubes to fully transistorized computers, an arena pioneered by Tsinghua researchers.[35] By the time of the US delegation's arrival in summer 1972, at the midpoint of the Cultural Revolution, China had already developed a computing industry capable of producing third-generation computers: a smaller, faster breed of machines based on integrated circuits. More surprising still, the PRC was producing these integrated circuits domestically as part of a semiconductor program dating back to the mid- and late 1960s.[36] In the midst of the Cultural Revolution, in fact, a Shanghai factory once dedicated to producing window handles was repurposed for the production of an integrated circuit digital computer. Still referred to as the "Window Handle Factory," the factory continued to employ many of the same women who had worked there before (figure 3.7).[37]

There was one key arena of computing that the delegation did not bear witness to: the computational processing of Chinese characters. Whether because of the era's dogged focus on mathematical and scientific computing, or because of lingering uncertainty as to the fate of Chinese-character script in an era of continued interest in wholesale Romanization, it was not until 1974 that mainland Chinese engineers began to dive seriously into the problem of Chinese-character information processing.[38] In October of that year, the PRC formally launched the "748 Program" focused on the question of computational Chinese information processing—with

Figure 3.7 Photographs from 1972 delegation of American Computer Scientists. Courtesy of Severo Ornstein

a particular focus on the research and design of Chinese interfaces. Spearheaded by the National Defense Science and Technology and the Chinese Academy of Sciences, the program helped catalyze a proliferation of experimental input devices and surfaces on the mainland.[39] Like their counterparts in the offices of Ideographix, moreover, mainland Chinese engineers focused on the development of custom-designed non-QWERTY interfaces. Although well aware of QWERTY, of course, they considered it to be the least promising of all potential avenues.

Peking University was one of the earliest and most important centers of Chinese interface research at the time. In 1975, a newly formed Chinese Character Information Processing Technology Research Office set out upon the goal of creating a "Chinese Character Information Processing and Input System" and a "Chinese Character Keyboard."[40] The group researched more than ten proposals for potential Chinese keyboard designs, some containing upward of 40 keys, others many hundreds.[41] The team evaluated three general directions for keyboard design: a "large keyboard" approach, which strove to provide one key for every commonly used character; a "small keyboard" approach, which referred to the QWERTY-style keyboard; and a "medium-sized keyboard" approach, which would attempt to tread a path between these two poles.

The team leveled two major criticisms against the QWERTY-style "small keyboard" approach. First, there were just too few keys. Chinese input sequences that relied on the "small keyboard" approach tended to be beleaguered by duplicate codes, they emphasized, with too many Chinese characters being assigned identical input sequences. QWERTY keyboards did a poor job of using keys to their full potential, the team also argued. For the most part, each key on a QWERTY keyboard was assigned only two symbols, one of which required the operator to depress and hold the "shift" key to access. A better approach, they argued, was a technique of "one key, many uses" (*yijian duoyong*)—that is, assigning each key a significantly larger number of symbols to make the most use of interface real estate (figure 3.8).[42]

The team also examined the "large-keyboard" approach to Chinese text input, in which two thousand or more commonly used Chinese characters were assigned to a large, tabletop-sized interface. At another center of early Chinese interface research, Nanjing's 734 Factory, engineers

Figure 3.8 Example of "large-keyboard" approach to Chinese keyboard interface

were working on two large-format prototypes: the RPH-2 model Pressure Sensitive Chinese Character Keyboard, and the RPH-3 model Electrostatic Chinese Keyboard.[43] The Wuhan External Device Research Institute and the Shenyang Computing Office were also collaborating on a large-format keyboard employing conductive rubber.[44] At the Yanshan Computer Applications Research Center, meanwhile, work was underway on the ZD-2000 Large-Format Keyboard.[45] Other centers of research on large-keyboard designs included the Xinhua News Agency.

Engineers at Peking University regarded the "large-keyboard" approach as excessive and unwieldy, however. The goal, they agreed, should be to exploit each key to its maximum potential, but also to keep the number of keys to a minimum.[46] After years of work, the team in Beijing ultimately settled upon a keyboard with 256 keys, which came to be called the "medium-sized" keyboard.[47] Of those 256 keys, 29 would be dedicated to various functions, such as carriage return, spacing, and so forth,

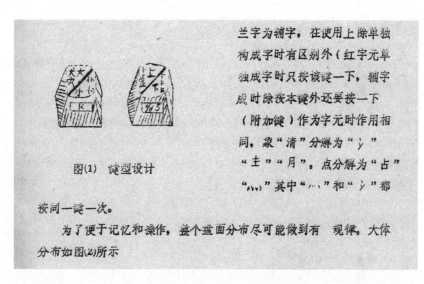

Figure 3.9 Sample button on the Peking University "medium-sized" keyboard

with the remaining 227 used to input text. The keyboard offered two "cases" ("upper" and "lower"), but not for the purposes of capitalization (again, there is no such thing as an upper- or lowercase Chinese character). Instead, the upper case was reserved for foreign scripts, such as Greek, Latin, and Cyrillic, along with punctuation marks and mathematical symbols; while the lower case was reserved for Chinese characters. The depression of each keystroke generated an 8-bit code, stored on punctured paper tape (hence the choice of 256, or 2^8, keys). These 8-bit codes were then translated into a 14-bit internal code, which the computer used to carry out the retrieval process.[48]

In their assignment of multiple characters to individual keys, the team's design was reminiscent of the IPX machine by Yeh and Ideographix. But there was a twist. Instead of assigning only full-bodied, stand-alone Chinese characters to each key, the team instead assigned a mixture of both Chinese characters and character components—that is, "radicals." Specifically, each key was outfitted with up to four symbols, divided among three varieties (figure 3.9):

full-sized Chinese characters (limited to no more than 2 per key)
partial Chinese character structures or radicals (no more than 3 per key)
the above-mentioned uppercase symbol (limited to 1 per key)[49]

In all, the keyboard contained 423 full-body Chinese characters, and 264 character components (sometimes referred to as "radicals").[50] When arranging these 264 character components on the keyboard, the Peking University team hit upon an elegant and ingenious way to help operators remember the location of each: they treated the keyboard as if it were a Chinese character itself, locating each of the 264 character components in whichever region of the keyboard that it tended to appear in actual Chinese characters. The grass-radical (艸 or ⁺⁺) for instance, was assigned to one of the keys along the top of the keyboard, because it is here where the grass-radical always appears in the structure of Chinese characters. The water-radical (氵), meanwhile, always appears on the left side of characters, and so was assigned to one of the left-most keys, for the same reason (figure 3.10).[51]

When considering the remarkable design of the Peking University keyboard, the same question emerges here as in our examination of Chan-hui Yeh's IPX system: Where did the idea come from for such an unconventional approach to human-computer interaction? As with IPX, the answer brings us back to the archives of modern Chinese information technology—in particular, an approach to Chinese printing and typewriting known as "divisible type," which harkens back to a series of experimental movable type Chinese fonts from the mid- and late nineteenth century, and later to a series of experimental mechanical Chinese typewriters from the 1910s through the 1940s.

"Divisible type" Chinese printing was developed in Europe and the United States beginning in the 1830s. Developed primarily by Samuel Dyer (1804–1843), Guillaume Pauthier (1801–1873), and Marcellin Legrand (fl. 1820–1860), the technique was predicated on decomposing Chinese characters into reoccurring, modular shapes (variations of the so-called Kangxi radicals). To print the single Chinese character 海 (hai, "sea"), for example, one in fact set two pieces of metal: one containing the left-side component 氵 (the "water radical"), and a second containing the right-side component 每 (mei, "every"). Meanwhile, to print 池 (chi, "pool"), the operator reused the left-side "water radical," but replaced 每 with the metal slug containing 也 (ye, a copular verb in Classical Chinese, or "also" in modern Chinese). Once forged into a moveable-type font, these modular pieces enabled the operator to print many tens

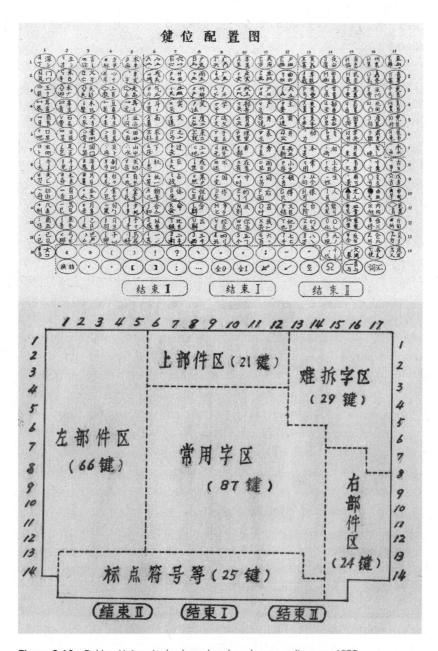

Figure 3.10 Peking University keyboard and explanatory diagram, 1975

of thousands of Chinese characters using around two thousand sorts (figure 3.11).[52]

"Divisible type" reappeared in the 1910s in the domain of Chinese typewriting. One of the earliest experimental Chinese typewriters in history—built by a young overseas Chinese student Qi Xuan during his time at New York University—was premised upon the same idea of subdividing Chinese characters into their component parts.[53] On Qi Xuan's machine, one set of highly common Chinese characters were, just like the Peking University keyboard, included in full-body form, the judgment again being that the frequent occurrence of such characters merited their inclusion on the machine "as is." Qi Xuan's second set of graphs, however, were partial *components* of Chinese characters, once again anticipating the strategy employed by Peking University. Using this set of character components, an operator would be able to "shape-spell" Chinese characters by imprinting one piece of the character on the page at a time. Typing the characters 芋 (*yu*, "taro") or 营 (Ying, a historic placename), for example, required the "grass radical" (⁺⁺) first, followed by either 于 (*yu*, "in, on, to") or 吕 (Lü, a historic placename), respectively. Meanwhile, typing the characters 宇 (*yu*, meaning "space") or 宫 (*gong*, "palace"), only required replacing the "grass radical" with the "roof radical" (⌐). In other words, by breaking Chinese characters into components, and then assigning these components to different type slugs, the operator could mix and match them to create many thousands of characters. More specifically, with a set of 4,200 full-bodied characters and 1,327 modular components, Qi Xuan's typewriter was capable of producing tens of thousands of Chinese characters (figure 3.11).

This strategy resurfaced yet again in the 1930s and 1940s with Lin Yutang and his MingKwai ("Fast and Clear") experimental Chinese typewriter—the first Chinese typewriter to feature a keyboard. Once again, the machine included two sets of Chinese glyphs: one comprising high-frequency Chinese characters, in full-body form; and another comprising character components with which the operator could "build up" their desired (lower-frequency) characters on the page (figure 3.11).

Whether knowingly or not, the engineers who built the Peking University machine drew directly on this legacy of Chinese technolinguistics. The major adjustment they made, however, was to employ Chinese

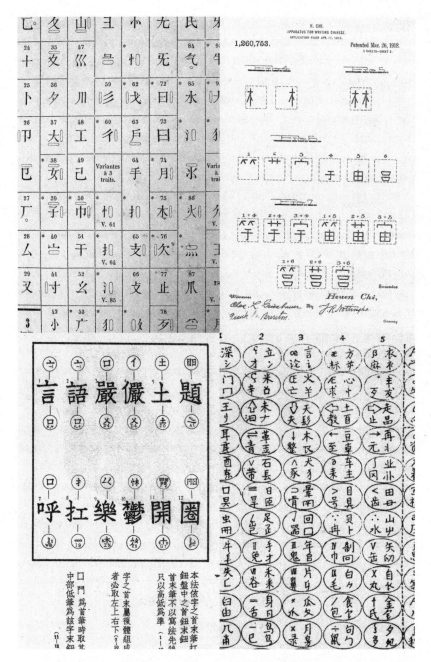

Figure 3.11 Comparison of "divisible type" Chinese printing, the Chinese typewriter prototype by Qi Xuan, the MingKwai Chinese typewriter by Lin Yutang, and the Peking University "medium-sized" keyboard

character components as a means of governing a computational system of information *retrieval*, rather than a process of mechanical *inscription*. In Marcellin Legrand's "divisible type" font from the late nineteenth century, Qi Xuan's experimental typewriter from the 1910s, and Lin's Ming-Kwai typewriter from the 1940s, operators built up Chinese characters on the page, piece-by-piece. But in the case of "medium-sized" keyboards from the 1970s, they used Chinese character components to retrieve their desired character from memory. Once a character was retrieved and displayed on screen, however, it was always "full-body."[54]

In its final design, the Peking University medium-sized keyboard was capable of inputting a total of 7,282 Chinese characters, which in the team's estimation would account for more than 90 percent of all characters encountered on an average day. Within this character set, the 423 most common characters could be produced via 1 keystroke; 2,930 characters could be produced using 2 keystrokes; and a further 3,106 characters could be produced using 3 keystrokes. Characters requiring 4- and 5-key input sequences numbered only 823.[55]

In terms of speed, the Peking University medium-sized keyboard boasted an overall average of 2.6 keystrokes per Chinese character, making it competitive with Anglophone text input where the average length of English words has been estimated by some at approximately 4.5 letters.[56]

The Peking University keyboard was just one of many medium-sized designs of the era. IBM created its own 256-key keyboard for Chinese and Japanese. In a design reminiscent of the IPX system, this 1970s-era keyboard included a 12-digit keypad with which the operator could "shift" between the 12 full-body Chinese characters outfitted on each key (for a total number of 3,072 characters in all).[57] In 1980, the Chinese University of Hong Kong professor Loh Shiu-chang developed what he called "Loh's keyboard" (*Le shi jianpan* 樂氏鍵盤), also featuring 256 keys (figure 3.12).[58]

IDEO-MATIC 66

On a winter day in 1976, a young boy in Cambridge, England, searched high and low for his beloved Meccano set. A predecessor of the American Erector set, the popular British toy offered aspiring engineers hours of

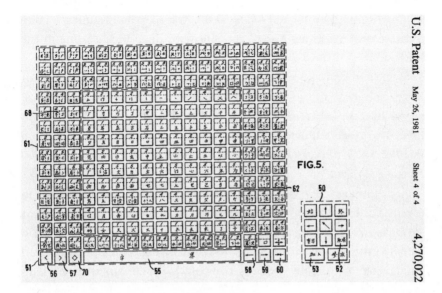

Figure 3.12 Loh's keyboard, by Loh Shiu-chang

inexhaustible modular possibility. Andrew had played with the gears, axles, and metal plates recently, but today they were nowhere to be found.[59]

Wandering into the kitchen, he caught the thief red-handed: his father, the Cambridge University researcher and former Royal Air Force wing commander, Robert Sloss. For three straight days and nights, Sloss commandeered his son's toy, along with the family's dining room table, engrossed in the creation of some kind of peculiar gadget, cylindrical and rotating. It riveted the young boy's attention—and then the attention of the *Telegraph-Herald* who dispatched journalists to see it firsthand. Ultimately, it attracted the attention and financial backing of one of the UK's most important telecommunications company, Cable and Wireless.[60]

Robert Sloss was building a Chinese computer.[61]

The elder Sloss was born in 1927 in the Scottish town of Dumbarton, a prime target for German bombers in World War II owing to its aircraft factory, shipyard, and gasworks.[62] Upon joining the Navy, he was subjected to a series of intelligence tests that revealed a proclivity for foreign languages. Stationed in Hong Kong in 1946 and 1947, he filled his off-hours writing film reviews for the *China Mail* newspaper, and running a radio program on ZBW Hong Kong, "the voice of British Broadcasting in the Far East."[63]

Sloss went on to join the civil service as a teacher and later, in the Air Force, became a noncommissioned officer.[64] Owing to his pedagogical experience, his knack for language, and his background in Asia, he was invited to teach Chinese at Cambridge and appointed to a lectureship in 1972. He went on to become head of Cambridge University's Chinese Language Project.[65]

At Cambridge, Sloss met a wunderkind named Peter Nancarrow (as an article in the *New York Times* characterized him). Nancarrow, twelve years Sloss's junior, originally trained as a physicist but later found work as a patent agent. The "bearded, 38-year-old" then taught himself Norwegian and Russian as a "hobby" before setting out to tackle computational Chinese translation.[66]

Chinese-English mechanical translation became the core of Nancarrow and Sloss's collaboration within the Cambridge Language Project.[67] The pair focused their energies on the automated production of Chinese character indices and concordances, and the computational translation of Chinese scientific texts into English—efforts that caught the attention of their Cambridge colleagues. "I can really think of nothing more important than the work to which Nancarrow and Sloss have set themselves for years past," emphasized the towering Sinologist, historian of science, and biochemist Joseph Needham in a private letter to the renowned Sinologist Michael Loewe.

The conjunction of computing, apparatus, numerisation, scanning techniques and television display with the character script of ideographic and syllabic languages is something which has been on the horizon for a long time past; and in the future it will be a vital progress in the realm of human intercommunication with all that that implies for mutual understanding and world peace.[68]

Success hinged on the creation of a digitized Chinese dictionary: a means of converting and digitally storing in a machine-readable format the tens of thousands of paper-bound Chinese vocabulary index cards that the duo had created.[69] The major choke point, however, was character input: namely, how to get all of these hand-written Chinese characters, definitions, and syntax data into a computer, both accurately and efficiently.

Over the following two years, therefore, Sloss and Nancarrow dedicated their energy to designing a custom-built Chinese computer interface.[70] It was this project that turned Sloss to his brief stint as a thief, tinkering through the prototyping phase with the help of his son's stolen toy.[71] By 1976, Sloss's tinkering bore fruit: a working prototype the duo called the "Binary Signal Generator for Encoding Chinese Characters into Machine-compatible form"—known alternately as the "Ideo-Matic Encoder" and the "Ideo-Matic 66" (named after the machine's 66-by-66 grid of characters).[72]

Each cell in the machine's grid was assigned a binary code corresponding to the X-column and the Y-row value.[73] In terms of total space, each cell was 7mm squared, with 3,500 of the 4,356 cells dedicated to Chinese characters. The rest were assigned to Japanese syllables or left blank.[74] The distinguishing feature of Sloss and Nancarrow's interface was not the grid, however, but its cylindrical design. Rather than arranging their 4,356 cells across a rectangular interface like those seen in Silicon Valley, Beijing, Hong Kong, and elsewhere, the pair decided to wrap the grid around a rotating, tubular structure. By using one hand to rotate the cylindrical grid, and another to move a cursor left and right, the machine then determined which of the 4,356 cells was being identified. Finally, the depression of a button produced a binary signal that corresponded to the selected Chinese character or other symbol (figure 3.13).

As novel as it might seem, the core design elements of the Ideo-Matic 66 once again marked a direct continuity with mechanical Chinese typewriting from decades prior. Amid all of the machines examined in this chapter, in fact, Sloss and Nancarrow's device harkened back to the deepest recesses of Chinese typewriting history, knowingly or not. In the mid-1910s, a prototype designed by Zhou Houkun—an overseas Chinese student at MIT—was premised upon precisely the same kind of character cylinder, wrapping a set of roughly 3,000 Chinese characters around a rotating cylinder—just like Sloss and Nancarrow. The Cambridge machine's resemblance to early Japanese kanji typewriters from the 1930s and 1940s was even more striking. On these machines, as on the Ideo-Matic, the operator directly manipulated a rotating cylinder of characters, initiating the typing mechanism once the desired kanji graph was in position (figure 3.14).

Figure 3.13 Ideo-Matic Encoder

The Ideo-Matic Encoder was completed and delivered to Cable and Wireless in the closing years of the 1970s.[75] Weighing in at 7 kilograms, and measuring some 26.8 inches wide, 22.4 inches deep, and 8.9 inches tall, the machine quickly garnered industry and media attention.[76] Cable and Wireless, the British telecommunications giant, purchased rights to the machine in hopes of mass-manufacturing it for the East Asian market. Both *The Times* and the *New York Times* covered the story as well, the former declaring that the "Chinese puzzle has now been cracked by the improbable team of two Cambridge lexicographers using a son's Meccano set, sealing wax, and string." "It was absurdly simple," the article continued, "but its consequences are profound. They now expect to publish their dictionary next year instead of next century."[77]

THE RETURN OF INPUT

Few letters housed in the Needham Research Institute in Cambridge University strike as furious a tone as the one ensconced in a manila folder

Figure 3.14 Comparison of the Zhou Houkun's first Chinese typewriter prototype, the Toshiba Japanese typewriter, and the Ideo-Matic 66 by Sloss and Nancarrow. Photo by SFO Museum

labeled "中Computerisation戰"—Joseph Needham's idiosyncratic shorthand we can translate as "Chinese Computerisation War."

"I write to dissociate myself completely from the account given in your article on Dr. Yeh Ch'en-Hui, Mr. Robert Sloss, Mr. Nancarrow and myself, with regard to the computerisation of Chinese."[78] "That I 'judged the two computers' is a mere fiction," he wrote to the editor of the *Kung Sheung Evening News* of Hong Kong. "If this rebuttal is not printed, and a copy duly received by us, advice will be taken on whether the article now complained of is actionable."

What had so infuriated the septuagenarian giant of Sinology that it brought him in 1978 to the brink of legal action against a newspaper based in Hong Kong?[79] What was the "Chinese Computerisation War," and who—if anyone—won?

The episode began back with Chan-hui Yeh and an article in January 25, 1978, issue of the *New York Times* titled "Two Britons Devise a Computer That Can Communicate in Chinese."[80] The article recounted the story of how Sloss and Nancarrow designed a "Chinese computer" capable of processing a total of 4,356 Chinese characters. Calling it "one of the neatest tricks in gadgetry of the decade," the *New York Times* reporter declared that the invention would make it "possible for a Chinese-speaking computer operator to communicate directly with his computer in Chinese."

Something about the article enflamed Yeh's sensibilities. Perhaps it was the cavalier tone by which Sloss and Nancarrow spoke of their invention, taking "a quiet delight in the sheer amateurism of their achievement," as the reporter phrased it. "Our view," Sloss and Nancarrow told him, "was that somebody had solved this problem, and we were simply too provincial to know about it. It was ridiculous to assume otherwise, but nobody had worked it out." Perhaps it was the unbelievable speed by which the team reportedly had solved the puzzle of Chinese computing: a matter of days, according to the article, as compared to many years for Yeh. Or perhaps it was the unflattering terms that Sloss used to describe a Chinese trade delegation who traveled to Cambridge to see the new invention: "When the Chinese came to visit, they reacted as if they had seen an electronic talking dog."

One thing was certain: Yeh needed to meet Sloss and Nancarrow, and to see the Ideo-Matic 66 for himself. He needed to visit Cambridge University.

From here the story grows increasingly cloudy. "Chinese Computers: Dr. Yeh Chen-hui Won the Day in London," an article in the *Kung Sheung Evening News*, reads like a hero's quest, tracing Yeh's passage to the UK and the lofty objective that motivated his trip: "to compare and decide which one of two Chinese computers is superior, the one invented by two British scientists or the one invented by him."

So dazzling was Yeh's machine, the *Kung Sheung* reporter continued, that "Sloss and Nancarrow immediately conveyed their apologies and said it was all a misunderstanding. They admitted that they were not experts in computers, nor did they know beforehand that Dr. Yeh had already had his computers patented in [the] UK." The reporter emphasized that "the judge was none other than the great British academician—Joseph Needham, author of *Science and Civilization in China*. The result: the Chinese scientist knocked the British out and won the day." "Dr. Joseph Needham went so far as to say that Yeh's invention might perhaps change the course of Chinese history."[81]

More than thirty years after this article was published, Yeh recounted the moment in largely the same light. As Yeh explained to me during our conversations, Needham went out of his way to be a gracious host. He "even climbed up the stairs—at that time he was 75 years old—and brought his tea pot and boiled tea for me."[82] Needham and Yeh spoke at great length, Yeh recalled, the late morning tea-time spilling into the afternoon—and then into the early evening. Learning of Yeh's system, Needham called in Sloss and Nancarrow to learn about it. "Obviously, you're way behind," Needham reportedly said to the pair, impressed by Yeh's more advanced work.[83]

Needham's recollection of the event—housed in that manila folder in Cambridge—was utterly different. The tea-time discussion did indeed happen, Needham acknowledged, but it was not scheduled especially for Yeh—Yeh's inclusion was purely a matter of convenience. "To save time," Needham wrote to the paper in Hong Kong, "he visited this Library for tea, when all our members usually meet together."[84] Needham was not even present when Yeh, Sloss, and Nancarrow met, he wrote to the newspaper. Mounting a point-by-point refutation of the article, which Needham dismissed as a "mass of terminological inexactitudes," he inveighed against the account on letterhead from the Science and Civilisation in China Project. "Sloss and Nancarrow," Needham underscored, "are not 'professors' at Cambridge University, nor 'Head of the Chinese Department,' but members of the teaching and research staff; and I am not a 'member of the Royal Academy.'"

What exactly happened, then, during that August meeting in Cambridge between Yeh, Needham, Sloss, and Nancarrow? Did Joseph Needham stand in awe of the IPX, pronouncing it superior to the Ideo-Matic 66? Did the Taiwanese engineer stand victorious, his British opponents slinking away in cowed, shameful defeat? Or was the entire meeting an act of courtesy, conducted in haste? Here, the archival record is silent.

The one unmistakable truth about that day, perhaps, was one that no party involved could have foreseen: that the IPX, the Ideo-Matic 66, and indeed all of the other custom-built devices examined in this chapter would soon meet with exactly the same fate—oblivion. Peking University's medium-sized keyboards will disappear quite soon from our story, as we will see in chapter 5, and so will all of the "large-scale" devices. What is more, not a single one of these devices has surfaced in any major museum collection (although one hopes that they still reside somewhere).[85] As for Sloss and Nancarrow, only a few Ideo-Matic Encoders are known to survive: one in a museum found quite literally at the "ends of the earth" (Land's End, in the southernmost, wind-beaten tip of the British mainland); another on the uppermost shelf of a high-density, closed-stacked storage inside one of the many libraries at Cambridge University.

Chan-hui Yeh's IPX met with the same fate. "We ran ourselves out of business," Yeh told me.[86] Ideographix was an instant success from practically the moment it became a public-facing company, he continued, that it was effectively unnecessary to invest in marketing, at least at the outset. Almost overnight, the Taiwanese market was saturated, Ideographix having offered such profound labor-saving technology to *Lianhebao*, the tax bureau, and other customers, that there was little need for those customers to seek further advancements. To keep up this kind of whiplash speed, Ideographix desperately needed to break into the mainland Chinese market, and yet the PRC was recalcitrant. For four or five years, Yeh explained, he courted the *People's Daily*, but to no avail. They had strung him along, Yeh explained to me, worried by the fact that Yeh was Taiwanese, and eager to build their systems for themselves. They did not want to lose control of their propaganda machine, he felt convinced.

There were other changes afoot as well. The era of custom-designed Chinese text processing systems was coming to an end. A new era was taking shape, one that Chan-hui Yeh, Ideographix, and indeed a good number of other major corporations, entrepreneurs, and inventors were largely unprepared for. This new age has come to be known by many names: to some, the software revolution; to others, the personal computing revolution; and less rosily, the death of hardware.

In an era when software was becoming the main commodity, the software was free for Ideographix. In an era when hardware costs were plummeting, with no floor in sight, for Ideographix it was the hardware that customers paid for (and which, naturally, they began to wonder: Why are we paying so much?). New devices were appearing on the scene as well, blindsiding inventors like Chan-hui Yeh: Japanese-built fax machines, for just a few hundred bucks a pop, and microcomputer "kits" flexing muscles once reserved for mainframes. Meanwhile, Yeh's system remained what it had always been: something akin to a telephone switchboard system, as compared to consumer-facing television set.[87]

The old offices of IPX in Sunnyvale, which Chan-hui Yeh walked me through on a March day in 2010, captured this oblivion most vividly. "We used to have 20,000 square feet," he explained to me, our footsteps echoing as we made our way.[88] The room looked as if, one day, dozens of workers had simply stopped what they were doing and walked out all at once—a kind of anti-Pompei. If the citizens of that ill-fated town were encased in a bath of molten rock, here in the offices of Ideographix, it was as if the staff had simply evaporated. Many desks were still weighed down with technical bulletins and product binders, some still open, as if its reader might return at any moment, a cup of coffee in hand.[89]

None of this dampened Yeh's enthusiasm, however, nor his healthy sense of self. Unlike Lois Lew, who we met in chapter one, Yeh was decidedly unsurprised when I found him. He was waiting for me, in fact—or someone like me, at least. Indeed, the very first time we spoke, he opened the phone call by saying that Ideographix was the "best kept secret in Silicon Valley." "I always knew someone would call."

4

THE INPUT WARS: ZHI BINGYI AND THE RETURN OF HYPOGRAPHY

In 1979, Zhi Bingyi stood in the luxurious Jinjiang Hotel banquet hall, steps from where Zhou Enlai and Richard Nixon had issued the Shanghai Communiqué seven years earlier. To Zhi's left was Roy Hofheinz Jr., the director of Harvard University's Fairbank Center, one of the world's preeminent institutes of Asian Studies. Seated before him were Chen Zhiyuan, director of the Shanghai Instruments and Meters Institute, and William Garth IV, president of the Graphic Arts Research Foundation—the same organization where, two decades prior, Samuel Caldwell had invented the Sinotype. It was time for Zhi to make history: to usher China into the age of minicomputing.

What a difference a decade makes (figure 4.1). It's hard to believe that just ten years earlier, in 1969, Zhi Bingyi was released from prison where he had been languishing for nearly three years, condemned to solitary confinement by his Red Guard captors. As China descended into the Cultural Revolution, radicals in his hometown of Shanghai had pledged undying loyalty to Mao Zedong, roaming the city in "seek and destroy" raids bent on purging the country of all vestiges of "Old China."[1]

Traumatic as they were, those days of anxiety and terror in the Mao-era "ox pen" played a central role in Zhi's newfound success. Unsure if he would ever see his wife again, and with no work to occupy his mind, Zhi filled those long hours in his six-square-meter cell with a single-minded obsession, one he and his wife later credited for saving his sanity. He set out to encode Chinese characters using the letters of the Latin alphabet. Summoning up images of Chinese characters in his mind's eye, he exploded them into fragments, over and over again, trying to find patterns he could exploit in the design of his system. He turned his mind into

汉字进入了計算机

GARF
OCT 79

Figure 4.1   Photograph of Zhi Bingyi

a particle accelerator, smashing apart the constituent elements of the Chinese universe, in hopes of discerning its subtler, subatomic regularities.

To make matters more complicated, Zhi had no paper. (The guards didn't permit it, save for when he was forced to write and rewrite political self-criticisms and confessions.) He did manage to steal a pen, though, using it to write secretly on the lid of a teacup, one of the few objects afforded him by his keepers. When turned over, Zhi discovered, the lid was large enough to fit a few dozen Latin letters. Not only could he keep his writing a secret, but he also could erase the letters and start again. Like a school child in ancient Greece, he kept his makeshift wax tablet close at hand, writing, erasing, and rewriting, for months and years on end.

Zhi's hallucinations laid the groundwork for a new Chinese input method, known alternately as "See the Character, Know the Code" (*Jianzi shima*), "On-Site Coding" (OSCO), or simply "Zhi Code" (*Zhima*).[2] This code, as we'll see, put Zhi Bingyi on the radar of engineers and technologists in the PRC and two foreign organizations—the Olympia Werke company in Germany and GARF in the United States. Zhi's code, many felt, might finally make it possible for the Chinese writing system to hold its own in the global computing age.[3]

The story of Zhi Bingyi encapsulates two of the most significant transitions in Chinese computing during the 1970s and 1980s. First, this period marked the re-ascendance of the QWERTY keyboard within Chinese computing—the return, that is, of human-computer interaction governed by both the Western-built keyboard and the Latin alphabet, rather than the many bespoke devices we examined in chapter 3. Despite this growing dominance of QWERTY and Latin script, however, Chinese input in this period did not use Latin letters to "spell" the phonetic sounds of Chinese words. As we will see, Zhi Bingyi's input system used the Latin alphabet to describe the *shape and structure* of Chinese characters, *not their phonetic values*. This marked a departure from the history of Latinization or Romanization as it took place—or had been attempted—in other parts of the world in the modern period. Conventionally, to Latinize or Romanize a script was to represent the pronunciation of that script in the Latin alphabet, as in the Romanization of Turkish or Vietnamese (along with many abortive efforts focused on Arabic, Bengali, Hebrew, Persian, Uzbek, and more). In a seeming paradox, then, Chinese computing from the mid-1970s through the 1980s marked a peculiar case of "Romanization without phoneticization": representing Chinese characters with the letters of the Latin alphabet (and sometimes Arabic numerals as well) but doing so *nonphonetically*.[4]

But Zhi was far from alone in these efforts. OSCO was one of dozens of input systems—and soon hundreds—invented during the 1970s and 1980s, each advancing a novel way to use Latin letters and Arabic numerals to get Chinese character information into a computer using a QWERTY keyboard. Like OSCO, the overwhelming majority of these systems were structure-based—which is to say, nonphonetic. So unpopular were pinyin-based input methods, indeed, that some speculated as to whether the further development of Chinese computing might spell the death of pinyin overall. "Never popular to begin with," one account from the era emphasized, "the various pinyin systems could die out if left behind by the computer revolution."[5]

The second transition was equally important, namely the transition away from computers and word processors capable of handling only one input system, to those capable of handling many. Unlike the story of Caldwell's Sinotype, or the IBM typewriter before it, the rise of mini- and personal computing opened up the possibility of computers capable of handling a variety of different IMEs, leaving it up to the user to select one based on personal preference. As we will see, Zhi was one of the few to survive this transition (at least for a time), working first on a computational system that featured his code exclusively, to another venture in which "Zhi Code" was one of a variety of different input systems that operators had the choice of using. The era of "one machine, one input system" was coming to an end.

THE RETURN OF QWERTY: ZHI BINGYI'S JOURNEY FROM CHARACTER RETRIEVAL TO CHARACTER INPUT

When the Cultural Revolution erupted in 1966, Zhi became a marked man. Having been named a "reactionary academic authority"—one of the era's many monikers for those condemned as enemies of the revolution—he was captured and confined in one of the era's infamous "ox pens."[6] On the wall of his cell, an eight-character passage made a chilling assurance to him and anyone unfortunate enough to set their eyes upon it:

Leniency for Those Who Confess, Severity for Those Who Resist (*tanbai congkuan, kangju congyan*) 坦白从宽，抗拒从严

The message was clear: We have the authority to destroy your life (if you resist). Or to make your imprisonment somewhat more tolerable (if you confess).

Without the comfort of family or work, this terrifying couplet was all Zhi had to occupy his time. And so he read it, over and over, for days, weeks, and months on end. And something began to happen—something that reminds us of the inherent strangeness of language.

No matter one's mother tongue, the process of becoming fluent in a language or literate in a script is in many ways a process of forgetting that language and writing are a form of arbitrary code. There is nothing inherently "candid, frank, or open" about the character 坦 (*tan*), for example, nor "white, blank, or clear" about the character 白 (*bai*). As with any young child, Zhi in his juvenile years would have looked upon these symbols as random assemblages of pen strokes, born out of a complex web of conventions whose origins we will never be able to reconstruct in full. But over the course of innumerable repetitions, something happens to us: the sounds and sights of language begin to approach, and then achieve, a condition of invisibility and givenness: 白 (*bai*) no longer "stands in" for *whiteness* by dint of painstaking study and memorization, but merges with it effortlessly. This merger is what we might call the fruition of every child's struggle to speak, read, and write: the struggle to make inroads into their family's and community's semiotic universe, transforming it from an indecipherable code to a medium of expression. To become fluent is to become, in a sense, code-unconscious.

While most experience this transformation as a one-way process, it can be reversed on occasion. A sound or symbol made second nature can be *denatured*—defamiliarized and queered, returning perhaps not to that original state of nonknowledge, but at least to a bifocal view in which we are somehow able to tap into the original meaninglessness of our mother tongue, even as we continues to hear, see, and speak it fluently.

This is what happened to Zhi Bingyi. As he whiled away in prison, mulling over these eight characters (seven, if we account for one character that is repeated), this act of repetition restored to them their inherent arbitrariness. By the hundredth reading—perhaps the thousandth, we cannot know—Zhi began to decompose these characters in his mind, into a variety of elements and constellations. The first character (坦), for example, could be readily divided into two distinct parts: 土 and 旦, and then further still into 十 and 一 (making up the component 土) and 日 and 一 (making up 旦). The second character 白 could be subdivided, as well, perhaps into 日, with a small stroke on top. Then 从, being the concatenation of two 人 graphs, and so forth. Even in this short, eight-character passage, the possibilities of decomposition were abundant.

Zhi extended this analysis to Chinese writing more broadly. For nearly three years, he decomposed characters into elemental parts, pairing each element with one or another letter of the Latin alphabet. (Zhi had no dictionary or reference material at hand, of course. Save for the Chinese characters on the wall of his cell, the thousands of other characters he analyzed had to be derived exclusively from memory.) His goal became that of developing Latin alphabetic encodings or "spellings" for every Chinese character.[7]

Zhi's interest in Latin alphabetic encodings had no connection to computing at this early point. Instead, his goal during his imprisonment (beyond maintaining his sanity and dignity) was to develop a more efficient, rational, and scientific way of organizing Chinese dictionaries, phone books, card catalog cases, name lists, and other bodies of information encoded in Chinese character script. Zhi wanted to discover the Chinese equivalent of "alphabetic order."[8]

In September 1969, when Zhi was released from prison, he rejoined his wife and family at their apartment on South Urumqi Road in Shanghai to live under prolonged house arrest.[9] His wife hoped that Zhi's release might mark a return to some kind of normalcy—a chance to rebuild. This was not to be, however.[10] Zhi Bingyi continued to obsess over his code, memories of his imprisonment never far from his mind.

On one occasion, he fell ill, registering a dangerously high fever. Instead of resting, though, Zhi puzzled over one of the lingering problems in his code: how to differentiate between characters that, according to his system, had the same "spelling" (for example, 吉, 台, and 古). His wife grew concerned. *Why did he need to work on it so hard*, she wondered? *Why couldn't he try and relax and recover?* "I'm not tired," Zhi tried to assure her. "Coding makes me happy." On another occasion, Zhi sat on a bus for hours, lost inside of a daydream, only to be woken up by the driver at the route terminus. Like he had during his days in the ox pen, he fell into a waking hallucination of sorts, reworking his character encoding again and again. Zhi's wife could hardly protest, however. As she readily acknowledged, Zhi's code had saved her husband's sanity, and perhaps his life.[11]

Shortly after Zhi's release, China's relationship with the world began to change dramatically. In 1971, the United Nations recognized Beijing as the official representative of China, granting the PRC a seat on the Security Council. In 1972, Nixon shocked the world by sending the first US presidential delegation to Communist China.

Profound geopolitical shifts were matched by equally dramatic shifts in the arenas of computing and information processing. China's recently completed DJS-6 computer boasted a speed of 100,000 operations per second (OPS), with up to a 32,000 word capacity memory.[12] The 111 machine, completed by the Institute for Computing Techniques in Beijing, clocked in at 180,000 OPS, with a 64,000 word core memory.[13] Then, in August 1974, Tsinghua University completed work on the DJS 130 Small-Scale Multi-Function Computer, patterned after the Nova 1200 by Data General.[14] In the domain of the global computing industry, meanwhile, an explosion in the consumer technology market, as well as the personal computing market, translated into an ever-increasing flood of mass-produced, QWERTY-based systems into the global market. All these outfits had China in their crosshairs, which remained frustratingly beyond reach during the giddy and extremist days of Maoism.

Zhi began to look at his code in a new light, his focus shifting away from dictionaries and phone books, and toward computers—that is, to input. Rather than being merely a system for reorganizing Chinese library card catalogs and the like, Zhi reasoned, perhaps its true impact would be in giving structure to a new era of QWERTY-based Chinese computing. After all, despite their growing familiarity in China, all of these varied Western-built computers were practically useless for anyone seeking to input and output Chinese-character text. The day was still not here when an average computer user could sit down with an IBM machine—or later an Apple—and use it for Chinese information processing.

Over the course of 1974 and 1975, the Shanghai engineer began to recast "Zhi-code" as a method explicitly designed for computational input, traveling to Beijing to introduce OSCO to the First Ministry of Machine Building. The following year, he shared a broader vision for Chinese information, one in which Zhi Code would take center stage.[15] Thinking far beyond the personal computer, indeed, Zhi began to fantasize about Chinese "information" itself. "Let's say you want to know which essays by Lu Xun discuss the reform of Chinese characters," he speculated. "All you have to do is give the library a call. A few minutes later, a computer would rapidly and accurately display the materials you need on your TV screen."[16]

In 1976, the pace of both geopolitical and technological change began to accelerate. Mao Zedong's death set in motion a profound sweep of political, economic, and social transformations. In 1979, normalization of relations with the United States abruptly swung open China's gates, introducing the mainland Chinese market to mass-manufactured personal computers, predominantly by companies based in the United States—all of which, of course, featured one or another version of the QWERTY keyboard. Instead of trying to beat back the tide of QWERTY-style devices—a move which would have forced China into a virtually autarkic position—one inventor after the next started to employ QWERTY as their "starting point," asking in effect: What can I do to render the QWERTY keyboard usable for Chinese character input?

This influx of low-cost, mass-produced, QWERTY-based systems spelled doom for the era of bespoke Chinese interfaces we examined in chapter 3. Slowly at first, then rapidly, systems such as Loh's keyboard, the medium-sized keyboards built at Peking University and elsewhere, the IPX by Ideographix, and the Ideo-Matic 66 began to disappear. A new generation of machines were being designed with QWERTY as their starting point, and input as their foundation. By the early 1980s, never again would another non-QWERTY keyboard or non-keyboard input surface ever seriously compete with the QWERTY keyboard in the Sinophone world.

HYPOGRAPHIC SEMIOTICS: WHAT IS AN ALPHABET THAT DOESN'T SPELL?

By the time Zhi Bingyi began to present publicly on his Chinese input system, the stage was already getting crowded. More than three dozen input designers took part in the 1978 National Academic Conference on Chinese Character Encoding, for example, held in the city of Qingdao.[17] Alongside OSCO, competing systems included the "Chinese Character Hierarchical Decomposition Input Method," the "Sound-Shape Four-Bit Floating Encoding Method," the "Three-Key Encoding Method," the "Three-Letter Encoding System," the "SYX Chinese Character Encoding

System," and the "Chinese Character Shape-Letters Encoding Method," just to give some sense of the diversity of proposals on the table.[18]

Zooming out to the broader Sinophone computing world, the number of competing Chinese input designs grows even larger. At the First International Symposium on Computer and Chinese Input/Output System," held in Taiwan in August 1973, about a dozen Chinese input systems were presented, including the "Chiao-Tung Radical System," the "Upper-Right Corner Indexing System," the "New Alphameric Code for Chinese Ideographs," and the "SINCO" (or "Synthetic Index Nomenclature for Chinese Orthography") system, among others.[19]

Like Zhi, many of these designers began their work, not with computers in mind, but dictionaries, filing cabinets, and phonebooks. In other words, their work was focused on character *retrieval* methods rather than character *input* methods—a history dating back to the "Character Retrieval Crisis" of the 1920s and 1930s, when Chinese linguists, publishers, educators, and entrepreneurs fought over the best way to organize Chinese character information environments. They were trying to find China's "equivalent" of Latin alphabetic order, in other words.[20] These debates spilled into the Communist period when, in 1961, the Ministry of Culture of the People's Republic of China formed the "Character Retrieval Working Group," its goal being to survey and judge China's existing character retrieval methods. A staggering 315 different retrieval methods were tallied in all, more than three times as many as in the 1930s.[21]

With the rise of microcomputing in the 1970s, China's long-standing "character retrieval problem" morphed rapidly into the "character encoding problem" and the "character input problem." For Zhi and his competitors, the operative question suddenly became *How should we retrieve Chinese characters from computer memory?* rather than from filing cabinets or phone books. Practically overnight, half-century-old systems of dictionary organization—such as the Four-Corner Retrieval System and the Five-Stroke Retrieval System (*Wubi jianzifa*)—were revived and repurposed as Four-Corner input and Five-Stroke input.[22]

Input techniques varied greatly across these systems. To input the Chinese character 电 (*dian* "electricity") using OSCO, for example, the input sequence was "DDDD." If using the Zheng Code input method, however, the input sequence was "KZVV"; "NY," in the case of Taiji Code; "JNV," in the case of Wubi input; "RGD," in the case of Yi input; "BL," in the case of Shape-Meaning Three Letter Code, and the list goes on. Some input systems were premised exclusively on numbers. To enter the character *dian* using Stroke-Shape Code, for example, the user would key in the sequence "6-0-1"; for Telegraph Code input it was "7-1-9-3"; and "6-0" in the case of the Chinese Character Stroke-Shape Look-Up Encoding Method.[23]

But how exactly do the Latin alphabetic letters "NY" lead to the Chinese character *dian*? What is more, how can the user *also* get to the Chinese character *dian* by way of "KZVV," "JNV," "RGD," "BL," "601," and so forth? If letters of the Latin alphabet were not being used to "spell" Chinese within these input systems—since none of these input sequences were based on the phonetic transliteration of the sound of Chinese characters—how should we understand their function?[24] Is the choice of letters and numerals entirely random, or do input system designers abide by any guidelines or conventions when trying to link alphanumeric sequences to the Chinese characters they are meant to retrieve? Is there a semiotics of structure-based hypography, in other words?

There were three general approaches to alphanumeric input during the 1970s and 1980s. First, the letters of the Latin alphabet (alongside Arabic numerals) were used as surrogates or *variables*, standing in for one or another structural feature of a given Chinese character. Second, letters and numerals were used as *measurements*, used to take *samples* of specific features or characteristics of any given Chinese character. While related to the variables approach, this method was not premised upon describing the overall structure of characters, but rather taking "readings" thereof, then using letters and numbers to record the measurements taken (we will clarify this momentarily). Finally, and least commonly, letters were used as *isomorphs*: a technique in which Latin alphabetic letters were used to encode structural features of a given Chinese character based on the perceived graphical resemblance between a given letter and a specific component of Chinese script. (Think of the English expression "U-turn," only where "U" is meant to represent the shape of a Chinese character component rather than the curving path of a vehicle.)

Zhi Bingyi's OSCO system was a prime illustration of the first of these approaches. To input the Chinese character *fu* (幅 "width") using OSCO, a user would enter the four-letter sequence: J-I-T-K. The first letter in this sequence (J) corresponded, not to any phonetic property of the character, but to a structural component located in the left-most side of the character itself: the component 巾 that, when seen in isolation, is pronounced *jin*. The code symbol "J," in other words, was derived from the first letter of the Hanyu pinyin pronunciation of the component.

The remaining letters in the sequence—I, T, and K—followed the same logic, each letter representing one or another structural component in the character, assigning a specific letter based on the pinyin pronunciation of that structural component when encountered in isolation. "I" was set equal to the component/character *yi* (一); "K" to *kou* (口); and "T" to *tian* (田).[25] Other examples from the OSCO code include:

D = the structure 刀 (with "D" being derived from *dao*, the pronunciation of this character when seen in isolation)

L = 力 (based on the pinyin pronunciation *li*)

R = 人 (based on the pinyin pronunciation *ren*)

X = 夕 (based on the pinyin pronunciation *xi*)[26]

Designers who adopted the variable approach enjoyed many degrees of freedom. To begin, they could choose how to decompose Chinese characters into modular pieces. In the case of OSCO, Zhi Bingyi piggy-backed on a centuries- if not millennia-old taxonomic system. Specifically, it was based upon structural components known in English as "radicals"— "classifiers" (*bushou*) in Chinese—a technique of decomposing Chinese characters dating all the way back to the Han dynasty text, the *Shuowen jiezi*, considered by some to be the first Chinese dictionary. More proximately, OSCO relied upon the Kangxi Radical-Stroke system, named after the Qing dynasty–era *Kangxi Dictionary*, which organized characters according to 214 *bushou*.

Although Zhi opted to employ radicals from China's imperial past, there was nothing to prevent other input designers from exploring other options. Other inventors might choose to decompose characters into "fundamental strokes" rather than radicals. Still others might decide to invent their own, proprietary systems of decomposition, based upon entirely made-up components of their own design.[27] While there were conventions, to be sure, there were no hard and fast limits to the variable approach.

Inventors also had leeway in deciding how to assign Latin letters (and perhaps Arabic numerals) to each of these elemental units. Zhi opted to exploit Hanyu pinyin phonetics—naming each of his entities after the pinyin pronunciation of those structures when appearing in isolation—but other approaches were viable as well. Within Cangjie input, for example—a system developed in 1976 by the Taiwanese engineer Chu Bong-Foo—some of Chu's character components were identical to those found in OSCO, and yet the assignment of these elemental units to Latin letters (that is, which elements would be assigned to the "A" key, versus "B," "C," etc.) followed a different pattern.[28] Ultimately, hundreds of input designers used the "variable" approach during the 1970s and 1980s, each of their decisions leading to sometimes stark, sometimes subtle, differences. Within the variable approach, then, the possibilities were effectively infinite.

The second approach—letters as samples—is best understood through the analogy of filtration. Imagine a piece of paper with three holes cut out of it. Placing this piece of paper over top of a Chinese character, all parts of that character would be filtered out except in those three zones. Now imagine a chart outlining every possible feature or characteristic we might encounter in those three "sample" zones—stroke-shapes, angles, and so forth—with each of these features being assigned a unique alphabetic, numeric, or alphanumeric code. By focusing exclusively on those three zones, it becomes possible to assign a specific value to any Chinese character by matching each of its features to a code unit.[29]

Three-Corner Coding—an input system developed in Taiwan by Yuan-wei Chang, Li-ren Hu, and Jack Kai-tung Huang—is a prime example of the measurement approach.[30] As suggested by the name, the system requires operators to analyze three corners (*jiao*) of a Chinese character, converting the structural features found in those zones into unique two-digit numeric codes based upon a look-up table the user had memorized. By entering the resulting six digits (three zones, with a two-digit code for each, that is), the Three-Corner Coding input system was thus capable of retrieving the user's desired character from memory (or presenting "candidate" characters, in the event that a six-digit code corresponds to more than one possible graph) (figure 4.2). In the case of the character 鄭, for example, the "three corners" refer to the top-left, bottom-left, and

三角編號 法規則

漢字係方塊形，通常共有四個角，此編號方法取其三個角的基本筆形，按照基本符號（99個主符號201個副符號）編碼，每個基本符號的代碼係二位數，全字號碼共計六位。

本編選法係根據取角和取形二原則，兩相配合處理。取角原則乃注重顧序和地位，取形原則在選取基本符號，以確定號碼。

［I］ 取角原則

取角原則的基本順序，是從左到右，由上而下，圖示如下：

1. 左上角
2. 右上角
3. 左下角
4. 右下角

因爲漢字結構繁複，在取角規則中應按字形不同，加以變動，以求得簡單而達到高度的編號效果。舉例如下：

(1) 若一字的四角均屬各個基本符號，只取前三角，最後一角略而不用。

(2) 若一字可由一個或二個基本符號組成，最後四位或二位號碼，應用 00 30 或 00 補足（因每字號碼必須爲六位）。

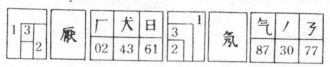

(3) 若一字中相鄰二角屬於一個基本符號，這二角應併作成一角，即該字只有三個角，因此恰合三角編號法原則，其編號取角有下列各種形狀：

(4) 若二個基本符號，完全佔去一字的四個角（這四角可稱外角），最後二位號碼，應取字中剩餘的最左上方的基本符號編號，這是第三角（這角可稱爲內一角）。

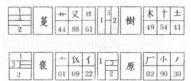

(5) 若一個基本符號，完全佔去了一字的四個外角，如口、冂、門等，最後四位號碼應取剩餘內部的基本符號作爲其他二角。這二個內角的編號順序，仍須按基本取角原則和下述取形原則，繼續依次編號。

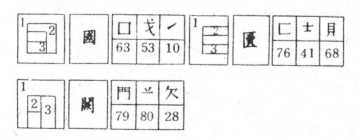

Figure 4.2 The Three-Corner Coding look-up table and sample encoding

top-right zones of the character. According to the coding system, this would result in the input code 801542—within which 80, 15, and 42 each referred to the modules found in those locations.[31]

As in the case of the variable approach, the sample approach afforded its designer practically infinite freedom. Instead of focusing on three parts or zones of Chinese characters, a designer could just as easily focus on two, four, five, or otherwise. Likewise, the precise location of these zones—the places where a user was supposed to take measurements—was up for grabs, with some inventors focusing on "corners" of characters, others on "tops and bottoms," and so forth. Designers could also decide which features a user would be required to identify, creating in effect their own proprietary alternative to the table seen in figure 4.2. Stroke size, shape, directionality, and more were all possible qualities that a designer might wish to focus on. Finally, the coding system itself—the pairing of particular features with particular alphanumeric codes—was completely open to interpretation. As with the variable approach, then, it was possible to have a theoretically infinite variety of input systems, all based on this same approach.

The isomorphic approach was the most curious, if least common, technique. Here it is helpful to familiarize oneself with so-called homograph attacks, a practice wherein malicious actors take advantage of the surface-level resemblances that exist between the letters of different writing systems in order to achieve one or another nefarious goal. In homographic "spoofing" attacks, for example, a computer user might be duped into clicking on a URL link marked "HM.com" or "HBO.com," only to find themselves on a malicious third-party website, rather than those of a well-known clothing retailer or the cable television outfit. In this form of homograph attack, the "H" in these links is not in fact the Latin alphabetic letter "H" (/h/) but the look-alike Cyrillic letter "H" (/n/)—enabling the attacker to route one to an entirely different URL. Other homographic pairs—also referred to as "confusables"—include 0 (the numeral) and O (the capital letter), B (Cyrillic letter /v/) and B (Latin letter /b/), P (Cyrillic letter /r/) and P (Latin letter /p/), among dozens of other examples.[32]

Transletteration, as this phenomenon is also called, need not be malicious. It has also been used in times when technological barriers prevent a person from using their desired orthography, or when the affordances of a given system allow users to "play around" with language. Examples include Faux Cyrillic, Volapük encoding, Arabic Chat Alphabet, or other one-time acts—such as using the Latin letter 'x' to stand in as a Hebrew aleph (א).[33] Other examples come from digital countercultures such as Leetspeak (13375p34k), a practice of writing that interchanges letters and numerals based upon perceived visual similarities, and "calculator spelling," where one uses the numeric output of an LCD calculator to produce short text messages (enter "77345663" on your calculator, for example, and you will find the word "eggshell" when you turn the display upside down).

During the 1970s and 1980s, a surprising number of inventors exploited the principles of transletteration to create full-fledged Chinese input methods. A key example is the work of H. C. Tien and a system he referred to alternately as the "Chinese Transalphabet" and the "radical transalphabet."[34] Tien created isomorphic equivalences for every radical in Chinese, using one or more Latin alphabetic letters to stand in for specific character components. To represent the "grass radical" [艹], for example, Tien's system used the letter sequence "tt"—positing a mutual resemblance. Other isomorphic pairings included "b" and 白, "p" and 尸, "pq" and 门, "pttq" and 鬥, "bb" and 比, "mm" and 马, and "bv" and 身, among many dozens of others (figure 4.3).

To grasp how Tien's Chinese Transalphabet worked in practice, we can return to the example of *dian* ("electricity") (this time in the "traditional" form of 電). The first part of Tien's code was D-I-A-N, making it one of the few systems at this time to make extensive use of Hanyu pinyin. The second half of Tien's code was reserved for his structure-based Transalphabet, however, which Tien used to disambiguate between the large number of Chinese homophones that share the same pronunciation—such as 點 (*dian*, "point, dot"), 典 (*dian*, "canon" or "classic"), 滇 (*dian*, toponym signifying "Yunnan"), among many others.

The second half of Tien's code sequence for 電 was T-M-V-V (figure 4.4). To understand the meanings of these letters requires a three-step process. First, within the structural component found on top of the character 電—namely, the "rain" radical 雨—Tien isolated the horizontal and vertical lines therein and likened the shape to an uppercase letter "T." The

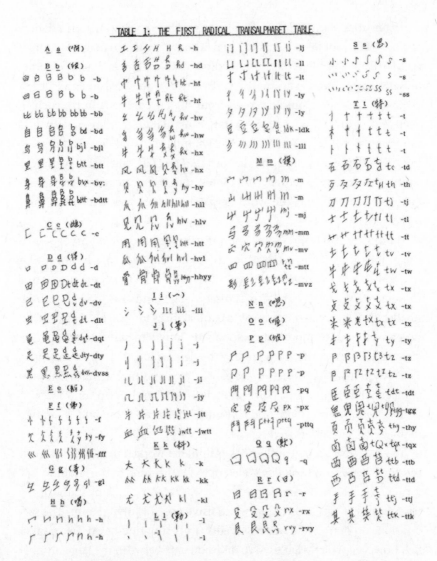

Figure 4.3 "The First Radical Transalphabet Table" by H. C. Tien (samples)

Character Component   Isomorphic Equivalent

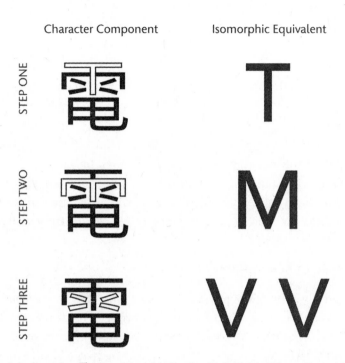

Figure 4.4   Encoding 電 (*dian*, "electricity") using H. C. Tien's Chinese Transalphabet

second letter—the uppercase letter "M"—was chosen for its resemblance to the cluster of strokes located in the middle section of the rain-radical. Finally, a pair of "V" letters was set equal to the smaller strokes located within this M-shaped structure. Taken together, then, the full "transalphabetic" code for the rain-radical 雨 was T-M-V-V. When combined with the first part of Tien's code, then, the full input sequence for 電 was thus D-I-A-N-T-M-V-V.

As unconventional as Tien's system may sound, this strategy was employed by multiple inventors during the 1970s, 1980s, and beyond.[35] In the "Chinese Character Shape-Letter Encoding Method," for example, each letter of the Latin alphabet was, once again, paired with one or more Chinese radicals, strokes, or other structural components with which it bore an isomorphic likeness.[36] In another input method—"Three-Key" input, developed in 1979—the letter "A" was used to represent the Chinese character/radical 人, for no other reason than that the two diagonally tapering strokes of the character matched the two diagonal strokes of the capital "A." Following this same logic, the uppercase "B" within Three-Key input represented the radicals 阝 and 阝; the lowercase "b" represented all but the top-most stroke of the character 五; the uppercase "C" represented 匚 and 匸; the uppercase "D" represented both 力 and 刀; the uppercase "E" represented 彐, the uppercase "F" represented 下, and so forth.[37]

With dozens—and soon hundreds—of input systems in existence, it is important not to get mired in detail, however. As with the history of writing more broadly—wherein the study of alphabets, syllabaries, abugida, and more could easily occupy multiple lifetimes—hypography is an immense terrain of possibility, as semiotically vast as writing itself.

THE HYPOGRAPHY MARKET

Why did Zhi's input system catch the attention of computer manufacturers, while so many others failed? In contrast to those who presented alongside Zhi at the 1978 conference in Qingdao, he went on to establish a manufacturing partnership with one of Europe's foremost precision engineering firms—the West German Olympia Werke company—which installed Zhi's OSCO input system on their new Chinese word processor, the 1011. Why was OSCO chosen, as opposed to the "Chinese Character Hierarchical Decomposition Input Method," the "Sound-Shape Four-Bit Floating Encoding Method," the "Three-Key Encoding Method," or otherwise? Phrased more broadly, what accounts for the success or failure of any individual method?

The merits of a particular input system were only part of the story. Equally if not more important were contingent factors that placed certain input methods on the radars of prominent companies and firms, as well as the boundless entrepreneurial energy that certain inventors invested in trying to proselytize on behalf of their systems. Pure accident—and in other cases relentless self-promotion—played as large a role as merit.

Circa 1978, there was little to suggest that Zhi Bingyi should have been the one to secure a high-profile manufacturing contract with a major foreign firm. He was a relatively low-profile engineer in Chinese computing circles, after all, and utterly unknown to foreigners (for a time, at least). "He is an elderly man," Harvard's Roy Hofheinz later wrote of his first

encounter with Zhi, "who is virtually unknown in Peking."[38] "I asked
several people in the Academy of Sciences about him before leaving for
Shanghai," Hofheinz added, "and got an ignorant response."[39] Nevertheless, it was Zhi—this relatively obscure, nearly retirement-age engineer
from Shanghai who would go on to secure a relationship with Olympia
Werke—as well as with the Graphic Arts Research Foundation a few years
thereafter.

To understand this unexpected outcome, we need to delve a bit deeper
into the engineer's early biography—particularly his uncommonly global
pathway through life. Zhi was born in Taizhou, in Jiangsu province, on
China's eastern coast. The month after his birth, in October 1911, a rebellion some 400 miles away in the city of Wuhan would end up toppling
the Qing dynasty.

He was part of the generation of Chinese students whose goal—
mandate, even—was to contribute to the modernization of their country
by training in a variety of applied disciplines.[40] To this end, Zhi completed his undergraduate education in 1935, receiving a degree in electrical engineering from Zhejiang University.[41] Upon graduating, he enjoyed
a rare privilege in his day and age: the chance to continue his studies
overseas, in Europe. He moved to Germany in 1936, in pursuit of a PhD
in physics, going on to receive his doctorate in 1944 from the University
of Leipzig.[42] Zhi spent nearly eleven years in Germany, becoming fluent in the language, and marrying a German woman. After spending her
entire adult life in China, she would come to be known by her Chinese
name: Zhi Aidi.[43]

Upon the couple's return to China in 1946, Zhi held a variety of distinguished posts, including director of the Electronics Laboratory at the
National Bureau of Industrial Research and faculty positions at Zhejiang
University and Tongji University. He also became a member of the Chinese
Academy of Sciences.[44] Nonetheless, his long-time experience overseas—
notably in Nazi-controlled Germany—made him suspect in the eyes of
the still-nascent Chinese Communist regime following the revolution of
1949. Ultimately, the price Zhi paid was solitary confinement.

Following release from prison, however, these once suspicion-inducing
connections to Germany began to open doors. Whether thanks to inter-
personal connections, or relationships he maintained from his time in

Leipzig, the first foreign company to become aware of Zhi Bingyi's work was German: the Olympia Werke company, a towering presence in the history of German precision engineering.

The timing could hardly have been better for Zhi. By the late 1970s, Olympia Werke was determined to reenter a Chinese market where, surprisingly enough, it already boasted a long presence—one dating back to the era of mechanical Chinese typewriting. During the 1950s, long before their encounter with Zhi, executives and engineers at the company collaborated with Chinese manufacturers to develop a line of mechanical Chinese typewriters: the Optima (*Eputima*) Chinese typewriter.[45] With the advent of word processing in the 1970s, Olympia wanted back in, this time pinning its hopes on the development of a computational system based upon a QWERTY interface. To do so, however, required an input system of some kind—the search for which brought them to Zhi.

Zhi and Olympia established a joint venture called Sinotype Systems (not to be confused with the Sinotype built by Samuel Caldwell, or the later iterations we will meet in this chapter).[46] The plan would begin, the parties agreed, with an initial production of five thousand machines, manufactured in Germany, and retailed at a price of US$7,500 (a price tag that placed it out of reach for everyday Chinese consumers, but not for government offices or other large institutions). The system would be known as the Olympia 1011, and the ultimate goal of the joint venture would be to shift the center of manufacturing from Germany to Zhi's Shanghai Instruments Research plant.[47] This unlikely collaboration represented a double reunion of sorts: Zhi's with the world of German engineering, and Olympia's with the domain of Chinese information technology (figure 4.5).

Chance encounters like this were more common than we might expect. One of the other rare success stories of the era was the above-mentioned Three-Corner input (*Sanjiao*) system, which captured the interest of Wang Laboratories in the United States.[48] In May 1978, company president An Wang formed a new "ideographic product development" branch, focused on the development of the Ideographic Word Processor. Jenny Chuang, a brilliant, Taiwanese-born engineer, was appointed to head the branch and quickly set about staffing her team.[49] Chuang focused her recruitment efforts locally, especially among individuals she knew via the Chinese

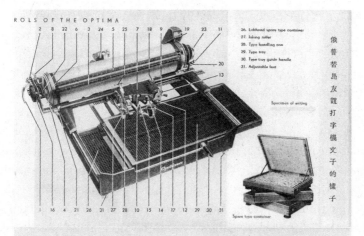

俄普蒂鳥友誼打字機文子的樣子

The typing of Chinese with the aid of modern technology – Olympia 1011

Olympia 1011 – a bi-lingual electronic memory typewriter with an astonishing range of features and versatile applications in the fields of science and technology, industry and commerce as well as culture and education.

A single keyboard with bi-lingual capabilities
The Olympia 1011 shows itself as a stand-alone desktop unit, about the same size as an electronic typewriter. It has a standard keyboard with 48 typing keys (including those of 26 Latin letters, 10 numerals and punctuation marks), 28 function keys and 8 switches.

One may wonder why no keys of the elements of Chinese characters are found in the keyboard as in other Chinese electronic typewriters. The answer is one of the unique features of the Olympia 1011 – a Chinese character generator (CCG), which can generate Chinese characters in a simplified or traditional form.

A unique encoding method accessible to both Chinese and foreigners
Everyone may know the typing of Latin letters well by operating the keys, but how are Chinese characters printed out by using the Latin-letter keys?

Modern-day technology has solved this problem by using a novel encoding method, by which the proper code for each Chinese character consists of 2, 3 or 4 Latin letters. This encoding method, developed by the Chinese scientist Dr. Zhi Bingyi of Shanghai, divides a character into 2, 3 or 4 segments

Figure 4.5 Olympia 1011 by Olympia Werke

Bible Church of Greater Boston.[50] By the time the project team was built, nearly every top-level member was of Taiwanese descent: lead programmer Ching-Wu Chen, a Taiwanese-born programmer who Chuang lured away from the Digital Equipment Corporation (DEC); Ke-Chieh (K. C.) Chu, also Taiwanese-born; and George Chuang (Jenny Chuang's brother-in-law, who worked on the printer), among others.[51]

The Taiwanese heritage of the Ideographic Word Processor team exerted an unmistakable influence on the team's selection of an input system. Simply put: Three-Corner input was popular in Taiwanese library circles at this time, and so the team decided to implement it.[52] Neither Chuang's team nor Wang Laboratories more broadly conducted anything approaching a systematic survey of the input marketplace, Chuang explained to me, or a rigorous stress-test of potential candidates. The decision-making process was a far more human, idiosyncratic, chance-laden affair.[53]

The success of a given system was not always a matter of chance, of course. Tireless self-promotion and entrepreneurial savvy played a part as well. The rise of Wubi or "Five Stroke" input—invented by Wang Yong-ming, and later adopted by the Digital Equipment Corporation (DEC)—is a prime example of a system whose success was propelled in no small part by relentless marketing efforts. Far from assuming that the intrinsic merits of his system would win the day, Wang personally bankrolled the publication of two periodicals: *Wang Code Computer* (*Wangma diannao*) and *Chinese Computer* (*Zhongwen diannao*).[54] While ostensibly focused on Chinese computing in general, *Chinese Computer* was effectively a prolonged infomercial for Wubi input, masquerading as a trade publication (of the five lead articles in the inaugural issue, for example, three were explicitly about Wubi input).[55] Wang also made masterful use of international organizations, as well, tapping into intellectual property rights regimes newly available to mainland Chinese inventors following China's normalization of relations with the United States and other countries. Wubi was patented, not only in China, but also in the United States (in 1986) and the United Kingdom (in 1987).[56]

Aggressive promotional campaigns were another mainstay for Wang, who targeted government offices, ministries, and state-owned companies within the PRC. In addition to debuting his system on March 29, 1984, at the Capital Iron and Steel Corporation, Wang successfully lobbied the country's Ministry of Post and Telecommunications to implement Wubi input alongside the standard four-digit Chinese telegraph code.[57] Incessant self-promotion bore fruit. By one estimate, Wubi input became the most dominant system in mainland China during the 1980s, its users accounting for perhaps 85 percent or more of all computer users in China.[58] (Then again, it is difficult to know whether one can trust such statistics—as chances are fair that it might be based on Wang's own claims.)

OSCO, Three-Corner, and Five-Stroke input were not the only input systems to gain traction during the era—Cangjie, invented by Chu Bong-Foo, was taken up by IBM and implemented on their Multistation 5550 series, for example. Overall, however, the fate of most input systems was to linger in obscurity, with perhaps only one or two out of every hundred gaining market traction, or catching the eye of software or hardware manufacturers.[59] Despite the slim odds of achieving commercial success, however, nothing stopped—or even slowed—the invention of an ever-increasing number of input methods. A survey conducted in 1985 revealed around 400 different input systems, for example. By 1986, that number reached 450.[60]

ONE COMPUTER, MANY SYSTEMS

In the thick of his collaboration on the Olympia 1011 project, Zhi Bingyi received an out-of-the-blue letter in November 1978. It came from the Graphics Arts Research Foundation—the birthplace of Sinotype two decades prior.[61]

Like Olympia, GARF had been out of the Chinese computing game for years and wanted to try again. Samuel Caldwell's untimely death in 1959 effectively brought an end to the foundation's involvement with Sinotype, yet US military interest in the machine didn't wane. Ted S. Bonczyk, one of the original members of the Quartermaster Research and Development Facility who collaborated with Caldwell, continued to work on the machine with military associates. Eventually, the Army established a new relationship with the Radio Corporation of America (RCA), establishing a $656,000 contract administered by the US Army Research and Development Engineering Command in Natick, Massachusetts.[62]

At RCA, a reformulated Sinotype team was led by Fred Shashoua, a talented engineer and language prodigy fluent in Persian, Arabic, French, and English.[63] In RCA's renovation of Caldwell's Sinotype, the original keyboard was preserved, as was the viewfinder. RCA abandoned Caldwell's electromechanical design, however, moving to one based on fiber optics and television. The end goal remained the exact same, however: a QWERTY keyboard-controlled system designed to retrieve Chinese characters from memory and register them photographically, ultimately with the goal of fashioning lithographic plates for use in offset printing (figure 4.6).[64]

Even with RCA's improvements, Sinotype continued to pose many of the same challenges that had foiled its highly anticipated debut during the Eisenhower administration.[65] From July 1967 to February 1969, the Electronics and Special Warfare Board of the US Army Airborne subjected the machine to a battery of tests.[66] For the machine to serve military purposes, it had to be "capable of making approximately 6,000 Chinese characters most used in newspaper and leaflet composition."[67] It needed to operate in a wide temperature range, from 40 degrees to 90 degrees Fahrenheit, in conditions of relative humidity up to 70 percent. It also had to be simple and fast to load, transport, and offload, without need for spare parts.

The machine's report card was checkered. On the one hand, the 2,300-pound device proved surprisingly agile, with a crew loading it on and off a USAF C-130 aircraft in the span of just two and a half hours, all without a platform dolly. It was immediately put into operation in just one and a half hours, without need for spare parts—all promising signs.[68]

The prototype fell short in other areas, however. It could not run as long as desired, without requiring maintenance. Textual output left something to be desired, as well, with the rows and columns of Chinese characters falling slightly out of alignment. It was also prone to electrical failures.[69] "The Ideographic Composing Machines tested are not suitable for Army use," wrote Walter N. Mott, a US Army major. It "may be suitable for Army use when deficiencies listed below are corrected," he added. Once again, it seemed, the US military was growing dubious regarding the battle-readiness of the device.

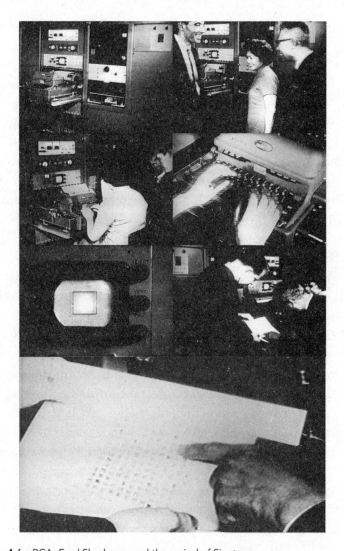

Figure 4.6   RCA, Fred Shashoua, and the revival of Sinotype

Ultimately, Sinotype's sojourn through the ranks of RCA and the American military brought it back to GARF.[70] This homecoming was due in large part to one man: Louis Rosenblum, who authored that November 1978 letter to Zhi Bingyi. Born in 1921 in New York City, he was yet another member of the MIT family in our story, graduating in 1942 with an undergraduate degree in applied math. Rosenblum took a job at Polaroid immediately following graduation, working on World War II–era military projects, studying under Harold Edgerton—the world-renowned professor of electrical engineering (and photographer of the famous "milk drop coronet" in the 1930s)—and later with Edwin Land on the development of instant photography. In 1954, he left Polaroid for Photon, where he concentrated on the photocomposition of non-Latin writing systems. Later, he worked for the Itek Corporation, another short-lived inheritor of the Sinotype project after Caldwell's death.[71] Deeply familiar with the late Caldwell's pioneering work on Sinotype, Rosenblum effectively adopted the project, and revived it after he joined GARF as a consultant in the mid-1970s.

During the same period, Rosenblum began to contemplate the Sinotype project more seriously. Reconnecting with GARF, which since Caldwell's time had grown to include a new cast of characters, Rosenblum helped assemble a core team.[72] Principal among them was Roy Hofheinz Jr., a man best known in Sinological circles as a scholar of Chinese Communist Party politics and rural administration, and as one-time director of Harvard's Fairbank Center.[73] Hofheinz also boasted a familiarity with computers, a skill set that made him the ideal candidate for Rosenblum's project.

Under Hofheinz's guidance, GARF began to work on a new generation of Sinotype, christened "Sinotype II."[74] The revised system would be based upon a Data General Nova 1200 with 32K of core memory—a system built primarily by John Forster, who served on the project as a "consulting engineer."[75] Sinotype II was supplemented with a Diablo 33F Moving Hard Disc with a disk cartridge patterned after the IBM 2315, formatted to hold 2.5 megabytes of data. For the purposes of display, the team used an industrial 15khz bandwidth television monitor, hooked to the main system by way of a Lexidata 200-D Video Image Processor interface. The screen would contain a 512-by-512 matrix, with each Chinese character occupying a 16-by-16 grid.[76] The interface would be a standard, 44-key, 2-shift QWERTY keyboard. Hofheinz helped develop the software.

If Sinotype II was going to succeed, advanced hardware wasn't going to be enough. It was going to need an input system. But Rosenblum and Hofheinz knew that the world of Chinese input had changed fundamentally since Caldwell's death. The number of competing input systems mushroomed—with hundreds now available, as opposed to the one or two possibilities circa 1959. Advances in hardware and software also made possible novel approaches that had been beyond Caldwell's reach.

In particular, the era of "one machine, one input system" was coming to an end—the era of Chung-Chin Kao and IBM, Caldwell and Sinotype, Three-Corner input and Wang Labs, Cangjie and the IBM 5550, and even Zhi Bingyi and the Olympia 1011. The growing power and affordability of minicomputers offered up the possibility of outfitting a single machine with multiple input systems—thereby enabling a user to toggle between different input systems, based on preference.[77]

For manufacturers, this "one computer, many input systems" framework presented obvious benefits. Betting everything on just one input system had placed companies in a vulnerable and volatile position, after all, rendering them susceptible to risks like those we saw in the case of Chung-Chin Kao and his four-digit cipher. In the one-machine-one-input framework, every commercial venture was an all-or-nothing affair: either the market adopted and embraced the input system, or it didn't. Not even the most advanced electrical engineering could change that fact.

Input designers saw benefits in this diversified arrangement, as well. With manufacturers moving away from exclusive agreements, input designers became "free agents," decoupling their systems from any single company, and instead trying to develop licensing agreements with an ever-increasing range.

It was this search for a new input system that brought Louis Rosenblum, Roy Hofheinz Jr., and GARF into contact with Zhi Bingyi—albeit by complete chance. By sheer coincidence, a short essay by Zhi was selected for translation and publication in mainland China's *New China News Agency English Language Report*. The August 13, 1978 report boasted of Zhi's successful partnership with Olympia, and shared limited details about the system. By some means, this English-language precis came to the attention of Rosenblum, who considered it interesting enough to pursue further. Later that same month, Louis Rosenblum got hold of a longer essay about Zhi, this one translated from the original Chinese by Hof-heinz himself (presumably at Rosenblum's request). The article, which appeared in the October 1979 issue of *Nature Magazine* (*Ziran zazhi*), was entitled "An Introduction to 'On Sight Encoding of Characters.'"[78]

From what he could deduce from the article, Hofheinz was clearly intrigued by the OSCO system. While reserving judgment, he noted to Rosenblum that the "basic concept of relying on ordinary spoken lan-guage means of describing the written language is an ingenious one."[79]

Rosenblum agreed, placing Zhi Bingyi squarely on GARF's radar.[80] In 1979, Zhi, Hofheinz, and Rosenblum communicated regularly, with dis-cussions quickly centering around an exhilarating idea: a GARF delega-tion, traveling to China with Sinotype II in tow, offering a demonstration of the machine to relevant parties in the PRC. Most intriguingly for Zhi, GARF recommended that they reprogram the machine so that, besides handling Caldwell's original input system, as well as Standard Telegraph Code input, it would be able to handle OSCO as well.[81]

After months of planning, GARF received a formal invitation from the PRC on September 13, 1979. The invitation was jointly issued by three parties: the General Bureau of Instruments and Meters, the First Ministry of Machine Building, and the *People's Daily*. The invitation requested that Hofheinz and his team arrive around November 5.[82]

The delegation from GARF comprised eight people: Richard (Dick) Solomon, the Sinologist, RAND corporation employee, and one-time aid for Henry Kissinger during "Ping-Pong diplomacy"; G. Prescott Low, pub-lisher of the *Patriot Ledger* newspaper in Quincy, Massachusetts, GARF's chairman of the board; William Garth IV, GARF's president; Louis Rosen-blum; William Shoenenberger, vice president of sales of the Optronics Corporation; John Forster; Susanne Mroczek, an administrative assistant at GARF; and Hofheinz. Their plan was to arrive in China on November 3 and stay through November 10—with the potential for extension.[83]

Zhi also tried to manage Hofheinz's expectations, however. The situ-ation in mainland China was changing rapidly, not only as a result of inroads made by foreign computing concerns, but also because of the rise of new domestic centers of computing. Zhi seemed worried, not only for Hofheinz and Sinotype, but also for himself. "I have to mention that recently competition for printing systems has increased," Zhi stressed.

"The British firm Monotype will soon have a demonstration of its products here, and Peking University has also developed a system. In spite of this I hope that GARF's demonstration will be successful and that a satisfactory cooperation contract will be materialized."[84]

Sinotype II was originally intended to drive a photocomposition machine, such as the one Optronics had built in the past.[85] The system was preprogrammed with a set of more than 8000 Chinese characters, all of which had been assigned four-digit codes according to the Chinese Telegraph Code (referred to as "STC" or "Standard Telegraph Code" numbers). Of these 8,000-plus characters, a subset of the 1,024 most common characters were also assigned OSCO codes, according to Zhi Bingyi's specifications. Prior to the trip, GARF had also produced 16-by-16 bitmaps for characters, in order for them to be displayed on the television monitor.[86]

The visit got off to an auspicious start. On the evening of November 3, a Saturday, a small team unpacked Sinotype II in front of Zhi Bingyi and his assistants. Unboxing the system, they connected the cables, and had the system up and running in less than 1.5 hours, all without a hitch. It was an "unplanned but rather effective display of the ruggedness and ease of use," Hofheinz glowed. From the moment it was unboxed, Sinotype II functioned with few interruptions from November 3 to 9 in Beijing, and then from November 10 to 13 in Shanghai. It "was further evidence of the manufacturability and reliability of the production sub-systems," Hofheinz continued.[87]

Demonstrations of the machine were even more impressive. Zhi and his team had prepared thoroughly, his typists reaching speeds of around 2,000 characters per hour, or about 50 characters per minute.[88] The team's inclusion of Standard Telegraph Code input—effectively a descendant of Chung-Chin Kao's four-digit code from the 1940s—scored an unexpected public relations coup, moreover, when it was discovered that *People's Daily* editor An Wenyi was, in effect, "fluent" in STC code. During the team's November 7 demonstration, performed in front of more than one hundred people in a conference hall at the *People's Daily*, An Wenyi took a seat at the console and proceeded to enter an unbroken string of Chinese characters using STC input. "Mr. An had been a radio/telegraph operator many years ago," a later GARF report recounted, "and had achieved great proficiency using a 'blind keyboard.' . . . He had memorized the more

Figure 4.7 Photograph of An Wenyi, deputy secretary of the *People's Daily*, entering characters on the Sinotype using Standard Telegraph Code (STC) input

than 8,000 STC numbers when he had been . . . a young soldier during the war of liberation."[89] "His elation at seeing ideographs echoed immediately on the video display terminal as he rapidly input four-digit codes is evidenced in one of our trip color photographs" (figure 4.7).

In mid-November, the delegation departed in positively stratospheric spirits, having signed a Joint Protocol in Shanghai on November 13.[90] Heralded as the "culmination of a ten-day tour," the protocol inaugurated the "joint development of modern computer equipment and systems for the typesetting of Chinese ideographs." The team left behind more than 800 sets of Chinese text specimens, prepared in advance of the trip on a Pagitron by Bill Shoenenberger's Optronics International. A bold set of plans had already begun to take shape, moreover, including a "Five-Year Plan" whose centerpiece was to be a "Joint Chinese-American Research Center" to be located somewhere in the Boston area.[91]

HALLUCINATIONS MADE REAL

In late October 1980, members of the Graphics Arts Research Foundation touched down on mainland Chinese soil for a second time.[92] Memories of the 1979 Joint Protocol were fresh in their memory, and hopes were high. This optimism was fueled over the intervening months. Photographs from the 1979 visit were passed back and forth. Letters of thanks and well wishes were exchanged. A bond had clearly been formed. Roy Hofheinz even had a chance to pay a visit to Zhi Bingyi in the interim, the pair meeting in July 1980 to hammer out draft details for a bold, $150,000 collaboration. The stage was set for something momentous.

But the China they found in 1980 was a surprisingly different place. Zhi and the Chinese side had become patently uninterested in GARF's Five-Year Plan. They now preferred something more flexible—to be able to order off the menu, so to speak, rather than committing to GARF's twelve-course banquet. Even before the delegation arrived, in fact, evidence of this shift was already becoming apparent to anyone at GARF who might have looked. In a letter directed to Louis Rosenblum on March 3, 1980, for example, Zhi began to downplay the idea of the Five-Year Plan, speaking instead about a "small-scale phototypesetting system proposal" that promised a more immediate pay-off. Rather than treating Sinotype as part of "an intelligent terminal" used to control a wide and far-ranging system—a key vision that the GARF team had advanced during its 1979 visit—Zhi sketched out what he described as a "conception for the project of a simple photocomposing machine," one that for GARF representatives sounded more or less like existing Pagitron machines: an input terminal, a laser phototypesetting unit, floppy disk output, a printer, a CRT display, and a 32-by-32 dot-matrix font.[93]

China was growing more self-confident, it seemed, in no small part thanks to its newfound bargaining power. Beginning in the early 1980s, computer manufacturers across the world were beginning to stumble over each other in an attempt to gain a toehold in the PRC (a topic we will explore in more depth in the next chapter). Why, then, would the Chinese side wish to lock itself in a five-year relationship with GARF—a relatively minor player in the field—when it had a world of options at its fingertips?

Other changes were diminishing GARF's chances further—notably in the arena of computer hardware. Suddenly, the Chinese side regarded GARF's choice of backend—the 16-bit Data General Nova 1200 minicomputer—as a complete nonstarter, insofar as the country's high-level ministers had already decided to invest in copying the DEC PDP-11. Even more than this, however, the fate of minicomputers in general was being called into question, in light of the meteoric rise of their nimbler cousins: the microcomputer.[94] In a few short years, the most popular imports to the PRC included the Apple II, the Cromemco CS-1, and the Cromemco CS-2 (named after the dormitory where the founders lived while students at Stanford University, Crothers Memorial Hall).[95] A photograph from the era captures the giddiness of the China trade, showing Cromemco founders Harry Garland and Roger Melen grinning ear-to-ear as they stood before a truck packed to the gills with PCs bound for the PRC (figure 4.8).[96]

The more money that poured into China's computing circles, moreover, the greater the forces of professionalization, formalization, and bureaucratization became. Back in the era of five- and six-figure partnerships, there was still considerable room for entrepreneurial individuals like Zhi Bingyi, Chu Bong-Foo, Wang Yongming and others to establish more or less personal ties to manufacturing concerns—even ones with the size and clout of IBM, Olympia Werke, DEC, and more. But as China's

Figure 4.8 Photograph of Harry Garland and Roger Melen shipping "Cromemco" computers to the People's Republic of China

computing sector grew in size and potential—to seven, eight, and nine figures, at least in the aggregate—computing technology steadily became the purview of Chinese bureaus and ministries, whether the Bureau of Machine Import and Export, the Ministry of Electronics Industry National Computer Industrial Bureau, or otherwise. We see this transformation play out most clearly in the progressive sidelining of Zhi Bingyi, for example, who went from being a relatively central player in the 1979 GARF visit, to a marginal player by 1980 (all despite Zhi's election to the Chinese Academy of Sciences, and his promotion to chief engineer of the Shanghai Instruments Research Institute, both in 1980).[97] Despite the oral agreement that Zhi and Hofheinz had come to during their July 1980 conversation, then, a follow-up meeting—this one privy exclusively to Chinese ministry officials, without Zhi in attendance—effectively took control over discussions with GARF, changing many of the terms of the tentative agreement.[98]

The GARF delegation boarded their outbound plane empty-handed. Negotiations had broken down entirely, sending the team into a scramble to pick up the pieces of a once-promising relationship. Hosting a Chinese delegation to the United States, exploring the possibility of outfitting Sinotype II with additional Chinese input systems, and maintaining close contact with their counterparts in Shanghai, they held out hope of salvaging the Joint Protocol.[99]

But it was not to be. The dynamics witnessed in the 1980s constituted the new normal, rather than an anomaly. "In the past two months," Louis Rosenblum wrote to Dick Solomon in 1982, in advance of Solomon's upcoming return visit to China, "we have become aware of three more micro-processor-based systems for Chinese ideographs. . . . Neither these nor any of the earlier systems that we know of has the combination of capacity and speed that are part of the Sinotype," Rosenblum assured Solomon—but it was cold comfort. Solomon's prognosis was grim. "My own conclusions about the current state of Chinese language word processing and possible interest in GARF are as follows: The Chinese are actively pursuing the development of their own systems and are unlikely to want to put their own resources into a 'foreign' system, unless it has unique characteristics that they feel they cannot reproduce on their own (let's say within a five-year time framework)."[100]

Solomon's return from China in 1983 brought only more bad news for GARF. He relayed word of an upcoming conference in China to be held in August 1983, wherein presentations were to be made on more than 200 different Chinese input systems—20 of which had already been implemented on machines. Solomon's interlocutor in China went on to explain to him that, by this point, "some of the most efficient Chinese input systems now take approximately one week to learn and can operate at [an] approximately 80-character-per-minute inputting speed with a reasonably experienced keyboard operator." When Solomon shared a sample of Sinotype's Chinese character printout with two of China's leading editors—perhaps in an effort to show the relative strengths of the system—the "reaction was a polite, 'Oh, now we can generate characters of this sort in a 32 × 32 matrix.'"[101]

GARF's efforts with Sinotype II ended in failure.[102] "I only regret that the longstanding connection with GARF came to such an end," Zhi concluded in a letter to William Garth IV.[103] The sentiment was clearly genuine, as Zhi's hopes along with GARF's were dashed on the very same rocks.[104]

As disappointing as this failed partnership must have been for Zhi, it would have been difficult for the Shanghai engineer not to stand in awe at what Chinese computing had become in the span of just a few years. A decade prior, sitting in his darkened cell, with long stretches of boredom punctuated by moments of dread, he traced out ephemeral alphabetic codes on the underside of a teacup, and hallucinated about a fully mature Chinese language information environment. Now, it seemed as if his hallucinations were becoming real.

In 1984, Zhi put forth a proposition that would have once been dismissed as hyperbole, if not absurdity: It was now within the realm of technological possibility for the computational input of Chinese characters, not only to match the speed, accuracy, and efficiency of English, but to surpass it.[105] In the near future, Zhi asserted, Chinese might very well become the fastest writing system in the digital age.

5

THE SEARCH FOR MODDING CHINA: PRINTERS, SCREENS, AND THE POLITICS OF PERIPHERALS

When the guests in Room 426 stepped out for the morning, housekeepers at the Great Wall Hotel set about their work. They weren't emptying ash trays for jet-lagged diplomats, however, or refilling hot water thermoses for tourists. Today, Room 426 was the makeshift office of IBM in China, one of the many foreign companies tripping over themselves to get in on China's personal computing "buying binge."[1]

It was 1985, and the epicenter of Beijing's PC revolution—"Electronics Street"—was humming. Along Haidian Road, in the city's Zhongguancun neighborhood, newly opened computing shops brimmed with curious students, teachers, scientists, and even Traditional Chinese Medicine practitioners, all jockeying to try out the latest devices.[2]

Foreign companies rushed to get in on the bonanza. Room 5063 at the Beijing Hotel became the provisional office of the Sony Corporation. Room 2201 at the Minzu Hotel: NEC. Room 1339 at the Friendship Hotel: PerkinElmer. Rooms 921 and 1022 at the Xiyuan Hotel: DEC and Toshiba. By one estimate, firms were racking up US$5,000 each month in room charges alone to house dozens of company representatives at a time.[3]

Their enthusiasm is simple enough to understand. Microcomputer imports to the PRC grew from a paltry 600 in 1980 to an anticipated 130,000 in 1985—and from there, who knew how high the number might grow? Fueling the fire, Deng Xiaoping's Reform Era government publicly committed an immense chunk of its closely guarded hard-currency reserves to purchasing computers and other advanced technologies—no less than US$1 billion in 1984 alone.[4]

But two formidable obstacles stood in the way of large-scale, across-the-board computerization. Cost was the first. At the Shanghai Computer

Store, for instance, customers could in theory walk out with a brand-new Apple II with 8K of RAM—if only they could stomach the thought of parting with three- or four-years' worth of wages. When machines like this did sell, it was mainly to Chinese work units, rather than individual consumers.[5]

Compatibility, the second obstacle, was the bigger problem. No matter how much expendable income a Chinese customer might have enjoyed, none of these Western-built machines could handle Chinese character input or output—not in the early and mid-1980s, at least, and certainly not "out of the box." Western-built printers couldn't natively output Chinese characters, Western-built monitors couldn't display them, and Western-designed operating systems couldn't deal with the intricacies of Chinese input methods. Effectively, personal computers were useless at this point for anyone hoping to work in Chinese.

In this chapter, we shift our focus away from keyboards and input systems to a broader ecology of computing peripherals: monitors, printers, and more. Despite the marginal sound of their name, "peripherals" are at the very center of the history of early Chinese personal computing. Like the personal computer itself, Western-manufactured monitors, dot-matrix printers, graphic cards, and more were designed with the Latin alphabet in mind—a deep-seated bias that, as we will see, was sometimes literally baked in at the metallurgical level. At precisely the moment when engineers had begun to overcome the problem of keyboard-based Chinese input, then—chopping off the first monstrous head we might call "the QWERTY problem"—the hydra of alphabetic order cropped up an entirely new set of challenges, each threatening to exclude Chinese script once again from the rapidly unfolding computing revolution.

This chapter also seeks to distinguish between "modding" "copycatting" and "piracy" that appear in accounts of Chinese computing during this pivotal period in the late 1970s and 1980s.[6] When encountering programs such as "Chinese DOS," knee-jerk reactions in the Western world have been to treat them just as so many "knock-offs," lumped alongside other examples of Chinese intellectual property theft (from the humorous "Kennedy Fried Chicken" to finely produced Prada counterfeits).[7] While piracy was no doubt widespread during the era, and while it no doubt extended into the realm of computing, the modifications of programs and devices we'll examine in this chapter were not motivated primarily by cost (*How to get a given product, only less expensively*) but by overcoming exclusion (*How to take a wide ecology of computing, practically every part of which exiles Chinese, and tweak it to be interoperable with a script employed by more than one billion people*).[8] All of the "hacking" and "modding" to be discussed in this chapter, in other words, was essential for Chinese engineers trying to find inroads into an exclusionary global information order.

FEATS OF MEMORY

The first domain to examine is computer memory, where Latin alphabet-centrism was on conspicuous display. At the advent of Latin alphabetic computing and word processing, Western engineers and designers determined that a low-resolution (barely legible) digital font for English could be built upon a 5-by-7 bitmap grid—requiring only 5 bytes of memory per symbol. Although far from aesthetically pleasing, this grid offered sufficient resolution to render the letters of the Latin alphabet legibly on a computer terminal or a paper printout, as well as Arabic numerals, punctuation marks, and other necessary symbols. Using this grid, the storage of the 128 characters in standard ASCII (the American Standard Code for Information Interchange) required just 640 bytes of memory—a tiny fraction of, for example, the Apple II's then 48K of motherboard memory.

To achieve comparable, bare-minimum legibility for Chinese characters, a 5-by-7 grid was far too small. For Chinese, engineers had no choice but to increase the size of the grid geometrically, to 16-by-16 pixels or larger—requiring at least 32 bytes of memory *per Chinese character*. The total memory requirement to store just the bitmaps (in either simplified or traditional form, but not both, and with no accompanying metadata) would equal approximately 256K for the 8,000 most commonly used Chinese characters, or four times the total capacity of most off-the-shelf personal computers in the early 1980s.[9] (All this, even before accounting for the RAM requirements for the operating system and application software.) In concrete terms, it meant that storing just 20 low-resolution Chinese characters demanded the same amount of memory as a full, low-resolution ASCII character set (figure 5.1).

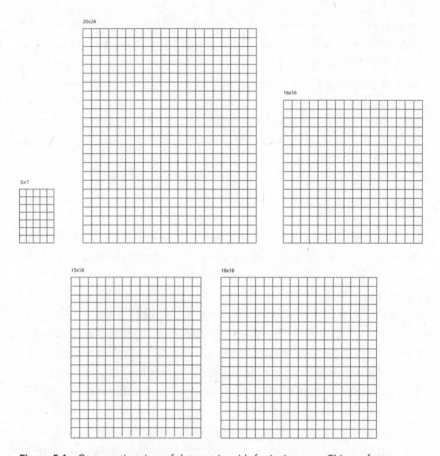

Figure 5.1   Comparative sizes of dot-matrix grids for Latin versus Chinese fonts

Chinese obviously contains more than 20 characters. Were we to imagine a font with perhaps 4,000 of the most used Chinese characters, the
total memory requirements to store just the bitmaps would equal approximately 128K, or twice as much as store-bought personal computers in
the early 1980s could handle.[10]

At higher resolutions, the problem of memory became even more acute.
By expanding this bitmap grid even slightly—from a bare-minimum 16-
by-16 to a modestly better 24-by-24—corresponding memory requirements for *each* Chinese character would jump to 72 bytes, or around
288K for the entire set of 4,000 characters (table 5.1).[11]

Table 5.1 Memory requirements of English and Chinese bitmaps at different resolutions, based on a 128-character set for English, and a 4,000-character set for Chinese

English (5-by-7)	640B
Chinese (16-by-16)	~128K (or 200 times the size of the English bitmap font)
Chinese (24-by-24)	~288K (or 400 times the size of the English bitmap font)

This was a problem that no one in Silicon Valley—or, indeed, any other center of commercial computing—had a solution to. For many, indeed, it had barely registered as a problem.

The number of installation disks that came with Chinese-compatible disk operating systems in the late 1980s and early 1990s serves as a testament to this challenge.[12] When customers purchased UCDOS 90, for example, the box contained a total of ten floppy disks. Of the ten, however, only two were dedicated to the operating system. The third and fourth disks were dedicated to a single low-resolution, 24-by-24 printer font.[13] The remaining six floppies were filled up with just three optional fonts (figure 5.2).[14]

Given these profound challenges, developers of early Chinese personal computers explored every available option in their effort to juice as much memory out of their systems as possible. We will explore two strategies in particular, sometimes employed in isolation, but often in concert: Adaptive Memory and Chinese Character Cards.

ADAPTIVE MEMORY

To examine the first of these strategies, we return once again to the offices of the Graphic Arts Research Foundation and its aborted efforts to develop the Sinotype II with engineer Zhi Bingyi. Despite that setback GARF continued to work on the Sinotype project well into the early 1980s. By this point, the foundation's advisory board had grown to welcome another cohort of renowned scholars: the Harvard linguist Susumo Kuno; and Richard Solomon, then head of the Social Science Department at the RAND Corporation, but better known for his pivotal role in Richard Nixon's visit to the PRC in 1972.[15]

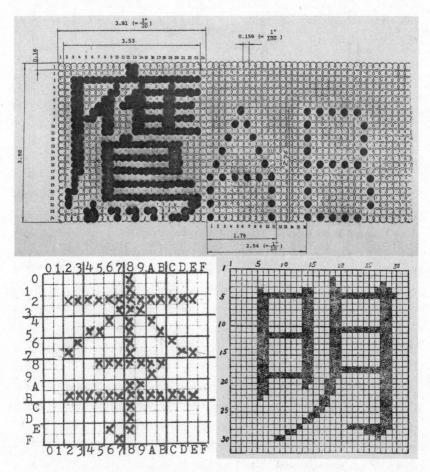

Figure 5.2 Examples of Chinese bitmaps

As stellar as this brain trust was, however, GARF's major breakthrough on the Sinotype project—the leap from a minicomputer-based system (Sinotype II) to one based on a microcomputer (Sinotype III)—was catalyzed by a college student whose only experience at GARF at that point was a brief, two-week gig working on data management for the Sinotype II project in 1979. He was Bruce Rosenblum, Louis Rosenblum's son.

As an undergraduate at the University of Pennsylvania and an aspiring photojournalist, Bruce was balancing time between coursework and his role as photo editor for the student-run newspaper *Daily Pennsylvanian*. The paper was remarkably advanced in terms of the equipment it operated, as well as the expertise of the students in charge. By the fall of Bruce's junior year, the paper's two Compugraphic typesetters were on their last legs. Along with three of his student colleagues at the paper, Bruce helped research potential replacements, exposing the team to state-of-the-art front-end systems for computational typesetting. They eventually settled on a combined $125,000 contract with two companies: Mycro-Tek in Wichita, Kansas, and Compugraphic in Wilmington, Massachusetts.

As for the Sinotype project—one that Bruce was aware of thanks to his father—a pivotal moment came in early May 1981. Bruce had just completed his final exams, and he stopped by the offices of the paper. His newspaper colleague Eric Jacobs was there, hard at work on a TRS-80 Model II personal computer from Radio Shack. Jacobs was contemplating how a microcomputer might be used to run the newspaper's business operations. Bruce observed for perhaps a half-hour before heading on with his day.

But those thirty minutes stuck with him. "It was the first time I'd ever seen anyone work on a microcomputer," Bruce recalled to me, "and those few minutes were the inspiration that triggered the whole Sinotype III project and eventually my career in computers."[16]

Bruce made an off-the-cuff remark in a phone call with his father later that same week. Referencing the immense cost of the Data General hardware used to build Sinotype II, Bruce conjectured that someone could probably program something equivalent or better on a microcomputer for a fraction of the cost—perhaps with as little as $10,000 worth of hardware, as compared to the more-than $100,000 price tag for the equipment GARF was currently using.

The elder Rosenblum was intrigued and asked Bruce if he himself might be up to the task of programming such a machine. Bruce boasted no formal training in computer science, although he had worked intensively with computers in high school and taught himself both PDP-8 assembly language and BASIC. "Sure," Bruce said, responding to his father's query with "the chutzpah of a newly minted graduate who had no immediate job prospects."

In June 1981, Bruce Rosenblum donned a three-piece suit and presented his Sinotype III proposal to Bill Garth, Prescott Low, and his father.[17] He tallied a total cost of just $12,500, split between hardware costs and programming labor. For the price, the proposal promised a Chinese word processor, running on an Apple II, delivered in approximately four months' time.

It would reduce the cost of Chinese computers by an order of magnitude—if it worked.[18]

Bruce got the job. He went on to program Sinotype III from June to November 1981, splitting his time between this and a full-time tour guide job at Independence National Historic Park in Philadelphia. During work breaks he would write out assembly code by hand, transcribing it at night. When Labor Day in 1981 came, and Bruce's tour guide job ended, he dedicated two months straight to finishing the code and delivered it to GARF.[19]

The Sinotype III system comprised five components: a Sanyo DM5012CM 12-inch monitor, an Epson MX-70 printer, a Corvus 10 MB "Rigid Disk Storage" to store the Chinese character bitmap database and the corresponding "descriptor codes," an Apple Disk Drive for the storage of text files, and the Apple II itself (figure 5.3)

Out of the box, the Apple II Bruce purchased came with 32K of RAM, extensible to 48K on the motherboard.[20] "We maxed that out even before the Apple II left the store," Bruce remarked by email to me. Since 48K of memory was still far too little for his purposes, however, Bruce opted for what, at the time, was a fully standard modification, commonly employed by so-called power users of the era: namely, to insert an additional 16K memory card in slot 0, thereby bringing the total available memory to 64K.

Even this was too little. "I needed more RAM to store a full encoding system," he wrote, "and also the 16-by-16 bitmaps for the 100 most frequent ideographs." He began to explore a "mod" of the Apple II that few if any others had tried before. "Somehow," Bruce recounted, "I figured out I could put a second 16k board in slot 2 of the Apple II, and that gave me a total of 80k. . . . Completely nonstandard," he continued, "but it worked with off-the-shelf components."

This modification, however, pushed the machine past its own limitations. The 6502 microprocessor on the Apple II was only capable of accessing 64K of memory directly—meaning that, even with the additional 16K

Figure 5.3 Photograph of the Sinotype III system

Bruce had managed to bootstrap in with the second memory board, there was simply no built-in way for the Apple II to simultaneously access these additional addresses in memory.[21]

To enable the Apple II to access 80K of memory, rather than just 64K, Bruce dispensed with the out-of-the-box operating system and programmed his own in assembly language. Key to his custom-designed program was the possibility of "selecting between two banks of 16K that overlap each other."[22] In other words, although only 64K worth of memory locations would be accessible at any one instant, by rapidly oscillating between the two memory expansion cards he could in effect trick the computer into accessing both at speeds that, from the perspective of the user, would be negligible. Bruce squeezed 25 percent more memory out of the system using this trick, enabling the inclusion of perhaps as many as 400 more Chinese characters in onboard memory. So "nonstandard" was this mod that, when he told an Apple engineer about it during one of their many conversations, the Apple engineer was shocked—he had never heard of, or thought of, doing such a thing.[23] The week before Thanksgiving, Bruce delivered the final code to GARF.

Even with Bruce's ingenious mod, however, he and his father estimated that a mere 600 to 1,000 Chinese characters would be able to fit in onboard memory.[24] When accounting for the size of Sinotype III's operating system, program applications, and the memory requirements of each Chinese character (approximately 40 bytes per graph, split between the bitmap and input codes), the vast majority of Chinese characters in the machine's lexicon would need to be stored somewhere else. Bruce briefly contemplated using PROM (programmable read-only memory) chips— but this idea quickly revealed itself to be a dead end. Circa 1981 and 1982, the largest PROM chips on the market maxed out at 2K of memory, which translated into a mere 28 to 51 Chinese characters.[25] In order to store 7,000 Chinese characters in this fashion, then, Bruce would have needed between 138 and 250 PROM chips. "That's a lot of chips," he remarked.[26] Floppy disks were a dead end as well, Bruce decided, not only because of the large number of disks it would have required, but also the slow access and retrieval speeds involved in fetching character bitmaps from floppy drive storage.

GARF opted for a third solution: to outfit Sinotype III with an external hard drive, which at the time was an almost unheard-of microcomputer accessory. To overcome the profound memory limitations, GARF would store lower-frequency Chinese characters "off-site" in the system's external hard drive: a 10 MB Corvus "Rigid Disk Storage." Many thousands of Chinese characters would be stored in this fashion.

But reliance on hard-drive storage still posed challenges for Sinotype III's operating speed. Hard drives relied on rigid magnetic disks— "platters"—that rotated within the device, and a head that read the contents of various "tracks" (not unlike the needle of a record player reading the grooves on an LP). Retrieval speeds depended therefore upon where the head was located at the moment of the retrieval request. If the user was fortunate, the head would be just about to arrive at the platter location of the desired data. But if the head had just passed that location, it would be like arriving at a bus stop only to watch the bus pull away—then waiting for the driver to complete the route and come back around.

Retrieval times for Chinese characters stored on the hard drive were ten times slower than those stored in RAM. As Louis Rosenblum explained in a June 30, 1981 letter to Zhi Bingyi, with whom GARF still maintained regular contact, the retrieval of Chinese characters stored in RAM could be achieved in approximately 100 milliseconds per character—a unit of time imperceptible by human cognition. As for the characters stored in external storage, however, as much as a full second would be needed to access and retrieve them—a unit of time well within the threshold of human perception.

A one-second input time would have proven devastatingly slow within the context of mid-1980s personal computing, where users in English-language contexts were quickly becoming accustomed to the sensation of real-time typing. Moreover, one second is (obviously) ten times as long as 100 milliseconds. The average user would be able to feel this differential each and every time he or she wished to input lower-frequency characters.[27]

To mitigate this problem, Louis Rosenblum hit upon an idea he referred to as "adaptive temporary storage" or "temporary adaptive memory." Sinotype III would be able to adjust the set of characters stored in RAM depending upon what the user had recently inputted. Upon initial boot-up, Sinotype III's onboard RAM would be outfitted only with a predetermined set of high-frequency characters. The inputting of any hard-drive-based infrequent character would take up to one second, as noted above. But "as each of the less frequent ideographs is keyboarded," Louis Rosenblum explained, "its code and dot matrix pattern will be noted in the random-access memory."[28] In other words, such characters would be temporarily copied from the hard drive to an onboard RAM cache, thereby reducing subsequent retrieval times to 100 milliseconds.[29]

"Adaptive memory" has antecedents in the history of analogue Chinese text technologies, a fact that the engineers at GARF knew (particularly because of their visit to Chinese newspaper plants during the

Sinotype II delegation, examined in chapter 4). In the physical layout of the Qing dynasty–era Wuying dian printing office, where printers moved peripatetically among the different cabinets of movable Chinese type, the relative frequency of these different Chinese characters—which numbered in the tens of thousands—informed the spatial logic by which they were arranged in the printing office. Frequent characters were placed as closely as possible to the press, and less frequent ones further away—not unlike the characters stored in Sinotype III's RAM versus the external hard drive.

Naturally, in absolute terms the computational speeds of Sinotype III were much faster than the act of physically fetching type from a cabinet. But the *relative* difference between high- and low-frequency character retrieval times—the 10-to-1 ratio between Sinotype III's hard drive and onboard memory, compared to distance between characters stored by a printer's side versus those stored, say, in a cabinet on the opposite end of the room—is commensurable.

"Adaptive memory" was also a core practice within mechanical Chinese typewriting, although not by this name. For Chinese typists, the most common characters were clustered as much as possible in the very center of the character tray bed, with less common characters located on the left and right flanks. The most infrequent characters were stored "offsite," in a wooden box of secondary usage characters—again comparable to the Corvus drive. To type a character from this set required the typist to use a pair of tweezers to retrieve the metal slug from the box, place it in the grid of their tray bed, and then type. In Chinese typewriting as in Chinese movable type, then, the 10-to-1 ratio between high- and low-frequency character printing speeds is again a reasonable estimate.

Even more than this, however, another technique was used by both Chinese printers and typists that bore uncanny resemblance to Sinotype III's adaptive memory. Namely, in order to reduce the time spent moving between type cabinets, compositors and typists often kept a small, additional set of Chinese characters close at hand—characters that, although infrequent in general, happened to appear very frequently in whatever text or typescript was being prepared on that particular day. Because these characters were what we might term "temporarily frequent," it made sense to keep them close-by until this particular job was completed—at which point the characters could be returned to their proper locations. Although they made no explicit reference to these late imperial and early twentieth-century Chinese precedents, Louis Rosenblum and GARF effectively replicated this approach, albeit in digital form.[30]

CHINESE-ON-A-CHIP

Toggling and adaptive memory helped, but there were still many thousands of characters that fell beyond the limits of such strategies. While high-frequency Chinese characters accounted for a large percentage of overall usage, the production of any kind of technical or specialist content would have certainly brought the user repeatedly into the off-site repository of Chinese characters. More of these "low-frequency" characters needed to be brought "on-site" if the experience of Chinese computing was ever going to approach the same feeling of instantaneity enjoyed by English-language counterparts.

Engineers in the early and mid-1980s began to explore more sophisticated and engineering-intensive hardware solutions, referred to alternatively as "Chinese Character Cards" (*Hanka*), "Chinese Cards" (*Zhongwenka*), "Chinese Character Generators," "Chinese Font Generators" (*Hanzi zimo fashengqi*), and most delightfully, "Chinese-on-a-Chip."[31] In effect, these stand-alone cards represented next-generation versions of the microprocessors found on the Ideographix IPX and Olympia 1011 systems explored in chapter 3. Rather than being hard-wired into the motherboard, however, these cards could be purchased independently and installed via motherboard expansion slots, akin to graphics cards and memory cards. No longer was it necessary to buy a stand-alone Chinese word processor, such as the Olympia 1011, to gain access to thousands of Chinese bitmaps and input encodings: the Chinese-on-a-Chip could be installed on the user's machine of choice.[32]

Among the earliest centers of Chinese computing to focus on Chinese Character Cards was Tsinghua University, where researchers developed an early model capable of storing approximately 6,000 Chinese bitmap patterns in 32-by-32 dot matrix format.[33] By the mid- and late 1980s, dozens of different *Hanka* entered the market, such as the Founder Chinese Card, Legend Chinese Card, Giant Dragon Chinese Card, and others manufactured and marketed by companies across China, Taiwan, Hong Kong, Japan, the United States and elsewhere.[34] By the closing years of the 1980s, practically all computers boasting Chinese- or Japanese-language capabilities featured a character card of one sort of another.

Thus, from the 1950s with Caldwell's Sinotype to the father-son Rosenblum team (with GARF) and Sinotype III in the 1980s, solving the memory problems associated with Chinese characters was the linchpin to opening the Chinese market to computing. Hacking computers to free up more memory, creating adaptive memory algorithms for prioritizing characters, and building dedicated hardware tailored to the particular needs of Chinese input/output (I/O) bridged the problem and helped catalyze the personal computing revolution in China.

But the problem remained: How to expand beyond the motherboard itself to everything that might connect to it? Our discussion continues, therefore, with a deep-dive into the challenges of "modding" early computer printers, monitors, and other peripherals originally incapable of handling Chinese text output.[35]

DOT-MATRIX PRINTING AND THE METALLURGICAL DEPTHS OF ALPHABETIC ORDER

Beyond the memory issues associated with storing and recalling Chinese fonts, further challenges confronted engineers seeking to render Western-manufactured printers compatible with the needs of Chinese script output. You cannot see what you cannot output, after all.

Commercial dot-matrix printing was yet another arena in which the needs of Chinese character I/O were not accounted for. This is witnessed most clearly in the then-dominant configuration of printer heads—specifically the 9-pin printer heads found in mass-manufactured dot-matrix printers during the 1970s. Using nine pins, these early dot-matrix printers were able to produce low-resolution Latin alphabet bitmaps with just one pass of the printer head. The choice of nine pins, in other words, was "tuned" to the needs of Latin alphabetic script.

These same printer heads were incapable of printing low-resolution Chinese character bitmaps using anything less than two full passes of the printer head, one below the other. Two-pass printing dramatically increased the time needed to print Chinese as compared to English, however, and introduced graphical inaccuracies, whether due to inconsistencies in the advancement of the platen or uneven ink registration (that is, characters with differing ink densities on their upper and lower halves).

Compounding these problems, Chinese characters printed in this way were twice the height of English words. This created comically distorted printouts in which English words appeared austere and economical, while Chinese characters appeared grotesquely oversized. Not only did this waste paper, but it left Chinese-language documents looking something like large-print children's books. When consumers in the Chinese-Japanese-Korean (CJK) world began to import Western-manufactured dot-matrix printers, then, they faced yet another facet of Latin alphabetic bias.

The politics of early dot-matrix printing—this embedded Latin-alphabet centrism—ran even deeper, as discovered in the early work of a man we met in chapter 3, Chan-hui Yeh, the developer of the 120-shift IPX device. When Yeh set out to digitize and print Chinese characters based on a bitmap grid of 18-by-22, his initial idea was an obvious one: to shrink the size of existing printer pins so as to pack more of them on the printer head. More pins, with smaller diameters, would in theory solve the problem. But Yeh's quickly discovered solution would not be so simply carried out.

The Latin alphabetic bias of impact printing, Yeh found, was baked into the very metallurgical properties of printer components themselves. Simply put, the metal alloys used to fabricate printer pins were themselves "tuned" to 9-pin Latin alphabetic printing; reducing the pins' diameters to the sizes needed would result in deformation or breakage.[36] To phrase this another way, the recipes for the metal alloys used to fabricate printer pins were painstakingly calibrated to the standard that worked for Latin alphabetic printing: 9-pin printer heads with each pin's diameter measuring 0.34mm. To use an alloy significantly more durable than this would have been a needless expense from the perspective of manufacturers who operated under the (likely unconscious) assumption of "A through Z."

To compensate, engineers in China, Taiwan, and elsewhere devised a series of workarounds. In one mod, some engineers tricked Western-built printers into fitting as many as 18 dots—two passes of a 9-pin printer, that is—within roughly the same amount of vertical space as nine conventionally spaced dots. So rather than being twice as tall as their Latin letter counterparts, Chinese characters would be the same height, but twice as dense in terms of pixel placement.

The technique was as ingenious as it was simple. First, an initial array of dots was laid down during the first pass of the printing head. But then, instead of laying down the second array of dots beneath the first, they reprogrammed the printer into registering this second batch of 9 dots *in between* the first set—like the teeth of a zipper, fastening together. Specifically, engineers rewrote printer drivers to modify the machine's paper advance mechanism, refining it so that it rotated at an extremely small interval so as to tuck the second set of 9 dots *inside* the first set (figure 5.4).[37]

Mods such as the ones described here were essential to early 1980s-era Chinese computing. Indeed, they formed a core and accepted part of the early economy of Chinese microcomputing—accepted in some cases even by the foreign companies whose products were being "hacked." A prime example here is the early history of the Sitong Group, known in English as Stone. Established in May 1984, Stone quickly rose to prominence as one of the most important players in early Chinese electronic typewriters and word processors. Many don't recall, however, that the company's first production and marketing strategy was focused squarely on modding—specifically on retrofitting the Japanese-built Brother 2024 printer for the Chinese market.

As the Stone employee Wang Jizhi recalled, there was only one 24-pin printer available on the mainland Chinese market circa 1984: the Toshiba 3070, imported by the Fourth Machinery Ministry. Given the staggering cost of this machine (with an import price of more than US$1,000), Wang and others at Stone saw an opportunity. Wang caught word that the Beijing Computer Technology Research Institute had gotten hold of a Brother 2024 24-pin printer. Although capable of printing Japanese kanji and kana, they learned, Chinese-language output came out garbled and unusable.[38]

Executives at Stone launched their new company effectively on the basis of "copycatting" the Brother 2024, rewriting the printer driver with the goal of rendering it compatible with Chinese. The team got in touch with a researcher at the Chinese Academy of Sciences, Cui Tienan, hiring

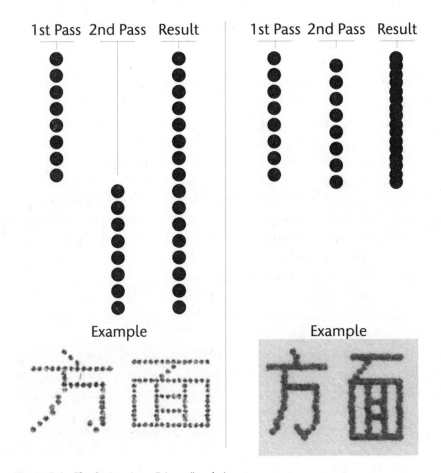

Figure 5.4 The 9-pin printer "zipper" technique

him and three of his colleagues. Cui and his crew managed to rewrite the printer driver in a mere eight hours, thereby opening the door to Stone. Acquiring the machines from abroad, the company would then release them on the Chinese market "after a little bit of computational opening-up work."[39] The plan worked, Stone's retrofitted 24-pin printer becoming a runaway success on the small but growing mainland Chinese market. (Meanwhile, Cui and his colleagues received a total of 200 RMB for their labor.)

The following year was even better. If 1984 was Sitong's year of (reprogramming) the Brother 2024, 1985 was the year of the ITOH-1570 color printer. This printer, also built in Japan, contained an onboard character generator (*Hanka*) capable of producing kanji characters. The encoding system depended upon the Japanese-standard JIS encoding, however, and so once again a retrofit was needed. Stone assigned Wang Jizhi with the task of "emulating" (*fangzao*) the Japanese Kanji Card in order to remake it as a *Chinese* Character Card.

Unlike reprogramming the Brother 2024 printer driver, however, modding the Japanese-built Kanji Card depended on hardware engineering experience Wang did not possess. Intriguingly, however, after Wang managed to get hold of the card itself, as well as the card's circuit diagram, he was also provided a copy of the firmware program *by the Oki company itself*. This Japanese company, in other words, was an active collaborator in the "copycatting" process, eager to help Chinese counterparts help them break into the Chinese-language market.

The gambit made good business sense for Oki. By the time Stone held its one-year anniversary celebration—a gala affair held at the Cultural Palace of Nationalities on May 16, 1985, attended by local government notables—the company was clocking annual sales of 31 million RMB (a three-fold increase from 1984). Instead of going to battle with the Stone copycatters, then, it was only logical to collaborate (and profit) with them instead.[40]

It was not until May 1986—two years into the life of the company, after it made its first small fortune by modding foreign dot-matrix printers—when Stone finally released the product for which it is best remembered today: the Stone MS-2400 Electric Chinese Typewriter (*Zhongwen dianzi daziji*) (figure 5.5).

POP-UP MODERNITY: CHINESE CHARACTER MONITORS

A third domain in which Chinese script butted up against the barriers of alphabetic order was that of mass-manufactured computer monitors. In certain respects, the politics of monitors were similar to those of printers, also involving the issue of character distortion. Unavoidably, even the lowest-resolution Chinese character bitmaps occupied upward of twice the vertical and horizontal space as Latin alphabetic letters, making Chinese appear comically oversized. Here, however, engineers could not take

Figure 5.5 The Stone Chinese word processor

advantage of the "zipper" technique used in printing. The grid of the monitor was, after all, completely fixed and unforgiving.

This had a profound impact on the user experience. Chinese language users could only see small portions of their texts at any one time, for example, for the simple reason that Western-manufactured computer monitors could only fit a far smaller number of Chinese characters on screen than Latin letters. On the Apple II, the monitor offered a total of 192 by 280 pixels, for example. Based on the lowest-possible resolution Chinese bitmaps—a grid of 16-by-16—this meant that a theoretical limit of 17 Chinese characters could fit on each line, with only ten rows of text fitting on the screen overall.[41] As if looking through a keyhole, users were forced to remember what text proceeded or followed the tiny sample shown on screen.

This precious real estate was imposed upon further by that unique feature of Chinese character display we examined earlier: the pop-up menu. Because of the inherently iterative process of Chinese input, in which users are constantly being presented with Chinese characters that fulfill the criteria provided by their keystrokes, an essential feature of Chinese computing is a "window"—whether software-based or hardware-based— that enables the user to review these Chinese character candidates.[42] On Caldwell's Sinotype, as we recall from chapter 2, a small cathode ray tube screen flashed images of Chinese character matches, allowing the operator to determine if a successful match had been made. On the machine formally known as Sinotype II—but which was, in fact, the third or fourth generation in that series—input strings appeared along the bottom edge of the screen, along with input candidates (figure 5.6).

With the advent of personal computers, mechanical windows such as those found on the Sinotype were integrated into the computer's main display. It became a software-governed "window" on the screen, rather than a separate, physical device. While this offered obvious advantages to consumers—such as having the pop-up menu as close as possible to the composition window, as well as dispensing with the need for additional devices or peripherals—there was also an important trade-off. The pop-up menu took up even more space on the screen, which in turn meant that even fewer Chinese characters could fit.

"Pop-up menu design," we might call it, became a critically important area of research and innovation within Chinese human-computer interaction and personal computing. Companies experimented with different menu styles, formats, and behaviors, attempting to strike a balance between the requirements of input, screen size, and the preferences of users. One approach was the "prompt line" (*tishi hang*), a rectangular bar along the bottom of the screen in which the recursive process criteria, candidacy, and confirmation could be visualized.[43] Later developers experimented with other shapes, locations, and layouts, however, as well as with enabling users to select which pop-up menu format they preferred from a set of options. On UCDOS 90, for example, pop-up menus were

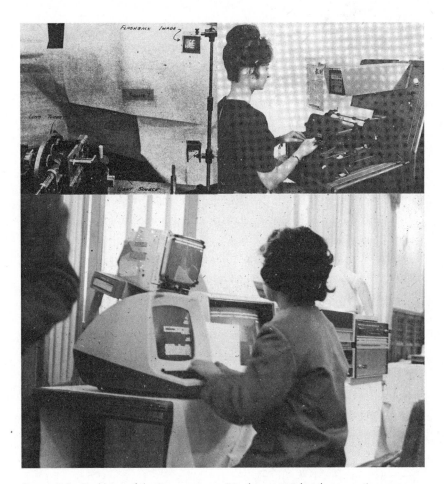

Figure 5.6 Evolution of the "pop-up menu" in the pre- and early computing eras

referred to as "Chinese input prompt user windows" (*Hanzi shuru tishi caiyong chuangkou*), and came in three different forms: a short, wide bar revealing up to 79 character candidates at a time; a narrower window, two rows high, capable of revealing 30 candidates at once; and a much smaller, portrait-oriented pop-up menu that was 12 rows tall and capable of revealing 6 candidates at once.[44]

There were pros and cons to each option. Displaying a large number of character candidates all at once increased the likelihood of finding the desired graph more quickly, for example, but came at the cost of screen space. Smaller windows were less intrusive, but in turn required users to scroll through multiple "pages" of candidates if their intended character was not found in the topmost recommendations.

When we tally up all these challenges—the small number of characters that could fit on the monitor, as well as the requirements of pop-up menu functionality—it is little wonder that Chinese engineers and firms were constantly on the lookout for next-generation monitors with higher resolutions. While it is perhaps natural for consumers to seek out higher-quality displays—eager to use more advanced CGA, EGA, or VGA monitors for business applications such as LOTUS 1-2-3 and AutoCAD, as well as for computer games—in the Sinophone computing world the primary motivation was far more basic at this point in time. Simply put, in upgrading from a CGA to an EGA monitor, the improvement in resolution (from a 640-by-200 display to one with 640-by-350 resolution) translated into an additional 9 lines of Chinese characters on the screen at any one time (from 12 lines to 21 lines—a vast improvement). An upgrade to a 640-by-480 resolution VGA monitor, meanwhile, would expand the text window even further—to 30 lines of displayable text.[45] Unlike personal computing in the Western world, where the hunt for improved graphics capabilities tended to be associated with domains of electronic gaming, or other graphics-intensive programs, in the context of Chinese computing, such improvements were necessary to fulfill much more basic text-processing applications.

INTERRUPTING MODERNITY

In the quest to mod their way to a Chinese-compatible personal computer, the final obstacle confronting engineers was in many ways the most challenging: How to get all these newly modded printers, monitors, Chinese Character Cards, and input systems to speak with the central processing units of Western-manufactured machines? Conventionally, the coordination of CPUs, monitors, printers, and keyboards was achieved thanks to a program known as the BIOS—the "Basic Input-Output software"—a program so essential to the functioning of a personal computer that it was the very first to be run when turning the computer on, even before the operating system. In addition to conducting critical self-checks to ensure that the system was in working order, notably the "Power-On Self-Test" or POST, the BIOS was responsible for activating all system hardware, including the keyboard and monitor, among other peripherals. Without the BIOS, nothing functioned.

The BIOS was biased, however. Like the monitors and printers examined above, the Western-designed BIOS was incompatible with the needs of Chinese computing. It was unequipped to handle the basics of Chinese input, for example, and the fundamentally recursive process of criteria-candidacy-confirmation. Neither could it handle the requirements of pop-up menus. The BIOS, too, would need to be hacked and modded. The final mod we will examine here—and in many ways the most influential one—was the "Sinicization" of the BIOS.

Typically speaking, the BIOS was a program unseen by—and, from a manufacturer's standpoint, ideally untouched by—average users. Stored in its own dedicated chip, it constituted a form of "firmware," baked into the architecture of the motherboard itself in Read-Only Memory (ROM). Only the most advanced users and computer enthusiasts ever customized the BIOS; the vast majority of everyday users left it alone—or were even unaware of its existence save for the fleeting few seconds at the beginning of the computer's boot-up process. If personal computers were ever going to serve a Chinese-language market, however, there was no choice but to design and build an add-on to the BIOS. They had to open up the hood.

One of the earliest mods of the Western-designed BIOS was CCDOS, a modification of MS-DOS premised upon the revision of the MS-DOS BIOS. CCDOS was programmed by Qian Peide, a graduate of the computing department at Nanjing University, who went on to take a position at Suzhou University. It was here where he began work on CCDOS, in collaboration with Institute Number 6 at the Ministry of Machine-Building Industries.[46]

To render MS-DOS compatible with Chinese character input and output, Qian targeted four BIOS subroutines, better known as "Interrupts" or "INTS": INT 5, INT 10, INT 16, and INT 17. Interrupts 5 and 17, which controlled printer functions, needed to be changed to permit for the output of Chinese character text. Interrupt 10 controlled video generation and needed to be reconfigured to enable out-of-the-box, Western-built systems to display Chinese on a computer monitor. Finally, Interrupt 16 provided information about the keyboard, including what keys were pressed and how they should be interpreted.[47] For Chinese input systems (*shurufa*) to function on mass-manufactured personal computers, INT 16 would also need to be rewritten, almost from scratch.[48]

Within the BIOS, monitor and printer controls posed similar problems for Chinese engineers. Conventionally, the Interrupts controlling the monitor display operated in "symbolic mode," also referred to as text mode, character mode, or alphanumeric mode. Effectively, this mode enhanced the efficiency of displaying Latin letters and numerals to a screen, first by dividing the screen into a preset grid of columns and rows, and then by using an onboard character generator to produce alphanumeric bitmaps more efficiently than if they were treated as graphical images.[49]

But symbolic mode could not be used in Chinese text processing. To display Chinese characters on a system running DOS, engineers had to redesign Interrupt 10 (INT 10) to change the default monitor mode from symbolic to graphical, and to bypass onboard character generators altogether (which were useless for the creation of Chinese bitmaps).[50] The same problems were found in INT 5 and INT 17, which governed printer functions. While ASCII printout data was handled symbolically, Chinese character bitmaps had to be treated as if they were raster images—that is, as if they were pictures rather than symbols.

For students of the Chinese language, the irony here will be apparent: the vast majority of Chinese characters are not pictographs, despite a longstanding trope in the Western world that portrays them as such.[51] In order for Chinese characters to function on early Western-built monitors and dot-matrix printers, however, it turned out that engineers had no choice but to treat Chinese characters *as if they were pictures*, without which it would have been impossible to establish compatibility between the Western-designed system and the Chinese writing system. To enter the world of computing, it seemed, Chinese would have to "dress the part," donning a costume and performing a peculiar act of digital orientalism that its Western hosts had so long assumed was a "traditional" part of Chinese culture.

A second phase of BIOS-hacking targeted keyboard operation. Although the QWERTY keyboard was handled via two Interrupts in the DOS BIOS— INT 9 and INT 16—Chinese engineers determined that only the second of these needed to be changed (INT 16). The former (INT 9) could be left as is, for the inputting of English.[52] The first thing that needed to be changed was the architecture of temporary storage, or "buffers." In order for the BIOS to handle the recursive processes essential to input, it needed more and larger memory buffers to store the user's alphanumeric sequences. Engineers developed three new buffers: an "Input Code Buffer Zone," in which keystrokes from the user would be captured and analyzed; a "Chinese Character Internal Code Buffer"; and a "Chinese Character Duplicate Code Buffer," in which a maximum of ten Chinese Character Internal Codes could be stored.[53]

Engineers then needed to instruct the BIOS on what to do with such input data. To convert a Chinese Character Input Code into its corresponding Chinese Character Internal Code, two routines were added. For numerical Chinese input codes, notably the Standard Telegraph Code and the Quwei code, the process was one of calculation, running the input code through a standard arithmetic process that resulted in the corresponding internal code. For all other input systems (which is to say, hundreds), the system used a lookup table method, which paired each input code to its corresponding internal code. Each input system of this sort needed to have its own dedicated lookup table in order to carry out this process.[54]

The duplicate code buffer was particularly important, since most Chinese input methods needed to provide users with multiple potential character matches for any given input code—notably pinyin phonetic input, in which pinyin values sometimes correlate to dozens of potential matches (an issue we will examine in depth in chapter 6). In cases when an input sequence corresponded to more than one potential internal code, a set of up to ten of these potential internal code matches were then stored in the duplicate code buffer. Thereafter, the depression of a numeric key by the operator—one through zero—indicated to the interrupt that it should return the value of whichever internal code occupied position 1, 2, 3, et cetera within the duplicate code buffer.[55] In other words, this was how users could choose which character candidate they wanted from the pop-up menu.

At the same time, depressing a different function key on the keyboard—either the page-up or page-down key—the modified INT 16 would interpret this as a command to empty the duplicate code buffer and repopulate it with the next set of up to ten internal codes. As technical as this operation may sound, the process described here was nothing more than the act of scrolling through the pop-up menu to find the desired character when it was not found in the uppermost list of candidates.[56]

Having rewritten each of these Interrupts, the final phase was to prepare the computer to access them when needed. As Chen explained, this last step of modding the BIOS was "to save the Chinese character processing modules in internal memory, and thereafter modify the vector of the corresponding Interrupt, assigning it to the entry address of the Chinese character processing program stored in memory."[57] The pointer for INT 1D, which helped manage the screen display, was rerouted to memory location 1888, the new CRT Initial Generation Parameter Storage (*chushi daicanshu cunfangchu*). The pointer for INT 1F, which managed the character font library, was redirected to memory location 2798, the address of the newly installed character font library. The pointer for INT 10, which also helped manage the computer monitor, was redirected to memory location 1848, the entry address for the "CCBIOS CRT Control Program." Finally, the pointer for INT 16, which managed keyboard input, was redirected to memory location 98B3, the entry address for the "CCBIOS Keyboard Control Program."[58] Finally, the program changed the monitor default resolution, setting it to the maximum resolution available: 200 by 640. It then employed INT 27H to "terminate and stay resident."[59] "No longer a program in ROM," Qian summarized, each modified Interrupt would become "a corresponding program in RAM."[60] "By modifying and expanding it in this way, the BIOS becomes CC-DOS's BIOS"[61]

This marked the completion of the process. "CC-DOS is an extension of PC-DOS," Qian was now in a position to declare.[62] Chen offered a somewhat more expansive summary: "As long as the keyboard management module, the monitor management module, and the printer management module are replaced with equivalent management modules capable of supporting Chinese character processing, the Sinification of the IBM-PC can be achieved."[63]

NO ESC

CCDOS was one of many Chinese-language operating systems designed to capture and mod the Western-designed BIOS.[64] Other operating systems were the focus of reprogramming efforts as well, including the CP/M operating system, PDP-11 Unix, and the CROMIX operating system designed for the Cromemco, among others. High-level programming languages were also modded, with Chinese programmers developing workarounds to permit Chinese character compatibility with ALGOL, FORTRAN, COBOL, PASCAL, PL/1, and more.[65] All told, then, there was hardly any domain of early personal computing—memory, printing, monitors, operating systems, and more—that was not the focus of intensive work by Chinese engineers seeking to render off-the-shelf systems compatible with the requirements of Chinese character input and output.[66]

But as brilliant and successful as many of these mods were, at the end of the day they remained just that: *modifications*. Although engineers managed to harness the unprecedented power of home computing on behalf of Chinese users, the autonomy and authority to create original systems—that is, systems that might address the needs of Chinese information processing from the ground up, without need for complex mods—was something that eluded them throughout the period in question. Modding was intrinsically precarious.

This precarity took a variety of forms. While the practice of modding helped in developing a wide array of systems, for example, it came at the expense of interoperability. The growing number of modded operating systems, high-level programming languages, and applications—each customized by a different outfit or company—were often incompatible with one another, further fragmenting a computational ecology that was already fractured into myriad character input systems, encoding standards, and more.[67]

Modding required constant vigilance and upkeep, moreover—no silver bullet or "set it and forget it" solution was possible. With every new computer program released on the market, programmers had to "debug" them line by line. The programs themselves often contained code that set display parameters for the computer monitor, for example, thus making it impossible to display Chinese character text. For most English-language word processing programs, a baseline assumption baked into many programs was a 25-by-80-character display format. This format was incompatible with Chinese character display, however, thereby requiring Chinese engineers to manually change each and every place in the program code where a 25-by-80 format was set. Fixes like this were often highly technical, requiring engineers to venture into the bowels of assembly code. (In Wordstar, for example, considered by some to be the program that helped launch the software revolution, the main culprits could be found at DS:0248 and CS:369E.)[68]

Operating systems and programs kept changing, moreover. Not long after the development of CCDOS and other DOS-focused mods, for example, IBM announced its move to a new operating system: the OS/2. "China and Chinese-language have been thrown into turmoil," one article from 1987 wrote, since no existing Chinese-language systems— whether in Taiwan or on the mainland—had yet to be adapted to it. "The race is on for developers to come up with the best match for IBM's MS/ DOS platform."[69] "Modders" were always vulnerable.

Modders were vulnerable to forms of historical erasure and misrecognition as well. In their own day and age, they tended to be characterized, at worst, as thieves and pirates, at best as uninspired reverse engineers. In a January 1987 issue of *PC Magazine*, for example, one cartoonist poked fun at Sinicized operating systems. "It runs on MSG-DOS," the cartoon's caption read.

History has been even less kind, rendering the work of early modders less visible by the day. Over the course of the late 1980s and early 1990s, the era of memory-scarce, low-resolution computing steadily came to an end. Advances in manufacturing and design gave rise to ever-diminishing costs for memory, denser bitmaps, ever-crisper printouts and screens, and ever-faster processing speeds. In the realm of printing, for example, a crop of new 24-pin dot-matrix impact printers began to be released on the commercial market, featuring pin diameters of 0.2 mm (as compared to 0.34mm on 9-pin printers). (Unsurprisingly, many of the leading manufacturers of these new printers were Japanese companies, such as Panasonic, NEC, Toshiba, Okidata, and others that also addressed the challenge of kanji output.[70]) In short order, these gave way to a new generation of ever-cheaper and ever-higher-quality ink jet and laser printers, all of which brought high-resolution Chinese printing further into the mainstream of office and home computing.[71] Moving into our current age of 4K resolution, dirt-cheap laser printers, and rock-bottom memory prices, moreover, it becomes that much harder even to imagine that problems like the ones outlined in this chapter even existed in the first place.[72]

But during the late 1980s and early 1990s, Western manufacturers steadily began to incorporate many of these hacks into the core architectures of their systems—the ability for the BIOS to handle Chinese Input Method Editors, for example, or the capacity to handle Chinese character output on monitors and printers. Soon even a "starter set" of Chinese IMEs was included on machines straight out of the box. In the wake of such "internationalization" and "localization," the work of early modders was effectively edited out of history—overwritten, as it were. Instead, the history of early Chinese personal computing has been retrospectively imagined as one in which the Western-built computer had always been language-agnostic, neutral, and welcoming. All but forgotten is the fact that so many of these changes were catalyzed, not by the universalism of Western personal computing, but by its foundational provincialism; and that most of these problems were first solved, not by the likes of IBM, Microsoft, and Apple, but by modders. Thanks to these modders, Western-manufactured personal computers became usable, and thus useful, to one-sixth of the global population.

6

CONNECTED THOUGHTS: CHINESE IN THE AGE OF PREDICTIVE TEXT

Thirty-nine floppy disks sheared out upon the desk: the latest version of CCDOS, released in the early 1990s. One disk contained the main application, with another two dedicated to IMEs and utility programs. Most of the rest were devoted to Chinese bitmap fonts in various resolutions. All in all, this Chinese disk operating system looked much the same as those from the preceding decade.

The input process was also relatively similar. Upon depressing the first set of QWERTY keystrokes, the IME would begin the process of comparing the user-provided search criteria against the system's onboard memory. It would present potential matches to the user for confirmation, and then display on screen whichever characters the user selected. With three input systems preloaded on the machine, and a dozen others available for installation, this system carried on the new tradition established in the 1980s, enabling users to toggle between their preferred IME.[1] So far, the story was the same.

There was one floppy disk, however, that set Chinese operating systems of this era apart from earlier versions: a disk labeled "Associative Character Compound Library" (*lianxiang ciku*), or more literally, the "Library of Connected Thoughts."[2] This disk—and the subtle yet powerful changes it helped make possible within Chinese IMEs—set Chinese computing down a path that has led to ever-more-powerful techniques of predictive text, and by now, an ever-narrowing gap between Chinese text input and artificial intelligence.

Glimmers of this future were already evident in the CCDOS 4.2 pop-up menu. Upon entering the first input sequence—L-I-A-N, perhaps—the IME pop-up menu would (as always) begin to offer potential matches. But when the user selected a desired character—联 (*lian* "to connect"), let's imagine—the IME would no longer take a moment of rest as it did in the past. Instead, it would immediately begin trying to figure out which character the user might want to input *next*. The IME pop-up menu would continue making suggestions: 合, for example, the selection of which would lead to 联合 or "united"; 邦 (leading to 联邦 or "federal"); 想 (leading to 联想, meaning "associative" or "connected thought"); 络 (leading to 联络 or "connection"), and so forth. If one of the IME's guesses proved prescient, the user could then choose one of these characters without the need to enter a new input sequence from scratch. All they needed to do was enter a number, 1 through 9 (figure 6.1).[3]

The "Library of Connected Thoughts" would not stop at the very next character but rather would keep on guessing. If the user selected 联

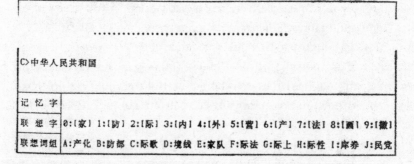

Figure 6.1 Early version of "Connected Thoughts" input pop-up menu

followed by 合, for example, the very next suggestion might include 国 (leading to the three-character Chinese word 联合国 ("United Nations"). If the user's selection was 联邦, by comparison, the very next "connected thought" suggestion might be the two-character compounds 德国 (*deguo* "Germany," giving us 联邦德国 or "Federal Republic of Germany") or 政府 (*zhengfu* "government," giving us 联邦政府 or "federal government"). Having conquered the problem of what we might term "short-phrase anticipation," Chinese hypography from the 1990s onward began to target long-phrase anticipation, attempting to accelerate the inputting of longer terms and passages predictively.

In this chapter, we examine Chinese computing in the final decade of the twentieth century and the opening decades of the twenty-first. We focus on two interlocking developments that emerged hand-in-hand during the period. The first was the growing popularity and eventual dominance of phonetic Chinese input: the inputting of Chinese using letters of the Latin alphabet to represent the pronunciation of characters, not their structure. This phenomenon took place much later in history than most realize, a full three decades after the invention of the first Chinese computer, and of Hanyu pinyin itself. And the late arrival of phonetic IMEs was not for lack of interest or efforts on the part of engineers, linguists, political leaders, and others. To the contrary, there was enormous pressure in favor of what we might call the "pinyin-ization" (*pinyinhua*) of Chinese information technology. This pressure was exerted both from abroad—in countless efforts by foreign missionaries to phoneticize Chinese—and from within China itself, up to and including Chairman Mao Zedong, who once called for the abandonment of Chinese characters and their replacement with a fully phonetic orthography.

The longstanding failure of pinyin input was, as we'll see, due to deep-seated problems with pinyin itself—its unattractiveness and, as some thought, unviability for the purposes of Chinese computing. When compared to structure-based input sequences—which often required only a few keystrokes to retrieve any given Chinese character—the notion of spelling out the full phonetic values of characters seemed to many like a great leap backward. Pinyin was bedeviled by the problem of homophones, moreover: the dozens of different Chinese characters that share the same pronunciation created profound challenges of disambiguation.

Pinyin input was late to the game because, in many ways, it was the least effective way of using Latin alphabetic letters to retrieve a Chinese character from computer memory.

But pinyin input was not a lost cause, which brings us to the second transformation of the era: the growing sophistication and widespread usage of predictive text and other powerful technologies of anticipation. As we will see, early experiments with pinyin input—undertaken at a time when structure-based approaches dominated—offered up a glimmer of hope of overcoming homophony, sequence length, and other challenges. There was a catch, however: in order for Hanyu pinyin to serve as an effective basis for Chinese input, it would need to be used *hypographically*—as a way of retrieving characters, that is, rather than "spelling" them per se. Simply put, the success of pinyin input depended on it violating the core principles of the Hanyu pinyin system upon which it was based, and embracing the principles of hypography we have examined at length thus far: what you type is *not* what you get, there is no essential or assumed relationship between keys and symbols, the primary transcript of input is ephemeral and disposable, and so forth. Pinyin input and Hanyu pinyin were not—and to this day are still not—one and the same thing. Pinyin input is a form of phonetic hypography, inseparable from but fundamentally different than Hanyu pinyin, a form of phonetic orthography.

Relying upon techniques drawn from probability theory and algorithm design, a new "input technocracy"—applied mathematicians, statisticians, and computer scientists, above all—began to design Hanyu pinyin–based input methods that relied heavily upon a sophisticated repertoire of predictive text technologies: autocompletion, abbreviation, contextual analysis, adaptive memory, character repositories, and more. By the 1990s, Chinese predictive text—paired with pinyin input—had in effect mastered the art of predicting a user's very next character, to the point where a new goal began to take shape: the anticipation and prediction of longer compounds and even passages. In the opening decades of the twenty-first century, as we will see, Chinese IMEs entered the Cloud, enabling predictive text suggestions to harness vast, dynamic, ever-growing, user-generated text corpora in order to deliver ever-longer and more accurate "connected thought" recommendations to users in the pop-up menu. Pinyin input is thus not only hypographic: it represents the most intensive and expansive engagement with the principle of hypographic "search writing" arguably in human history. So important was "connected thought" that the term was eventually enshrined as the name of one of China's most influential computing companies ever: *Lianxiang*, better known in the West as *Lenovo*—the firm that, in 2005, took over IBM's personal computing business.[4]

THE PROBLEM WITH PINYIN

By the 1980s, mainland China's officially designated system for transliterating Chinese into Latin characters—Hanyu pinyin—had been around for nearly three decades. Although formally promulgated by the PRC state in 1959, the system had failed to take root in mainland China, due in no small part to the tumult of the Cultural Revolution. Conditions were simply too unstable to carry out the patient, grinding work of inculcating Chinese students into the new system. It would not be until after the Maoist period that the real work of *pinyinhua* (pinyin-ization) could begin in earnest.[5]

In the post-Mao period, pinyin steadily became a feature of everyday life, serving as a kind of parallel writing system that ran alongside character-based Chinese writing. When Chinese toddlers learned to read and write Chinese characters, for example, they often learned pinyin first, to assist them with the memorization of standard, nondialect pronunciation. Meanwhile, as people navigated their everyday lives, the sight of pinyin became more commonplace, whether on street signs, bus schedules, book covers, and elsewhere.

The PRC government's promotion of Hanyu pinyin was also inseparable from China's longstanding efforts towards nation-formation—particularly with regard to the persistent question (and, from the state's perspective, problem) of Chinese linguistic diversity.[6] Hanyu pinyin is premised upon China's "standard dialect," a form of Chinese historically spoken in north China in the environs of Beijing. As a means of spelling the sounds of Chinese characters, pinyin is effectively indecipherable for monolingual speakers of Cantonese, Fujianese, and other "nonstandard" Chinese languages. The promotion of pinyin education in China's schooling system thereby doubled as a way of requiring speakers of Cantonese, Shanghainese, and other dialects to learn "standard Chinese."

Growing familiarity with Hanyu pinyin was not enough to make it a viable basis for Chinese input, however. Five intractable problems hindered its uptake within computation. The first was the length of the input string. When strictly following the syntax of Hanyu pinyin, the full phonetic spelling of Chinese characters leads to longer input strings on average than those found in structure-based input systems. Circa 1985, one study estimated an average of 3.5 pinyin letters per Chinese character: 0.5 units, or more than 15 percent, longer than typical input strings in many of the structure-based approaches we have seen.[7] Consider our earlier example of 电 (*dian* "electricity"). While the Hanyu pinyin spelling of this character contains four letters—D-I-A-N—many of the most popular structure-based input systems at the time required only three. In Five-Stroke input (*Wubi shurufa*), 电 could be inputted with the three-key sequence: J-N-V. The same was true for Cangjie input (L-W-U), Yi input (R-G-D), and dozens of others.

The second problem was homophones, which compounded the issue of long input sequences. If we revisit the example of 电, it would be inaccurate in fact to say that this character's pinyin input sequence was only four letters long. After all, the pinyin value *dian* corresponds, not only to the character 电, but to upward of two dozen other characters as well. Even when the user entered the pinyin input string D-I-A-N, the task remained to find and select the desired character within the pop-up menu—all of which required additional keystrokes. By comparison, structure-based input sequences tended to correspond to far fewer potential character candidates; and, in some cases, led to only one possibility. Even for structure-based input systems that might require a four- or even five-key sequence to input 电, then, the process was still faster than in pinyin.

The third problem with pinyin involved what we might call "nested codes": valid Hanyu pinyin spellings that contain other valid (but different) pinyin spellings inside them.[8] Consider the pinyin phoneme *ling*. On its own, this spelling corresponds to a host of Chinese characters with this same pronunciation (i.e., the homophone problem). In addition to this, however, *ling* also contains two other valid Hanyu pinyin spellings: *li* and *lin*, both of which correspond to still other Chinese characters. (Other examples include *shang*, which contains *sha* and *shan*; and *geng*, which contains *ge* and *gen*.) While nested, overlapping codes can be found in structure-based input systems, too—as we saw in our discussion of Caldwell's Sinotype in chapter 2—the scale of this problem within pinyin was far more formidable. All of this made pinyin algorithmically expensive, in other words, requiring the computer to perform a far more rigorous processes of disambiguation, as it vied to figure out whether a user had entered a complete sequence (*li*, for example), or if that user was still "en route" to another, longer one (*lin* or *ling*). In the case of structure-based input systems, situations like this were either far less common or nonexistent.

We can describe the fourth problem as taxonomic unevenness. In the case of structure-based input systems, designers were free to determine which structural elements of Chinese characters would be assigned to which keys on the QWERTY keyboard, and even how to define and categorize these structural elements in the first place. By tweaking and tuning their IMEs, often over the course of years, inventors were able to create "evenly distributed" taxonomic systems, wherein any given alphanumeric input sequence corresponded to as few Chinese characters as possible. Structure-based input systems had disambiguation built into their DNA, so to speak.

When it came to Hanyu pinyin, engineers enjoyed no such leeway to optimize, for the obvious reason that the sounds of Chinese characters represent the products of millennia of cultural change and evolution, not the brainchildren of input system designers. The extremely common Chinese surnames 李 (Li), 陈 (Chen), and 王 (Wang), to broach just three obvious examples, share pronunciations with dozens of other characters—and there was nothing a pinyin IME developer could do about that. For this simple reason, some pinyin spellings correspond to relatively few Chinese characters, while others correspond to a great many. Pinyin does not (and cannot) have disambiguation "baked in" at the level of input sequence.

If Hanyu pinyin were to be used as the basis for Chinese input, engineers would have no choice but to develop sophisticated disambiguation techniques to be applied "after the fact"—that is, after a user had entered an input string into the machine. Upon receiving the input sequence

"LI," for instance, the Input Method Editor would need to figure out which of the dozens of characters the user might want, whether by presenting Chinese character candidates to the operator for confirmation, or by undertaking algorithmically based forms of contextual analysis. All of this made pinyin input far more computationally demanding than structure-based alternatives.

Hanyu pinyin was also incapable of serving as its own character encoding back-end—that is, as the "addresses" in memory where any given character could be stored and accessed. Early in the history of Latin alphabetic computing, the American Standard Code for Information Interchange (ASCII) was established in the 1960s. Based upon a 7-bit architecture, its 128 addresses offered sufficient space for all the letters of the Latin alphabet, along with numerals, punctuation marks, and key functions. These codes, by virtue of their cross-platform standardization, also enabled information exchange between systems, ensuring that two computers using ASCII would be able to send and receive texts successfully.

Chinese internal character storage encoding was not standardized to any degree until the 1980s and 1990s, decades after the initial promulgation of Hanyu pinyin. Without a stable convention that could function as the Chinese counterpart of ASCII, engineers instead developed a hodgepodge of bootstrapped measures, some using the Chinese telegraph code and its four-digit ciphers; and some even using the *Mathews' Chinese–English Dictionary*—a reference work published in 1944 in which, for reasons not identified in the original work, assigned nonrepeating numerical codes to every single entry in the volume.[9] In the absence of a standard character encoding scheme, many structure-based input systems served "double duty": both as an input method and as a way of assigning nonrepeating addresses to Chinese characters in computer memory. In the case of Standard Telegraph Code input, for example, the very same four-digit codes that a user entered on the keyboard could also serve as a back-end address system within the computer's character database. This practice was so common, in fact, that many structure-based IMEs were referred to in Chinese both as "input methods" (*shurufa*) and as "encoding methods" (*bianma*).

Hanyu pinyin could never hope to perform this dual role. Due to the problems of homophony, self-nesting, and more, Pinyin input required some kind of separate, dedicated character encoding system to be operating in the background.

Despite all of these challenges, however, interest in Hanyu pinyin–based input methods did not wane. There was one immense advantage that pinyin enjoyed, after all, which set it apart from all of its structure-based counterparts: the full backing and budgetary support of the mainland Chinese government itself, which was investing unprecedented quantities of money, time, and labor toward the promotion of Hanyu pinyin throughout China. In effect, the support of the country's Ministry of Education had made it possible to teach hundreds of millions of Chinese people the basics of pinyin, at a time when even the most successful inventors of structure-based input systems could only hope to reach thousands—maybe tens or hundreds of thousands. Hanyu pinyin had problems, to be sure, but also vast technical and political potential.

FROM HANYU PINYIN TO PINYIN INPUT: EXPLOITING THE HYPOGRAPHIC POTENTIAL OF PHONETIC IMES

What would be the result of delinking pinyin input from Hanyu pinyin to explore its hypographic potential and abandon the idea of "spelling?" What impact might this have on the challenge of input length, homophony, and more? In this section, we examine two early explorations of what we might call "pinyin hypography": "Double Pinyin" (*shuangpin*), invented by Chen Jianwen, and "Character compound codes" (*cihuima*), developed by Zhi Bingyi, who we first met in chapter 4. These two systems represented attempts to solve key problems of pinyin input—notably string length and disambiguation—by exploiting pinyin as a hypographic rather than an orthographic system.

"Dual Romanization" or "Double Pinyin" was developed beginning in the late Maoist period by Chen Jianwen and Xue Shiquan, among others. Like many other input systems of the era, it originated outside the domain of computing, and was only later applied to the question of Chinese character processing (Chen and Xue were both librarians at the time, working at Nanjing University).[10] As the name "Double Pinyin" suggests, the system was premised upon using Latin alphabetic letters in two different ways at once: one orthodox, the other unorthodox. First, letters could be used to represent the phonetic value of a character, as they do in Hanyu pinyin. At the same time, however, Latin letters could be used *nonphonetically*, serving as substitutes for other (typically longer) Hanyu pinyin values.

Take, for example, the "Double Pinyin" encoding for the character 拆 (*chai*, to demolish). Within Chen Jianwen's early system, the operator did not write C-H-A-I but rather I-L—with the letter "I" representing the Hanyu pinyin value "ch," and "L" representing the Hanyu pinyin value "ai." To encode 商 (*shang*, merchant), meanwhile, the operator did not write S-H-A-N-G, but U-H—"U" representing the Hanyu pinyin value "sh," and "H" representing "ang."[11] In certain cases, the full phonetic value of a character could be represented using a single letter, as in the case of 安 (*an*, peaceful), which in Double Pinyin was simply "J."[12] Double Pinyin was a "pinyin" system, but one that used a highly unorthodox means of "compressing" the Latin alphabetic string.

Chen later transformed his system into one serviceable to the computational processing of Chinese characters. From June 1978 to June 1979, Chen completed his plan to encode the characters based on Double Pinyin and went on to codevelop an input system in the early and mid-1980s with help from Xue Shiquan and Qian Peide: the Double Pinyin Encoding Method (*Shuangpin bianma*), which would be preloaded on CCDOS.[13] The Double Pinyin pairings were as follows:

A: ia
B: uan
C: iao
D: ua, ia
E: ie
F: ong, iong
G: ian
H: eng, ng
I: y
J: an
K: in
L: ui, üe
M: uang, iang

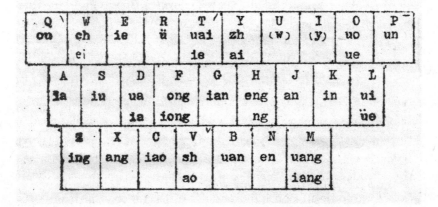

Figure 6.2 Double Pinyin (*Shuangpin*) system on a QWERTY keyboard

N: en

O: uo, ue

P: un

Q: ou

R: ü

S: iu

T: uai, ie

U: w

V: sh, ao

W: ch, ei

X: ang

Y: zh, ai

Z: ing

Double Pinyin offered up the possibility of resolving at least one of Hanyu pinyin's foremost challenges: long input strings. By using techniques of hypermediation, string lengths could be drastically shortened (from "JIANG" to "JM," for example, or "NENG" to "NH"), offering up the possibility of dramatically increased input speeds.

From the late 1970s through to the 1990s, Dual pinyin was a focus of intensive experimentation among a number of engineers. Circa 1978, for example, Zhang Tinghua developed a system he called "Double Pinyin Double Radical" (*Shuanpin shuanbu*), premised upon a combination of

Chen's Double Pinyin technique but paired with elements from structure-based approaches, as well.[14] In the 1980s, Zhang Guofang's "Fifty Element" input system treaded a similar path, pairing Double Pinyin phonetic encoding with yet another proprietary structure-based method.[15]

Double Pinyin was one of the earliest techniques to reveal the broader potential of Hanyu pinyin as a foundation for computational Chinese input. The only proviso, however, was that pinyin input would need to function as a hypographic system, drawing upon the resources of Hanyu pinyin, but using these resources for the purposes of *retrieving Chinese characters from computer memory*, not "spelling" them in full.

A second and even more important line of experimentation in phonetic hypography began to take shape around the same time. Here, designers studied how to use Hanyu pinyin–based input to enter longer Chinese terms, names, expressions, and phrases, rather than individual characters. Examples from present-day pinyin input might include the input sequences "ZG" or "JSJ," the first being a common pinyin input abbreviation for "Zhongguo" ("China" 中国), and the latter *jisuanji* ("computer" 计算机). Input designers began to experiment with ways of creating what might be thought of as Chinese "acronyms," stringing together the initial phonetic value of two or more Chinese characters in a row (*zhong-guo* becoming "ZG" and *ji-suan-ji* becoming "JSJ").

One of the earliest experiments with pinyin acronyms is found in the work of Zhi Bingyi. We learned in chapter 4 that Zhi's OSCO input system was structure-based, using letters of the Latin alphabet (in four-letter sequences) to describe the shapes rather than sounds of Chinese characters. These structure-based sequences were used to input one Chinese character at a time. At the same time, however, Zhi dedicated a small subset of his code to what he called "character compound codes" (*cihuima*)—also called "OSCO Quick-Codes." These were special-purpose two-letter sequences enabling a user to input, not single Chinese characters at a go, but a series of highly common two-character Chinese compounds. To enter the two-character Chinese word for "safety" (安全 *anquan*), for example, an OSCO user could type in "AQ," the "A" corresponding to *an* and "Q" to *quan*.

Unlike the rest of Zhi's code, which relied exclusively on structure-based sequences to input one Chinese character at a time, in the case of Quick-Codes, the OSCO system would know to retrieve this common two-character Chinese word from memory, thereby saving the operator time. To input the two-character compounds "extremely" (*feichang* 非常) and "revolution" (*geming* 革命), meanwhile, the Quick-Codes were "F-C" and "G-M," respectively (table 6.1).

The logic of Zhi's Quick-Codes epitomized a key difference between structure-based input and phonetic pinyin input. While pinyin was an inferior option when it came to inputting Chinese characters one-by-one, it offered major advantages when it came to inputting them in *groups*. Within modern vernacular Chinese, after all, an exceedingly large number of what English speakers would refer to as "words" comprise two Chinese characters, not just one (such as *anquan*, *feichang*, and *geming* above). In addition to compounds, moreover, there are many thousands of idiomatic expressions, terms, and names that comprise three or four characters. When it came to these multicharacter sequences, Zhi realized, it was more efficient to base at least a portion of his otherwise structure-based OSCO code on phonetics.

To underscore the key point here, Zhi's Quick-Codes clearly did not obey the syntactic laws of Hanyu pinyin. As with Double Pinyin, OSCO Quick-Codes employed Hanyu pinyin hypographically. That is, "AQ," "FC," "GM," and so forth are not valid or "legal" Hanyu pinyin spellings of *anquan*, *feichang*, and *geming*. When viewed *hypographically*, however, acronyms such as these are entirely viable ways of retrieving these two-character compounds from memory—not just viable, in fact, but far more effective. Insofar as GM, FC, and AQ cannot possibly be mistaken for any "real" Hanyu pinyin value, the Input Method Editor can treat them, with complete certainty, as an input sequence whose goal is to retrieve a two-character compound from memory.

These early experiments by Zhi, Chen, and others offered up a tantalizing promise: by employing pinyin as a hypographic system, one could take steps toward resolving the problems of homophones, string-length, and nested sequences that otherwise condemned Hanyu pinyin to being perhaps the least useful approach to computational Chinese input. Moreover, initial findings supported this optimism. Quick-Codes offered users a far more economical means of inputting at least some of the most common Chinese terms. While the inputting of any one of these

Table 6.1 Partial list of two-letter OSCO Quick-Codes

AP	安排	*anpai* "to arrange"
AQ	安全	*anquan* "safety"
BD	不但	*budan* "not only"
BF	部分	*bufen* "portion"
BG	报告	*baogao* "report"
BH	不会	*buhui* "would not, to not know how to"
BJ	北京	*Beijing*
BM	部门	*bumen* "department"
BN	不能	*buneng* "can't, be unable to"
BS	表示	*biaoshi* "to express, manifest"
BY	必要	*biyao* "essential"
BZ	帮助	*bangzhu* "to help"
DB	代表	*daibiao* "representative"
DS	但是	*danshi* "but, however"
DW	单位	*danwei* "work unit"
EQ	而且	*erqie* "furthermore"
EY	意义	*yiyi* "meaning"[1]
GB	干部	*ganbu* "cadre"
GM	革命	*geming* "revolution"
KX	科学	*kexue* "science"
LS	历史	*lishi* "history"
LX	路线	*luxian* "line"
MY	没有	*meiyou* "not to have, to not exist"
MZ	民主	*minzhu* "democracy"
RG	如果	*ruguo* "if"
SM	什么	*shenme* "what"
SX	思想	*sixiang* "thought"
TD	态度	*taidu* "attitude"

Table. 6.1 (continued)

TG	提高	*tigao* "to raise"
TY	同意	*tongyi* "agree"
WT	问题	*wenti* "problem"
ZX	主席	*zhuxi* "chairman"

1. Note that, in theory, the OSCO Quick-Code for this compound should be "YY," based on the pronunciation of the characters. This anomaly, along with the fact that no other Quick-Codes in the list seem to comprise two identical letters, suggests that there were further parameters or limits that Zhi Bingyi and Olympia Werke had to follow, disallowing Quick-Codes such as "AA," "BB," "CC," etc.

two-character compounds would normally require eight keystrokes using standard OSCO codes, Quick-Codes required only two. Through recourse to Quick-Codes, Zhi estimated that the average number of keystrokes-per-character could be reduced from 4 to approximately 2.6.

Most importantly, perhaps, these early experiments in pinyin hypography served as a proof of concept, throwing a spotlight on its deeper potentials. Thinking all the way back to Samuel Caldwell's Sinotype alphabet—and the many structure-based input systems that followed—by and large all structure-based systems had been fine-tuned to reduce the ambiguity of inputting *individual* Chinese characters. This disambiguation was "baked into" the system in the ways a designer decided to decompose Chinese characters into elemental parts, and how they assigned these elements to different letters of the Latin alphabet. Pinyin, by comparison, was ill-suited to the unambiguous inputting of individual Chinese characters, which in turn required that techniques of disambiguation be added *on top of it*, as a secondary process beyond the input sequence itself.

At first this might seem to disqualify pinyin input as hopelessly inferior to structure-based approaches, but in the longer view, engineers like Chen and others began to show that these "after-the-fact" disambiguations—although more computationally intensive than structure-based protocols—offered up *more powerful* techniques, in terms of the number of Chinese characters than could be unambiguously entered at once. Thinking back to Samuel Caldwell and his accidental discovery of "minimum spelling" (i.e., autocompletion), we might say that the developers of pinyin input decided to push this central axiom of Chinese computing—that adding layers of mediation can result in faster and more effective systems—to a new level. In a peculiar twist, then, pinyin's initial drawback—single-character input—became a driving force for what was to become its greatest strength: the inputting of two, three, or more characters at a time.

But such promises were still a distant notion circa the late 1970s and early 1980s. Each of Zhi's Quick-Codes corresponded to only one potential Chinese compound, for example—a miniscule fraction of the many thousands of two-character compounds in the Chinese lexicon. Even in situations when a given two-letter code could, in theory, have corresponded to more than one compound—Zhi assigned his Quick-Code "MZ" to *minzhu* (民主 "democracy"), but in theory he could have also assigned it to *minzu* (民族 "nationality"), *mingzi* (名字 "name"), *manzu* (满足 "satisfy"), and more than a dozen other two-character terms. Limitations of memory and processing prevented this, however. Only later would programmers begin to develop IMEs capable of handling pinyin "quick codes" capable of pointing to more than one possible match.

Yet even when accounting for these early limitations, one point is abundantly clear: for any engineer in the 1980s and 1990s seriously committed to bringing pinyin input on par with structure-based counterparts, there was an immense incentive to explore techniques of abbreviation, hypermediation, and (as we will see toward the close of this chapter) predictive text. Only by tackling this next-level challenge of input—inputting multicharacter passages—would phonetic input be able to hold on its own.

PINYIN AND THE NEW INPUT TECHNOCRACY

Beginning in the 1980s, Hanyu pinyin started to make inroads within mainland China's educational system, as the Chinese government finally followed through on a program it launched nearly three decades prior. Within a newly stabilized schooling system, students were trained in pinyin from a very early age, in tandem with or even before the study of Chinese characters themselves. Pinyin was becoming a commonplace sight for many in their day-to-day lives, moreover, whether in newspapers, street signs, or elsewhere.

While growing familiarity with Hanyu pinyin was undoubtedly an important factor in the recent rise of pinyin input, familiarity alone didn't resolve the computational problems outlined above: homophony, nested input sequences, and more. Character encoding is a case in point. As noted, pinyin input could not function as its own character-encoding set the way structure-based input systems could. By consequence, the long-standing absence of an agreed-upon Chinese character internal-encoding standard served as a major obstacle to the viability of pinyin input.

This situation began to change in the early 1980s. In 1981, mainland China's Association for Standardization enacted GB 2312-80, drafted by the North China Institute of Computing Technology for the Fourth Ministry of Machine Building (figure 6.3).[16] GB Code, as it is often referred to, was based upon a two-byte structure, with one byte providing the row address of a given character (referred to in Chinese as the *qu* or "area") and the second byte providing the column address (referred to as *wei* or "position").[17]

The formulation of a stable character encoding gave pinyin input designers a foundation on which they could work. More than this, GB Code was designed expressly with pinyin input in mind. Within the first layer of the code, which contained a set of the 3,755 most frequent characters, characters were sequenced according to their Hanyu pinyin phonetic values, with the very first character in GB Code—address 0176 0161 (or Hex B0A1)—being the character 啊 (*a*), an onomatopoetic character indicating surprise or exclamation. This was followed by the characters 阿 (*a*) at address 0176 0162 and 埃 (*ai*) at address 0176 0163, and so forth.

Organizing GB Code in this way provided pinyin input with two advantages. First, it simplified the process of sorting Chinese character corpora according to pinyin alphabetic order, insofar as characters—as long they form part of the first layer of high-frequency characters—need only be sorted according to the numeric value of their code points. This organizational logic also greatly facilitated the job of pinyin-based Input Method Editors. Upon depressing the first letter of a pinyin phonetic value—the letter "c," for example—table look-up and string-matching

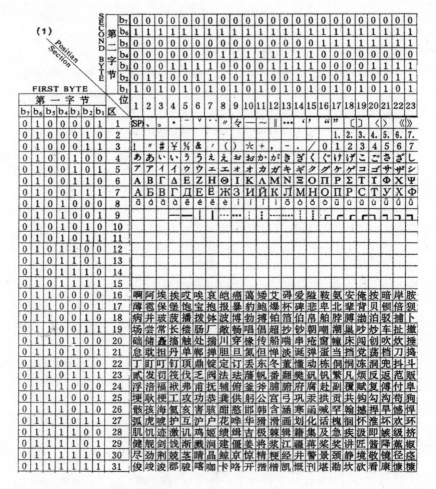

Figure 6.3 GB Code sample

processes could concentrate efforts within a single, contiguous zone containing all characters beginning with the pinyin initial letter "c" (beginning at address number 0178 0193 with 擦 / *ca* / "to rub" and extending to 0180 0237 / 错 / *cuo* / "mistake"). With the depression of another keystroke—"h," perhaps, bringing the overall sequence to "ch"—the string-matching algorithm could narrow its search even further, between code 0178 0229 (插 *cha* "insert") and 0180 0194 (绰 *chuo* "spacious").[18]

The standardization of character encoding was but one part of the story. To overcome the even more pronounced challenges of homophony, nested codes, and input length, pinyin input would need increasingly sophisticated techniques of disambiguation and shortcuts, expanding far beyond the rudimentary forms seen in the work of Zhi Bingyi and others. But there was a problem: while a structure-based input system could be designed, at least in theory, with little more than a philological mindset, ample time, and (ideally) a robust Chinese dictionary, the techniques needed for successful phonetic input extended into the realms of probability theory and applied mathematics. To develop a successful pinyin-based system was to be at home in a world, not of "radicals" and "strokes," but of "T-trees," "node arrays," Markov chains, stochastic algorithms, and information theory more broadly.[19] Put simply: no person or team without a robust foundation in statistics, probability, algorithm design, or programming had even the slightest chance of developing phonetic Chinese input systems with the sophistication and efficacy of Sogou pinyin, QQ pinyin, Google pinyin, or any of the other phonetic input systems one encounters on the market from the late 1980s onward.[20]

Fortunately for the domain of pinyin input research, the 1980s was a time of rapid institutionalization and professionalization in mainland China, with myriad research centers and programs founded with an express focus on Chinese information processing. These included a new center at Tsinghua University, the Yanshan Computer Applications and Research Center, the Computer Research Institute based at the Chinese Academy of Sciences, the Shanghai Printing Institute of Technology, the Beijing Computer Research Institute, and the Institute of Scientific and Technical Information of China, among many others.[21] The early 1980s also witnessed the launch of an ever-growing number of research societies and journals dedicated expressly to Chinese computing and information processing. In 1981, the Chinese Information Research Association was formed in Tianjin, followed by the Chinese Information Processing Society of China, the Chinese Language Computer Society, the China Instrument and Meters Society, the Chinese Computing Society, and the China Computer User Society, among many others.[22] While this period of institutionalization benefited all domains of Chinese computing— including that of structure-based input—pinyin input research benefitted disproportionately. A new pipeline of mathematicians, statisticians, and programmers—Reform Era China's new technocratic elite—brought the specific skillsets needed to tackle the computationally intensive domain of phonetic hypography.[23]

This was also a period of unprecedented global integration, with once disconnected pockets of Chinese computing research across the Sinophone and Chinese-Japanese-Korean (CJK) worlds being linked together for the first time in history. While there were a limited number of influential CJK computing conferences in the 1970s, the number of gatherings, trade shows, and delegations mushroomed during the 1980s.[24] In just the opening four years of the new decade, for example, people interested in Chinese computing would have had their pick of at least ten major industry conferences.[25] The same was true for international trade shows and delegations, with computer scientists and other technologists traveling back and forth between various parts of the Sinophone world, as well as to and from the United States, Japan, and Europe.[26]

Pinyin input researchers reaped immense benefit from this era of exchange. In particular, one of the most enduring impacts of this period of global exchange was the exposure it provided to researchers working in another key part of the CJK universe: the arena of phonetic Japanese input. Japanese computer engineers and linguists were ahead of their Chinese counterparts when it came to phonetic hypography, particularly the design of contextual analysis algorithms.[27] For the computational input of Japanese—a writing system that comprises both Chinese-character script (kanji) and phonetic syllabaries (hiragana and katakana, collectively known as kana)—one of the thorniest challenges was the development of algorithms capable of determining if a given input string should resolve to a kanji character, a sequence of phonetic kana syllables, or a combination of both. Consider the simple sentence, "I read a book" (*hon wo yomimashita* 本を読みました). Within this short passage, there are two kanji characters (*hon* 本, meaning "book" and *yo* 読, part of the past-tense verb *yomimashita* "I read"), and five Hiragana syllables (を, み, ま, し, and た). While simple to explain in prose, the distinction between kanji and kana posed a remarkably challenging puzzle for computational input. Upon depressing the first syllable *ho*, an input algorithm had to assess at the outset if the user intended to input kanji at all, and then determine which kanji character the user might be entering—perhaps 保

(*ho*, "guarantee"), 帆 (*ho*, "sail"), 穂 (*ho*, "ear"), 堡 (*ho* of *horui* 堡塁 "fort," or otherwise). Perhaps the user was en route to entering a phrase in Hiragana, however—such as ほとんど (*hotondo* "almost") or ほんとに (*hontoni* "really")—in which case the syllable should be left "as is," rather than converted to a kanji graph.[28]

One of the keys to successful Japanese input—and to what is sometimes referred to as kana-kanji conversion—was contextual analysis: the algorithmic analysis of an input string that dynamically reexamines with each keystroke the context of a user's input sequence to determine what they are trying to write. Although similar to some of the structure-based Chinese input systems we have examined thus far, what set Japanese kana-kanji conversion apart was the sophistication with which it disambiguated between a wide array of potential kanji homophones *and* determined which part of a given sequence should be resolved to what kind of Japanese orthography (that is, to kanji or to kana).

Kana-kanji conversion was not directly applicable to Chinese input (since Chinese does not have a syllabary like kana), and yet the underlying mathematics and statistical models worked out by Japanese researchers provided a boon to researchers working on pinyin disambiguation and contextual analysis. As in Japanese input, pinyin input could be made far more efficient if algorithms could be used to conduct real-time contextual analysis of pinyin input strings, offering probabilistic determinations as to which of the many potential Chinese characters the user was likely after.

It was not long before this strategy of contextual analysis began to be applied to phonetic Chinese input on a larger scale—a strategy that made immense strides in resolving problems of input string length, nested sequences, and homophones. In the HKT-100H system by Hua Ko Electronics, for example, the inputting of the pinyin syllable *ge*—which corresponds to the pronunciation of dozens of Chinese characters—was aided by a programmatic analysis of the lexical context in which the user entered the sequence. By analyzing preceding text, the program used co-occurrence analysis to provide a reported 98 percent accuracy rating in terms of the particular *ge* it suggested in the pop-up menu.[29]

Contextual analysis boasted a "learning" functionality as well. In absolute terms, the character 瑭 (*tang*, a rare form of jade), is a highly uncommon Chinese character, and thus unlikely to show up in the first batch of character suggestions in the pop-up menu. If one happened to be a dealer in rare forms of jade, however, then this character might be highly common. As such, when entering the pinyin phrase *tang* for the first time, the user may have to scroll through a series of pop-up menu pages to find 瑭; but upon the second or third time doing so, this character would be dynamically "promoted" within the IME, appearing towards the top of the IME menu upon subsequent entries of *tang*. Such functionality was of far more benefit to pinyin input than to structure-based approaches, although it was used in both.

PINYIN INPUT IS NOT HANYU PINYIN

By 1990, engineers working on Chinese information technologies were already in their fifth decade of concentrated work on a variety of anticipatory text technologies—some of them virtually unknown in the Anglophone world. Autocompletion functionality was baked into the very first computational system for Chinese in the 1950s (the Sinotype, see chapter 2). "Adaptive memory"—where the inventory of Chinese characters stored in onboard computer memory is dynamically updated, depending upon a user's most recent entries—was a focus of experimentation in the early 1980s (see the example of Sinotype III, chapter 5). It was these anticipatory text technologies, as well as even more powerful ones developed from the 1990s onward, that ultimately solved the longstanding problems of pinyin, making phonetic input a viable approach for the first time.[30]

While this commitment to predictive pinyin input is evident in many different arenas of research in the 1980s and 1990s, one of the clearest metrics we have was the creation and proliferation of digital Chinese character databases, known in Chinese as *Hanziku* or *ciku*. These were digital files containing thousands—and soon tens and hundreds of thousands—of multicharacter Chinese compounds, proper names, place names, technical terms, four-character idiomatic expressions, and more. The purpose of these repositories was to enhance predictive text and contextual analysis, thereby increasing accuracy and efficiency of Chinese Input Method Editors.

Beginning in the mid-1980s, Chinese input libraries mushroomed, with a host of institutes, companies, government bureaus, and everyday users working to create new ones and expand existing ones. Circa 1986, one of the largest digital lexicons was developed by the Beijing Institute for Aeronautics and Astronautics.[31] Meanwhile, a library of over 100,000 common words and terms from traditional Chinese medicine were preinstalled on the YYDOS Chinese operating system.[32] Even at this early stage, the scale and scope of such input libraries—and even more so the labor which went into their creation—was staggering. During the mid and late 1980s, Zhang Guofang—a doctor by training—spent years combing through a corpus of around 10 million characters, encompassing lower-school and middle-school textbooks, along with other materials. On this basis, he went on to produce what he referred to as a four-level lexicon or *ciku*. The first level included a set of 5,633 commonly used character compounds. Levels 2 through 4—which entailed 96,000 character compounds—drew upon research conducted at Chengde Medical School and People's University.[33]

By the second half of the 1980s, the subfield of pinyin input was reaching a point of maturity, with phonetic hypography approaching parity with structure-based alternatives in terms of speed and accuracy, and far surpassing them in terms of the number of potential users. Chinese computer scientists were becoming more confident by the day, some even beginning to offer projections about the general trajectory of Chinese predictive text advancements—a kind of "Moore's Law," if you will, but focused on predictive Chinese input.

In a 1986 essay by Qi Yuan, for example, the author outlined what he saw as four basic stages of Chinese-character text processing, with each stage defined in terms of a differently sized lexical unit that Chinese IMEs were able to handle effectively: the character (*zi*), the word or phrase (*ci*), the sentence (*ju*), and the paragraph or passage (*yu*). By Yuan's estimate, Sinophone computing circa 1986 was somewhere between the second and third of these stages—that is, between the effective inputting of multicharacter phrases, and the inputting of entire Chinese sentences.[34]

By 1989, some were offering even more confident diagnoses. "At present," Zhang Guofang wrote, "input speeds for modern Chinese have clearly already surpassed speeds for Western languages." Zhang continued:

Into the future, as character- and compound-based processing undergo even deeper research, and as even more ideal character libraries of even higher quality are created, we have every hope that users, with leisurely keystrokes, will be able to reach or greatly outpace the speed of human speech.[35]

In contrast, many in the West continued to offer starkly different outlooks, including one of the most renowned Chinese linguists: "In word processing," John DeFrancis wrote (also in 1989), "we normally input alphanumeric symbols and print out text expressed in the same symbolic code. This is very simple and highly efficient, especially when the operator is a skilled touch typist." As DeFrancis concluded:

There is not, and I believe never can be, as efficient a system for inputting and outputting Chinese characters.[36]

Audacious as Qi's and Zhang's prognostications were—and despite opposing views expressed by some in the West—by and large they have come true. By the 1990s, predictive text was so widespread that China's highest national standards bureau began to issue guidelines pertaining to which two-, three-, and more-character compounds needed to be included as part of any Chinese input system, text processing program, or computer. Known as the "General Word Set for Chinese Character Keyboard Input," this 110-page standard included 43,540 entries, divided into three ranks based on frequency.[37] These, and dozens of other shortcuts, allowed users to retrieve a set of commonly used compounds specially encoded in computer memory. Also in the early 1990s, three organizations—including the influential Chinese Information Society—joined forces to produce the "Compendium of Commonly Used Words for Use in Chinese Character Keyboard Input."[38]

Chinese disk operating systems from the era—such as CCDOS, UCDOS, and others—began to come preloaded with extensive Chinese input libraries, along with functionalities that enabled users to expand and personalize these libraries as needed.[39] Particularly fascinating was the terminology that began to stabilize around these predictive text approaches—specifically the Chinese term *lianxiang*, or "association" (also "associative" or, more literally, "connected thought"). In CCDOS 4.2, for example, one installation disk contained three Chinese character repositories known respectively as LXCK.CZ1 ("Large-Scale Associative Character Repository"), LXCK.CZ2 ("Medium-Scale Associative

Character Repository"), and LXCK.CZ3 ("Small-Scale Associative Character Repository").[40]

The proliferation of Chinese character libraries benefited pinyin input far more than structure-based input. As we have examined in this book, structure-based input systems have historically been fine-tuned to facilitate the inputting of individual Chinese characters. A host of disambiguation techniques are baked into the very syntax and logic of the input system, with the goal of making sure that each potential alphanumeric input string corresponds to the least possible number of potential character matches. Thus, while structure-based approaches did make use of predictive text, and thus of Chinese-character libraries, typically the suggestion of the "next most likely" character or characters came only after a successful inputting of one's current character.

Pinyin input, as we have seen, was historically ill-suited to the inputting of single characters, again for reasons of homophony. By virtue of necessity, then, pinyin input systems have tended to harness predictive text functionalities as early in the inputting process as possible, often even before a single character is fully inputted. In more concrete terms, the proliferation of ever-more-robust Chinese-character repositories enabled pinyin input to employ ever more (and longer) multicharacter abbreviations of the sort we saw in rudimentary form in our earlier discussion of Zhi Bingyi and his Quick-Codes. Pinyin input abbreviations only work to the extent that there exists a robust library of multiple-character compounds and phrases against which such abbreviations can be compared, however, and through which matches can be suggested and made. Thus, although both structure-based and pinyin-based input systems stood to benefit from these character libraries, what we might call the "ceiling" of potential benefit was higher for pinyin.

So common were these techniques of predictive text, co-occurrence analysis, Chinese input libraries, and more that special terms had to be invented to refer to pinyin input that did *not* employ one or another form of predictive abbreviation or double-mediation. Two such terms were *quanpin* and *chunpin*, which can be translated as "complete pinyin" and "pure pinyin," respectively. In these formulations, the adjectives "complete" and "pure" denote a mode of pinyin input "completely" and "purely" based upon pinyin phonetics. In UCDOS90, for example, six input systems came preloaded: Quwei, Wubi zixing, Wubi, "fast pinyin" (*kuaisu pinyin*), "pure pinyin" (*chun pinyin*), and ASCII (English).[41] (Notice here the distinction made between "pure" and "fast" forms.)

One of the earliest and most popular pinyin input methods, Tianhui, provides a helpful example. The original guidebook for this system suggests no fewer than five different ways to input the two-character name "Tianhui" (天汇) itself. One could use a setting called "complete pinyin" (*quanpin* 全拼), entering the sequence t-i-a-n-h-u-i. One could also use one of four other settings, however, none of which involved the full "spelling out" of the Chinese compound. Using the "simplified pinyin" (*jianpin* 简拼) setting, which required users to input only the initial pinyin phonetic value of each character, the correct input sequence was "t-h." Using "mixed pinyin" (*hunpin* 混拼)—a setting that allowed users to input *either* the full phonetic pinyin values of a character *or* just the initial value—the correct sequences were either "t-h-u-i" or "t-i-a-n-h." Finally, a setting called "simplified pinyin + stroke-shape" (*jianpin* 简拼 + *bixing* 笔形)—in which numerals and Latin letters could be used either to describe the phonetic pinyin value of a character, or as variables that represented one or another structural feature of the desired Chinese character—the correct sequence was "t-1-h-4." In other words, out of the five "pinyin" methods an operator could use to input just this one two-character compound, only one of them was based strictly on the syntax of Hanyu pinyin, in the sense of full phonetic spellings. The other four methods, while using pinyin as their underlying "platform," harnessed the letters of the Latin alphabet hypographically (table 6.2).[42]

Such approaches have persisted and proliferated in the 2000s, making it possible to input a single Chinese word—the term "typewriter" (*daziji* 打字机), for example—using at least four or five different alphanumeric sequences (all within the same pinyin IME) (figure 6.4).

In just the past few years, moreover, a new generation of Chinese predictive text has emerged: Cloud input (*yun shurufa* 云输入法). Unlike IMEs from the 1980s through the 2000s, where the entire input process took place inside the computer, systems by Sogou, Baidu, QQ, Microsoft, and others have begun to harness enormous, distributed, user-generated Chinese-language text corpora, as well as ever-more sophisticated natural

Table 6.2 Inputting 天汇 (*tianhui*) using Tianhui pinyin input

Pinyin Mode	Input Sequence Entered by User	Screen Output
"Full pinyin"	tianhui	天汇* * i.e. identical screen output for all input sequences
"Simplified pinyin 1"	th	
"Mixed pinyin 1"	thui	
"Mixed pinyin 2"	tianh	
"Pinyin + Stroke-Shape"	t1h4	

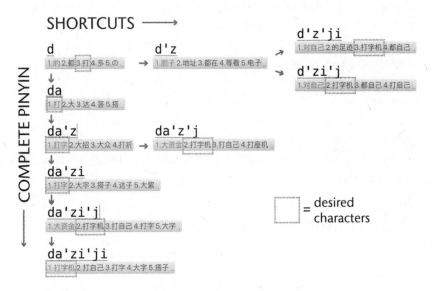

Figure 6.4 Pinyin IME shortcuts

language processing algorithms. These systems create immense, user-generated character repositories, growing and changing in real time, not unlike the repositories that help dynamically populate the Google search bar. In 2013, Microsoft researchers touted the growing power of its Chinese IME, while Sogou boasted far greater accuracy and performance for its cloud-based IME.[43] "Long sentence accuracy"—the ability for an IME to convert a long and complex sequence of alphabetic letters into an accurate, multicharacter Chinese passage—has grown from

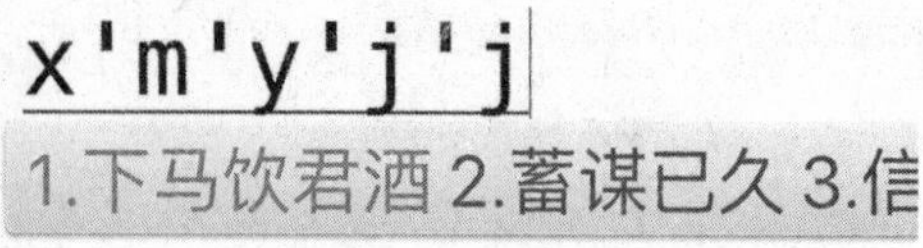

x'm'y'j'j

1.下马饮君酒 2.蓄谋已久 3.信

Figure 6.5   Inputting the poem "Parting" using Sogou pinyin input

a reported 62.5 percent on locally stored IMEs to 84 percent with cloud input, Sogou reported, while "short sentence accuracy" was reported to grow from 91.52 percent to 96 percent.[44]

Loading up a phonetic IME, for example, a user can enter the string *zhrmghg* and watch as it correctly suggests 中华人民共和国 (Zhonghua renmin gongheguo or "the People's Republic of China"). If a user prefers an example from deeper in antiquity, entering the input string *xmyjj* is a possibility. Chances are reasonable that the Cloud IME might recommend a stanza from "Parting" by famed Tang-dynasty poet Wang Wei: 下马饮君酒 (*xiama yin jun jiu* "I dismount from my horse and offer you wine . . .") (figure 6.5). Admittedly, this is one of the best known of all Tang poems, yet I invite the same user to switch their computer back to English-language mode and enter the string *sicttasdtamlamt*. Did your machine catch this comparably famous passage by Shakespeare? Chances are slim.

WHAT DOES PINYIN INPUT CHANGE?

From the 1990s onward, pinyin Input Method Editors have eclipsed structure-based approaches. What impact has the rise of pinyin input had? One question pertains to speed and efficiency: Is pinyin input superior to structure-based input systems? Was parity achieved? While it is difficult to determine this with certainty, there is reason to suspect that the answer is *no*. Time and again, Chinese input competitions of the kind first described in the introduction to this book have been won by contestants using one or another structure-based input system, rather than pinyin. Insofar as structure-based systems are far more effective at the unambiguous entry of individual characters; and because structure-based systems are also able to take advantage of anticipatory technologies such as predictive text; the case is very strong that, on the whole, pinyin input represents a step *backward* in terms of the overall speed of Chinese input.

What pinyin input might lack in speed, however, it more than makes up for in accessibility. This brings us to the second main point: the development of viable pinyin input methods has led to an unprecedented massification of hypography, with far greater numbers of Chinese speakers able to draw upon their knowledge of Hanyu pinyin to employ one or another pinyin input system on their computers or mobile devices. As compared to structure-based systems like Wubi, Cangjie, or others—all of which require users to learn a new syntax from scratch—pinyin-based IMEs have the ability to "piggyback" on the immense and growing number of Chinese people who have received extensive educations in Hanyu pinyin, often from a very young age. Using these pinyin-based input systems, along with a wider number of lesser-used nonphonetic Chinese Input Method Editors, hundreds of millions of Chinese computer and new media users have transformed China from a backwater of the global information infrastructure to one of its driving forces and most lucrative marketplaces. Parallel to the production of these information technologies for Chinese writing, writing itself has been fundamentally transformed.

But "massification" has not benefitted all Chinese speakers equally. Premised upon only one of China's many different spoken forms, Hanyu pinyin is not equally accessible to speakers of Cantonese or a host of other Chinese languages. Pinyin input inherits this bias. Unlike structure-based input systems, which were and are agnostic when it comes to non-standard Chinese languages, pinyin input leaves users little choice but to input characters according to "standard" Chinese pronunciation. Just as the Chinese government's original promotion of Hanyu pinyin has served its broader goals of nation-formation and linguistic standardization, so too does pinyin input require speakers of Cantonese, Shanghainese, and other dialects to "unify," to a certain extent. In this regard, pinyin input helps turn Chinese computing into another tool of state-led Chinese nationalism.[45]

The rise of pinyin hypography also changes the political economy of input entrepreneurialism. Although there are many different types of pinyin input systems, both in terms of proprietary software and underlying code, the ways in which these systems differ from one another is radically different from the ways that structure-based approaches do. To be more precise, in structure-based input, the inherent instability of hypography reveals itself in the form of a never-ending proliferation of new systems, each one advancing its own ontology of structural elements, its own syntax, and its own way of assigning elements to Latin letters and Arabic numerals. The diversity we find in structure-based input, in this regard, is *conspicuous* and *centrifugal*.

Diversification behaves differently in the arena of pinyin input. No matter how many different pinyin IMEs are proposed, prototyped, patented, and commercialized, all of them remain linked to, and in some sense reinforce, the underlying platform of Hanyu pinyin. Whether Sogou pinyin, QQ pinyin, Google pinyin, or otherwise, all of these systems normalize Hanyu pinyin, even as they compete with one another. What is more, the differences between these competing systems are subtler and less conspicuous as compared to the difference between, say, Wubi and Three-Corner, or OSCO and Cangjie. Most often, differences between competing pinyin IMEs come down to differences in the accuracy of each IME's predictive functions, the robustness of their character libraries, and even the aesthetics of their overall UX design. If structure-based input tends toward a conspicuous and centrifugal form of diversification, then, the diversification of phonetic input has tended to be centripetal and somewhat imperceptible. (But we should never misinterpret this tendency to mean that all pinyin Input Method Editors are the same. They are not.)

Even with the rise of a pinyin input, however, it is important to remember that Chinese input overall remains what it has been since the beginning of our history: a hypographic, retrieval-based system of textual production that relies on the same three-part process of criteria, candidacy, and confirmation that we have been examining throughout this book. Despite the uncanny resemblance, pinyin input and Hanyu pinyin are fundamentally different. Indeed, as we have seen, the success of pinyin input was itself a product of this fundamental difference, the immense hypographic potential of pinyin being what rescued it from the lowest rungs of the computational hierarchy. Pinyin input owes its success to an embrace of hypography.

CONCLUSION: WELCOME TO HYPOGRAPHY

In September 2013, at the very peak of the "character amnesia crisis," I sat down for tea in a neighborhood Starbucks with an octogenarian man and his granddaughter. It was the first time I'd met a Chinese veteran of World War II—the Eight-Year War of Resistance Against Japan, as the conflict is known to many in China.

The conversation came about thanks to a letter-writing campaign. My wife and I had slipped hundreds of letters into mailboxes all along Beijing's Second Ring Road, addressed to Chinese IME inventors I discovered in the PRC patent database.[1] Every one of these inventors—including this elderly war veteran, much to my surprise—laid claim to a novel Chinese input method and the certainty that their creation would at last solve the puzzle of computational Chinese text entry.[2]

Jiang Kun spent fifteen years working on his system—"Sound Code" (*Yinma*), as he called it.[3] The system was premised on the decomposition of Chinese characters into graphic elements, not unlike many of the systems developed from the 1950s through the 1980s. The system was interesting to read about, certainly—not least because of the inventor himself—but unlikely to reroute my research in new directions. It might make for a charming footnote, I reasoned; if not, I was still grateful to be there.

Then his granddaughter turned the wheel sharply on our placid Tuesday afternoon conversation and sent us careening down a cliff.

I think my input system is a bit easier to use.

She whispered the words just softly enough that her grandfather wouldn't hear.

I'm also working on a Chinese input system.

So is my father.

My language skills were clearly failing me, I thought. Three generations within the same family, working on three different input systems? She confirmed: Not only had her grandfather developed an input system for Chinese, but she and her father had as well. Her father had even quit his job to work on his input system full time, she underscored.[4]

After a pleasant exchange, Jiang Kun stood up briskly, smiled, and reached out to shake my hand. We snapped a photograph, and the conversation was over. I walked out of the café in a haze.

It was that day in fall 2013 when it hit me: The Chinese Input Wars of the 1980s and 1990s were still raging, and a peace settlement may never be brokered. Input may never settle down.

During the nineteenth century, and most of the twentieth, it was readily apparent why so many engineers, linguists, entrepreneurs, policymakers, and others would have committed themselves to the problem of Chinese in the information age. Chinese-language technologies were objectively slower and less effective than those for English and many other languages. Chinese telegraphy was slower. Chinese typewriting was slower. So was early Chinese computing. These deficits posed monumental problems for state-builders, moreover, along with diplomats, industrialists, researchers, entrepreneurs, military planners, financiers, educators . . . (the list is too long to enumerate). Beyond practical problems, there were also abstract, emotional motivations at play. On a conceptual level, "speed" and "efficiency" had served for so long as both a "mantra" and a "measure of men"—core metrics by which societies, languages, and technologies were compared and judged—that the markedly slower performance of Chinese raised disturbing questions about Chinese cultural and civilizational "fitness." Were Chinese characters compatible with the modern era?[5] Would China need to abandon them?

But the twenty-first century is an entirely different story. Chinese-language e-commerce platforms, apps, chat rooms, online literature sites, and social networks are the envy of tech firms and venture capitalists around the world.[6] Market valuations of some Chinese tech companies beggar belief.[7] Barely weeks after my exchange with Jiang Kun and his granddaughter in Starbucks, moreover, Huang Zhenyu—the man we met in the introduction of this book—stepped into an auditorium in Henan and bested his Chinese typing competitors with a speed of 221.9 characters per minute—3.7 Chinese characters every second. Based on any number of metrics, it would seem, Chinese has more than proven itself as one of the most successful writing systems of the digital age. What accounts, then, for this frenetic, irresolvable activity? Why does the number of Chinese input systems—already in excess of a thousand—continue to grow? Did the Jiang family—or, indeed, the dozens of inventors who submit IME patent applications each year—truly think they could outpace that? At 3.7 characters per second, hasn't Chinese emerged victorious in its more than century-long conflict with Latin alphabetic supremacy? Why, then, aren't Chinese engineers, inventors, companies, and users calming down and enjoying a well-deserved victory lap?[8] Hasn't alphabetic order at last been overthrown?

Two factors are essential to understand the continued restlessness of Chinese input; the first is technolinguistic. Within English-language typing, pushing the key marked "Q," then seeing the symbol "Q" appear on screen, is a taken-for-granted feature of human-computer interaction—so taken for granted that it has fostered a mythology in which alphabetic textual input is understood by many as an "immediate" act wherein "depression equals impression." I say "mythology" because, of course, in Anglophone as in Chinese computing there exists no pregiven relationship between any particular button and any particular symbol—to depress the key marked "Q" is simply to close a circuit that could actuate any one of an infinite number of potential operations. Printing "Q" to screen is just one such outcome.

Having operated under this framework for over a century, however—a convention that dates back to the era of mechanical typewriting—Anglophone computing makes it possible for users to be lulled into a deep and (for some) unshakeable complacency: the inability to summon to mind the fundamentally arbitrary nature of all key-symbol relationships. The arbitrariness, and thus plasticity, of all human-machine interaction sinks to the bottom of consciousness, like a pebble sinking to the bottom of a lake. All that remains is the assumption that Q *must equal Q.*

The situation in Chinese human-computer interaction is structurally different. The key-symbol relationship is just as arbitrary as in English, but in this case everyday users are, so to speak, constantly reminded of this arbitrariness. When there is no possible way to fit each Chinese character on its own dedicated key—when, by definition, what you type is *never* what you get—the *nonidentity* between keys and symbols becomes not just "possible," but the starting point. A kind of persistent, structural mismatch ensues that prevents hypography from settling into the same kind of complacency found in conventional models of human-machine interaction. When "Q" equals "Q," after all, a "keyboard interface" is nothing more or less than the shape of the keyboard itself, along with the precise allocation of symbols. Once that interface has stabilized into a particular shape and distribution—whether QWERTY or otherwise—then the "interface" has by definition "stabilized."

When "Q" doesn't equal "Q" (not necessarily, at least) then the *keyboard itself is not actually the interface.* The keys Q-W-E-R-T-Y all stay in the same places, and yet all one needs to do is change the IME to change the interface. In this way, a kind of buoyancy constantly pushes the intrinsic arbitrariness of language back to the surface of consciousness—no longer a pebble, but prosecco bubbles rising and bursting into the realm of critical awareness, over and over again.

This is not to say that everyday Chinese computer users are all armchair philosophers of language or would-be semioticians. Nor is it to say that input doesn't engender its own habits of mind and practice. It is to say, however, that no matter which Chinese IME users employ, there is no way they can look down at a QWERTY keyboard and pretend to see "Chinese characters" in any straightforward or self-evident way. To input Chinese using a QWERTY keyboard or a trackpad is always a matter of *invoking* or *retrieving* Chinese characters from memory. The A-B-Cs of the Latin alphabet are just a means of doing so, one among the infinite number of ways this act of retrieval could be achieved.

The buoyancy I speak of, then, has nothing to do with Confucianism, Communism, Buddhism, Daoism, or any other ill-conceived "predisposition" or "essence" of Chinese society or culture. It isn't a choice or a goal, even, so much as the historical byproduct of two centuries of technolinguistic inequality: a period that from the advent of electric telegraphy onward placed character-based Chinese script at a profound structural disadvantage—a disadvantage that engineers, linguists, and others sought to overcome.

The second reason for the continued reworking of Chinese IMEs is sociopolitical. We need to remember that the history examined in this book, and in *The Chinese Typewriter* before it, forms a nearly two-centuries-long, crucible-like period of civilizational anxiety.[9] This was a period in which China's position in the world was thrown into radical doubt within a new hierarchy of modernity and civilization dominated by the West. Through so much of this history, China found itself in a position of deficit and compensation—of compensating for shortcomings, whether real or perceived, through means self-imposed or imposed from without.

Chinese input is yet another part of this anxiety-ridden history of "compensation"; IMEs were invented, after all, solely for the purpose of circumventing the incalcitrant problem of "fitting" Chinese on a Western-style keyboard. Indeed, each and every one of the components in China's hypographic revolution originated as some kind of work-around, mod, patch, hack, make-do compromise, or "Plan B" designed to make the best of a bad situation. Even the engineers who pioneered Chinese input never held input up as a formal "alternative" to Western-style technological modernity; never did they exult hypography as a uniquely "Chinese" solution to challenges of the digital era. As much as figures like Chung-Chin Kao, Samuel Caldwell, Zhi Bingyi and others reveled in the ability of their systems to match, and perhaps even surpass, the speed of English-language typing, they never went so far as to suggest that English should perhaps steal a page from Chinese computational input and follow suit.

But then something peculiar happened. Since the turn of the twenty-first century, all of the many compensatory techniques and technologies examined in this book and in *The Chinese Typewriter*—ingenious work-arounds and hypermediations in the era of Chinese telegraphy, natural-language tray beds in the era of Chinese typewriting, and of course, Input Method Editors themselves—got *faster than the mode of textual production they were built to compensate for*: English and the longstanding model of one-key-one-symbol, what-you-type-is-what-you-get. Hypermediation, it turned out, boasted a much higher ceiling than "immediacy."[10]

To achieve 3.7 characters per second, however, is not the same thing as to shift a centuries-old paradigm. It does not, by itself, fully dislodge the legacies of Latin alphabetic dominance that permeate information technology at multiple registers simultaneously. Long after Chinese input has begun to surpass English typing in speed and accuracy, this longstanding sense of deficit—of somehow falling short of *one-symbol-one-key*, *what-you-type-is-what-you-get*, and *immediacy*—forms a belief system that will likely continue to exert influence on the minds of engineers, entrepreneurs, and amateurs alike. For how long, no one can say.

WHAT ARE THE LIMITS OF HYPOGRAPHY?

Is 3.7 characters per second the end of the story? Will hypography get even faster? Is hypography limitless? When it comes to English-language typing, there is a speed that human beings seem unlikely to exceed—200 words per minute, give or take. What about input? Is there a "speed limit" when it comes to hypography? If so, how do we calculate it?

Any attempt to answer this question will be far from straightforward. Thanks to the entangled relationship between hypography and predictive technologies—autocompletion, probabilistic retrieval algorithms, immense and ever-growing Chinese character libraries, cloud-based input, large language models (LLMs), and generative artificial intelligence— Chinese input has long since surpassed the stage wherein single Chinese characters can be anticipated before their full-length input sequences are keyed in. By now, anticipating two-, three-, and four-character strings is commonplace. Even ones extending to seven or more characters have been realized, as in the hyper-abbreviated pinyin input sequences *zhrmghg* (中华人民共和国) and *xmyjj* (下马饮君酒).

Imagine you are a literary scholar, employing a "connected thought" Chinese IME plug-in focused on Chinese poetry and literature. (If you're a medical professional, an aeronautical engineer, a physicist, a pharmacist . . . there are IME plug-ins for you, too.) As soon as you enter the first few characters of a well-known poem, let's say, your input system offers up a long passage from your desired text. Upon confirming, your composition window fills up with 10 characters—or perhaps 20 or 30. In a matter of a few seconds, you will have entered anywhere between 300 and 1,800 "characters per minute."

There is one speed limit at play, however, at least for the time being: the speed of human intention. All of the anticipatory technologies considered thus far presume that a writer already knows what they intend to write. The IME's job is merely to predict this preexisting intention. But what happens when capacities of "anticipation" exceed the speed by which there can exist something to anticipate? Could hypography begin to suggest words or passages the writer hadn't thought of—moving from feedback to "feed-forward"—and thereby surpass the speed of intention?[11]

Yes—and it's already happening. Whether in terms of automated journalism, grammar assistants, automatic translation, or most recently, generative AI technologies such as ChatGPT, nascent forms of such technologies already exist.[12] When I log into ChatGPT and enter a hypographic sequence—or what has since come to be referred to as a "prompt"—how to calculate the "words-per-minute" (WPM) of the text that results? Let's say I input the prompt "Produce a 1,000-word essay, complete with footnotes, condemning plagiarism committed by Ivy League academics" and receive a result within 10 seconds. In theory, this passage will have been produced at a speed of around *6,000 words-per-minute*. At this point, however, the very concept of WPM—this tried-and-true metric by which societies have been measuring productivity for over a century, as well as assessing the comparative advantages of one writing technology over another—begins to lose all meaning, collapsing under the weight of its own absurdity.

Keep in mind that this is only hypography's "incunabula" period: the first fifty years of an epochal change in text technology, akin to the opening half-century of the Gutenberg era.[13] There is every reason to expect that hypographic technologies will become more powerful, to the point where they are capable (or are treated by humans as being capable) of assessing the broader "meaning" of a draft in formation, and of making recommendations to the writer at the level of words, phrases, paragraphs, pages, or entire essays. Far beyond the "Gutenberg Galaxy," to borrow Marshall McLuhan's well-worn phase, we are now witnessing the making of the "hypographic human." Plus, there is nothing to say that hypographs must remain textual in nature—not in the conventional sense of the word, at least. Although the goal of hypography is textual production—the goal of *zhrmghg* is to create 中华人民共和国—hypographs themselves can also be gestural, facial, auditory, and olfactory. Anything that can be recorded and analyzed, in fact—the way a person walks, sits in a chair, even urinates—has the potential to serve as a hypograph.[14] What does "WPM" mean, then, if and when textual production becomes a full-bodied form of computationally augmented coauthorship? This is not to suggest that hypography is limitless—only that whatever its limits, they are unlikely to abide by the same rules as those seen in the domain of "immediate" human-computer interaction.

WHAT ABOUT ENGLISH?

"Does English have input?" A college student in Beijing posed this question to me at the end of a course I taught at People's University on the global history of information technology. "Are there IMEs for English?"

"Yes," I answered, thinking of systems such as Cirrin or ShapeWriter. "But virtually no one uses them."[15]

As I discussed in the introduction to this book, hypography is not and has never been limited to China or the Chinese language. Whether in the form of stenotyping, the Palm Pilot Graffiti Alphabet, T9 input, or otherwise, hypographic practices have been in use within English for over a century. What sets Chinese computing apart, then, is not the existence of hypography, but the scale and intensity of its usage. In English, hypography remains a hyperspecialized practice, reserved for specific domains of work (court stenography, for example), or in cases when practical limitations or physical abilities make the use of "conventional" QWERTY-style typing untenable or unattractive (as with the Palm Pilot and other small electronic devices). In Chinese, by contrast, hypography is ubiquitous.

A helpful analogy comes from music, or electronic music specifically— namely, MIDI, or musical instrument digital interface. With the advent of computer music in the 1960s, it became possible for musicians to play instruments that looked and felt like guitars, keyboards, flutes, and so forth but that—thanks to digital computing—produced the sound of drum kits, cellos, bagpipes, and more. What MIDI made possible for the instrument form, IMEs make possible for the QWERTY keyboard. Just as one could use a MIDI piano to play the cello, or a MIDI woodwind to play a drum kit, within the context of input, one could use a QWERTY keyboard and the Latin alphabet to "play" Chinese. For this reason, even when the "instrument form" remains consistent (in our case, the now dominant QWERTY keyboard within Sinophone computing), the number of different instruments that said controller can control is effectively unlimited.

The QWERTY keyboard as it functions in the Anglophone world is the same proverbial "MIDI controller" as in the non-Latin world. The only difference is: for a host of reasons, Anglophone computer users have been convinced that the only instrument they can play with this MIDI piano *is the piano*—and, furthermore, that the only proper way to play it is *exactly how they would play a conventional piano*. The idea of using QWERTY to "play English" in an alternate way—where "Q" might equal something other than "Q"—is a hard-to-swallow idea for all but the most aggressive of "early adopters." For mainstream computer users, moreover, the idea that it is possible to "play English" using some other kind of device entirely is undreamed-of (unless they're using their remote control to enter search terms in Netflix or entering an address in their car's GPS using a rotary dial). When it comes to a computer or a smartphone, after all, how many have heard of (let alone used) Cirrin or other "unistroke" text input systems? How many have outfitted their computers with "chorded keyboards" like Writerhander, BAT, TipTapSpeech, FrogWriter, Microwriter, SiWriting, or CyKey?

The *what-you-type-is-what-you-get* mode of human-computer interaction is pervasive, powerful, and enduring; it quietly airbrushes out certain parts of computing history that, in theory, should be difficult to ignore. Take "The Mother of All Demos," for instance, the famed 1968 presentation by computer visionary Douglas Engelbart. Heralded by so many for having predicted the future of computing—featuring prescient demonstrations of video conferencing, hyperlinks, graphical user interface, and the mouse long before their time—few seem to recall (or perhaps care) that Engelbart also demonstrated a hypography based "chorded keyboard." At the time he regarded it as on par with the computer mouse,

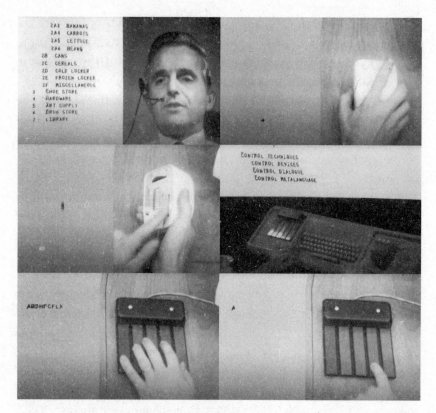

Figure C.1 Douglas Engelbart, the 1968 "Mother of All Demos," and the chorded keyboard

something that might forever transform the way human beings interacted with computational devices (figure 7.1).[16] (For this prediction, Engbelbart failed.)

It's not that hypography is unknown to Anglophone computing and new media, then—just that it's limited mainly to the fringe rather than the mainstream. While there has been no shortage of research on alternative HCI design, mainstream engineers and interface designers have, for the most part, done everything in their considerable power to craft computational experiences that mimic in many ways the act of typing on a 1924 Remington, only more ergonomically. They have done everything to *preserve*, that is, the core "logic" of mechanical typewriting from a century ago: one-symbol-one-key, what-you-type-is-what-you-get.

This is not surprising. After all, why would average users want to dedicate precious time and energy to learning complex, highly mediated systems of textual input when they already enjoy the "best of all possible worlds": immediacy? When presented with anything that falls outside this core—be it chording keyboards, autocompletion, predictive text algorithms, or otherwise—Anglophone computer users always have the option to regard these things as optional, auxiliary, or extra. As a result, much of the algorithmic and computational power of the microcomputer has been "left on the table"—at least when it comes to textual input.[17] In contrast, engineers focused on non-Latin computing have spent incalculable hours trying to figure out ways to *reimagine* human-computer interaction and *harness* the computational power of microcomputers to make their input experiences more intuitive or efficient—a pursuit driven by baseline necessity.

Ironically, some of the clearest illustrations of the enduring power of "normal" typing comes in Western criticisms of the QWERTY keyboard. In his famous 1985 essay, "Clio and the Economics of QWERTY," the economist Paul David posed a basic question: Given how inefficient the QWERTY keyboard is, in terms of how the letters of the Latin alphabet are arranged, why did it remain dominant? Why was it never displaced by other keyboard arrangements in history that were, he argued, demonstrably better? How did inefficiency win the day?[18] His answer became a mainstay of economic thought: economies are shaped, not merely by rational choice, but also by the sedimented accumulation of decisions from the past. Economic paths are "path dependent." (Many people have heard of the term "path dependency." Few realize it was coined in an article about QWERTY.)

David's essay placed a target on QWERTY's back, emboldening others to take up arms and follow in his footsteps. Jared Diamond (of *Guns, Germs, and Steel* fame) lambasted QWERTY as "unnecessarily tiring, slow, inaccurate, hard to learn, and hard to remember." It "condemns us to awkward finger sequences" so that "in a normal workday a good typist's fingers cover up to 20 miles on a QWERTY keyboard." He continued his withering critique by explaining that "QWERTY typing tends to degenerate into long one-handed strings of letters." QWERTY, in Diamond's opinion, is a "disaster."[19]

When we scratch at the surface of this exhilarating iconoclasm, however, things start to look a lot less revolutionary. It's true that a tiny minority of people in the Anglophone world began to question the sanctity of the QWERTY keyboard beginning in the 1980s, some even calling for its overthrow. But a far greater number of people outside of the Latin alphabetic world have been doing so for nearly one hundred years longer. It was in the 1880s, not the 1980s, when language reformers, technologists, state-builders, and others across modern-day East Asia, South Asia, the Middle East, North Africa, and elsewhere began to ask: How can we overcome QWERTY?[20]

The stakes of these non-Western critiques of QWERTY were also far more profound than those articulated in the writings of David and others. David lamented the loss of a few words-per-minute owing to the QWERTY layout. Diamond wanted to squeeze a bit more juice from his keyboard and ease the strain on his wrists. Meanwhile, Chinese critics were wrestling with whether QWERTY-style interfaces—alongside other Latin alphabet–centered information technologies—might exclude Chinese writing from global technolinguistic modernity altogether. Reformers in Japan, Korea, Egypt, Siam, India, and elsewhere worried about the fates of their writing cultures—the future of their countries.

The core irony of David's piece comes when the author explores ostensible "solutions" to QWERTY. For Anglophone critics of QWERTY, the answer we typically hear, with few exceptions, is *alternate keyboard layouts*—that is, by moving the keys to different locations on the board, we can liberate ourselves from the shackles of QWERTY. A pet layout for many has been the Dvorak Keyboard, designed by the University of Washington professor August Dvorak, who worked on his interface from roughly 1914 through the 1930s and beyond. Like many of those who have criticized QWERTY in the Western world, they vaunted the comparatively scientific disposition of letters on the Dvorák keyboard, heralding it as an emancipatory device that could free typists from the "conspiracy" of the "culprit" QWERTY.[21]

Although seemingly radical, when read within an Anglophone HCI context, such alternatives seem negligible—trifling, even—when compared to those emanating from the non-Western world. Chinese technologists and language reformers, along with many in the non-Western world at large, knew better than this. To "overcome" QWERTY was going to require far more thorough remedies than a mere rearrangement of letters. It wasn't even QWERTY per se that needed to be confronted and overcome, but the deeply embedded, interlocking web of assumptions within which QWERTY is just the most conspicuous feature. One key, one symbol. What-you-type-is-what-you-get. The myth of "immediate" writing. Writing that can be nothing else than an act of composition. These were the obstacles that ultimately needed to be overcome, and no amount of "rearrangement" was going to do the job, Dvorak or otherwise. The answer they arrived at, I argue in this book, was hypography.

Chinese is just one of the many non-Latin scripts to abandon conventional models of immediate HCI and venture headlong in the direction of hypography. The world over, QWERTY and QWERTY-style keyboards are ubiquitous. But "typing" in the conventional sense is not. For Chinese, Japanese, Arabic, Burmese, Devanagari, or any number of other non-Latin scripts, Input Method Editors—as well as a host of other "middleware" programs—are the rule rather than the exception, employed to solve one shared problem: the longstanding and deeply embedded bias of personal computing (and before that, telegraphy, typewriting, hot metal composing, and more), which operates to the advantage of the Latin alphabet, and to the disadvantage of most other writing systems.

Consider that:

- For the estimated 467 million Arabic speakers in the world, computational work-arounds are required to permit Arabic—a script in which the letters of the alphabet almost always connect as a rule, and change shape depending upon context—to interoperate with Western-built word processing programs in which letters are assumed not to touch, and not to change shape.
- For the more than 70 million speakers of Korean, computational work-arounds are needed to enable Hangul—an alphabet in which letters change both their size and position based upon contextual factors—to function within Western-built computing environments that, at the outset of personal computing, made no accommodations for this kind of script.

- For the estimated 588 million speakers of Hindi and Urdu, the approximately 250 million speakers of Bengali, and hundreds of millions of speakers of other Indic languages, still other computational workarounds had to be developed as well. Although there is a relatively small number of consonants in Devanagari script, for example, they often take the form of "conjuncts" whose shapes can vary dramatically from the original constituent parts (a feature of these writing systems that early, Western-built computers once again could not handle).

Even today, no fewer than seven different types of computational workarounds are necessary to render QWERTY or QWERTY-based personal computers compatible with Burmese, Bengali, Thai, Devanagari, Arabic, Urdu, and more (note that I do *not* write "to make Burmese . . . compatible with QWERTY"). These are:

1. Input Method Editor and recursive pop-up menus, required (as we have seen) for Chinese, but also Japanese, Korean, and many other non-Latin orthographies
2. Contextual shaping, essential for scripts such as Arabic and all Arabic-derived forms, as well as Burmese
3. Dynamic ligatures, also required for Arabic, but also Tamil, among others
4. Diacritic placement, required for scripts such as Thai that have stacking diacritics
5. Contextual reordering, in which the order of a letter or glyph changes depending on context (essential for Indic scripts such as Bengali and Devanagari in which consonants and following vowels join to form clusters)
6. Splitting, also essential for Indic scripts, in which a single letter or glyph appears in more than one position on the line at the same time
7. Bidirectionality, for Semitic scripts like Hebrew and Arabic that are written from right to left, but in which numerals are written left to right (figure 7.2)

When calculating the total population of those whose writing systems have been systematically excluded from the logics of Western-manufactured typewriters, linotype machines, monotype machines,

Contextual Shaping & Dynamic Ligatures

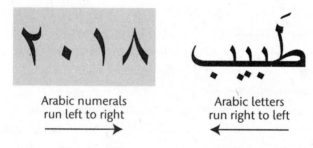

ب (b) in
medial position

ب (b) in
isolated position

Bidirectionality

Arabic numerals
run left to right
→

Arabic letters
run right to left
←

Reordering & Splitting

क + आ = का
क + इ = कि

When Hindi consonant क (k) combines with vowel आ (a),
vowel changes shape, but maintains position/order.

When same consonant combines with vowel इ (i),
vowel changes shape and position/order.

Figure C.2 Examples of contextual shaping, dynamic ligatures, reordering and splitting personal computers, and more—and which, as a result, have turned toward hypographic modes of HCI—the tally quickly exceeds 50 percent of the global population. In other words: at the beginning of the personal computing revolution, and for a good many years thereafter, the majority of people on earth were unable to use personal computers without "modding" them, whether by means of hardware or software solutions.

It would not be a stretch to say that the continued growth of computing and new media owes much of its fortune to this frenzy of hypographic "modding" in the non-Western and non-Latin world. For example: while still imperfect, Arabic is for the first time appearing on the screen and the printed page in its correct, connected forms; Korean Hangul is being presented with the *batchim* intact; and Indic conjuncts are appearing in correct formats, just to name a few. Collectively, Asia, Africa, and the Middle East are now home to some of the most vibrant and lucrative IT markets in the world, with Euro-American firms clamoring to make inroads. For those who cling to the myth of immediacy, such vistas might seem to confirm for them what they already knew to be true: that the magnificence of Western engineering and innovation once again bestowed its gifts upon the "rest" of the world.

But this is a falsehood. It was not the Western-designed computer that saved China and the non-Western world. It was China and the non-Western world that saved the Western-designed computer—saved it, that is, from its foundational limitations, both conceptual and material. Without Input Method Editors, contextual shaping, dynamic ligatures, rendering engines, layout engines, adaptive memory, contextual analysis, autocompletion, predictive text, the "modding" of the BIOS; the hacking of printer drivers, "Chinese-on-a-chip," and, above all, an embrace of hypography, no Western-built computer could have achieved a meaningful presence in the world beyond the Americas and Europe.[22] Today, hypography is the global norm. Hypography made global computing possible.

What becomes of our understanding of computing and new media, then, when the number of languages requiring "auxiliary" programs outnumber those for which computers work the way they "should?" When hypographic technologies become more widespread than supposedly "normal" orthographic ones, how does our understanding of electronic writing change?

Returning one final time to Huang Zhenyu and the 2013 Chinese input competition, we might ask: Would Huang Zhenyu have been able to write out President Hu Jintao's speech by hand, with just a pen and paper? And if he had proven incapable of doing so—if he, too, had "lifted his brush, but forgotten the character," would we really feel comfortable calling him amnesiac, aphasic, or illiterate?

Writing has changed. Our frameworks for understanding it must change as well.

ARCHIVAL ABBREVIATIONS

AS Academia Sinica (Taipei, Taiwan)

BMA Beijing Municipal Archives (Beijing, China)

CW Cable and Wireless Archives (Porthcurno, UK)

CSP Cambridge University Library Special Collections (Cambridge, UK)

CHM Computer History Museum (Mountain View, CA)

DNA Danish National Archives (Copenhagen)

DEPL Dwight D. Eisenhower Presidential Library (Abilene, KS)

HI Hoover Institute (Stanford University)

HML Hagley Museum and Library (Wilmington, DE)

HUBSSC Harvard University Business School Special Collections

IBM International Business Machines Corporate Archives (Somers, NY)

DK Donald Knuth Papers (Stanford University)

LOC Library of Congress (Washington, DC)

LR Louis Rosenblum Collection (Stanford University)

MBHT Museum of Business History and Technology (Wilmington, DE)

MLCR Mergenthaler Linotype Company Records, 1905–1993, Archives Center, National Museum of American History, Smithsonian Institution

MIT Caldwell Papers (Massachusetts Institute of Technology, Cambridge, MA)

MOP Museum of Printing History (North Andover, MA)

NARA-MD National Archives and Records Administration (College Park, MD)

NARA-SB National Archives and Records Administration (San Bruno, CA)

NCM National Cryptologic Museum (Annapolis Junction, MD)

NDL National Diet Library (Tokyo, Japan)

NI Needham Research Institute (Cambridge, UK)

OA Olivetti Archives (Ivrea, Italy)

PTM Porthcurno Telegraph Museum (Porthcurno, UK)

RHTC Rolf Heinen Technology Collection (Drolshagen, Germany)

SI Smithsonian Institute (Washington, DC)

SMA Shanghai Municipal Archives (Shanghai, China)

SPEC Stanford University Special Collections (Stanford University)

TMA Tianjin Municipal Archives (Tianjin, China)

TSM Thomas S. Mullaney East Asian Information Technology History Collection (Stanford University)

UW University of Washington Special Collections (Seattle, WA)

WL Wang Laboratories Corporate Papers (Harvard University Business School Special Collections)

YU Yale University Special Collections (New Haven, CT)

INTERVIEWS AND CORRESPONDENCES

Aled Tien, son of H. C. Tien, March 16, 2015

Andrew Sloss, son of Robert Sloss, February 21, 2011

Ann Welch, relative of Samuel Caldwell, September 5, 2013

Bruce Rosenblum, March 3, 2017

Chan-hui Yeh, March 21, 2010, and April 18, 2010

Jenny Chuang, September 3, 2013

Jenny Stagner, relative of Samuel Caldwell, September 4, 2013

Jiang Kun [蔣琨] and Jiang Wei [蔣薇], September 17, 2013

Jim Dowling, relative of Samuel Caldwell, August 20, 2013

Joseph Becker, January 18, 2013

Lee Collins, January 18, 2013

Lily Ling, September 25, 2017

Lois Lew, January 14, 2019

Louis Rosenblum, April 21, 2014

Patrick Finnigan, April 15, 2022

Peter Samson, September 5, 2023

Roger Melen, June 13, 2023

Severo Ornstein, June 30, 2016

Winston Kao, son of Chung-Chin Kao, April 23, 2014

Email from Jiang Wei [蔣薇], granddaughter of inventor Jiang Kun [蔣琨], to the author. September 15, 2013.

Letter addressed to Judy Ling from the offices of the Graphics Arts Research Foundation. June 9, 1981. Louis Rosenblum Collection. Stanford University.

Letter from A. P. Paine to J. T. Mackey. February 7, 1944. Mergenthaler Linotype Company Records, 1905–1993, Archives Center, National Museum of American History. Smithsonian Institution. Box 3628.

Letter from C. H. Griffith to George A. Kennedy. December 1, 1943. Mergenthaler Linotype Company Records, 1905–1993, Archives Center, National Museum of American History. Smithsonian Institution. Box 3628.

Letter from C. H. Griffith to George A. Kennedy. February 11, 1944. Mergenthaler Linotype Company Records, 1905–1993, Archives Center, National Museum of American History. Smithsonian Institution. Box 3628.

Letter from C. H. Griffith to J. T. Mackey. April 27, 1944. Mergenthaler Linotype Company Records, 1905–1993, Archives Center, National Museum of American History. Smithsonian Institution. Box 3628.

Letter from C. H. Griffith to J. T. Mackey. February 11, 1944. Mergenthaler Linotype Company Records, 1905–1993, Archives Center, National Museum of American History. Smithsonian Institution. Box 3628.

Letter from Chung-Chin Kao to A. P. Paine. February 7, 1944. Mergenthaler Linotype Company Records, 1905–1993, Archives Center, National Museum of American History. Smithsonian Institution. Box 3628.

Letter from Chung-Chin Kao to C. H. Griffith. April 29, 1944. Mergenthaler Linotype Company Records, 1905–1993, Archives Center, National Museum of American History. Smithsonian Institution. Box 3628.

Letter from Chung-Chin Kao to C. H. Griffith. February 7, 1944. Mergenthaler Linotype Company Records, 1905–1993, Archives Center, National Museum of American History. Smithsonian Institution. Box 3628.

Letter from Chung-Chin Kao to Griffith. August 21, 1946. Mergenthaler Linotype Company Records, 1905–1993, Archives Center, National Museum of American History. Smithsonian Institution. Box 3628.

Letter from Chung-Chin Kao to J. T. Mackey. February 18, 1944. Mergenthaler Linotype Company Records, 1905–1993, Archives Center, National Museum of American History. Smithsonian Institution. Box 3628.

Letter from Chung-Chin Kao to J. T. Mackey. October 1, 1946. Mergenthaler Linotype Company Records, 1905–1993, Archives Center, National Museum of American History. Smithsonian Institution. Box 3628.

Letter from Chung-Chin Kao to Mirovitch. May 13, 1946. Mergenthaler Linotype Company Records, 1905–1993, Archives Center, National Museum of American History. Smithsonian Institution. Box 3628.

Letter from Chung-Chin Kao to R .H. Turner. October 1, 1946. Mergenthaler Linotype Company Records, 1905–1993, Archives Center, National Museum of American History. Smithsonian Institution. Box 3628.

Letter from Chung-Chin Kao to Thomas S. Watson. February 2, 1946. IBM Corporate Archives. New York.

Letter from Eugene B. Mirovitch to J. T. Mackey. April 26, 1944. Mergenthaler Linotype Company Records, 1905–1993, Archives Center, National Museum of American History. Smithsonian Institution. Box 3628.

Letter from F.C. Frolander to R. R. Mead. June 11, 1946. Mergenthaler Linotype Company Records, 1905–1993, Archives Center, National Museum of American History. Smithsonian Institution. Box 3628.

Letter from Florence Anderson (Associate Secretary, Carnegie Corporation) to W.W. Garth Jr. (President, Graphic Arts Research Foundation). Mergenthaler Linotype Company Records, 1905–1993, Archives Center, National Museum of American History. Smithsonian Institution. Box 3628.

Letter from George A. Kennedy to C. H. Griffith. December 17, 1943. Mergenthaler Linotype Company Records, 1905–1993, Archives Center, National Museum of American History. Smithsonian Institution. Box 3628.

Letter from George A. Kennedy to C .H. Griffith. February 17, 1944. Mergenthaler Linotype Company Records, 1905–1993, Archives Center, National Museum of American History. Smithsonian Institution. Box 3628.

Letter from H. A. Burt to R. R. Mead. June 7, 1946. Mergenthaler Linotype Company Records, 1905–1993, Archives Center, National Museum of American History. Smithsonian Institution. Box 3628.

Letter from J. C. Plasteras to R. R. Mead. June 13, 1946. Mergenthaler Linotype Company Records, 1905–1993, Archives Center, National Museum of American History. Smithsonian Institution. Box 3628.

Letter from Jack Kai-tung Huang to Louis Rosenblum. April 16, 1980. Louis Rosenblum Collection. Stanford University.

Letter from John Forster to Hugh. April 15, 1980. Louis Rosenblum Collection. Stanford University.

Letter from Joseph Needham to the *Kung Sheung Wan Pao* (September 25, 1978). In "Correspondence and other documents relating to an inaccurate report published in 'Kung Sheung Evening News,' Hong Kong, about the computerisation of Chinese." Reference SCC2/24/1. Creator R.P. Sloss, Joseph Needham, Lee Tsung-Ying. Dates 9/25/1978 to 10/10/78. Needham Research Institute.

Letter from K. E. Eckland to Pardee Lowe. N.d. (August 1959, sometime after August 10, 1959). Pardee Lowe Papers. Accession No. 98055–16.370/376. Box No. 193. Hoover Institute Archives.

Letter from Li Liuping to Roy Hofheinz. September 13, 1979. Louis Rosenblum Collection. Stanford University.

Letter from Louis Rosenblum to Richard Solomon. June 21, 1982. Louis Rosenblum Collection. Stanford University.

Letter from Louis Rosenblum to Volta Torrey. October 13, 1960. Louis Rosenblum Collection. Stanford University.

Letter from Louis Rosenblum to Zhi Bingyi. June 30, 1981. Louis Rosenblum Collection. Stanford University.

Letter from Louis Rosenblum to Zhi Bingyi. November 1, 1978. Louis Rosenblum Collection. Stanford University.

Letter from Mirovitch to Mackey. April 30, 1946. Mergenthaler Linotype Company Records, 1905–1993, Archives Center, National Museum of American History. Smithsonian Institution. Box 3628.

Letter from Mirovitch to Mackey. June 19, 1946. Mergenthaler Linotype Company Records, 1905–1993, Archives Center, National Museum of American History. Smithsonian Institution. Box 3628.

Letter from Needham to Michael Loewe. September 26, 1979. SCC2/23/65. Needham Research Institute.

Letter from R. G. Crockett to T. S. Bonczyk of the Quartermaster Research and Development Section. June 18, 1958. Subject: "Contract No. DA19–129-QM-458 Quarterly Progress Report of June 18, 1958." Louis Rosenblum Collection. Stanford University.

Letter from R.G. Crockett to T.S. Bonczyk. June 18, 1958. Crockett listed as Project Administration.

Letter from Richard Solomon to William Garth IV and Louis Rosenblum. August 2, 1982. Rosenblum Papers. Stanford University.

Letter from Roy Hofheinz to Li Liuping. September 27, 1979. Louis Rosenblum Collection. Stanford University.

Letter from Roy Hofheinz to Louis Rosenblum. December 7, 1978. Louis Rosenblum Collection. Stanford University.

Letter from Roy Hofheinz to William Garth IV. January 16, 1980. Louis Rosenblum Collection. Stanford University.

Letter from Vannevar Bush to W. W. Garth Jr. August 19, 1954. Louis Rosenblum Collection. Stanford University.

Letter from Vannevar Bush to W. W. Garth Jr. May 9, 1955. Louis Rosenblum Collection. Stanford University.

Letter from W. W. Garth Jr. to Vannevar Bush. December 14, 1953. Louis Rosenblum Collection. Stanford University.

Letter from William Garth IV and Louis Rosenblum to Roy Hofheinz. January 14, 1980. Louis Rosenblum Collection. Stanford University.

Letter from Zhi Bingyi to Louis Rosenblum. March 3, 1980. Louis Rosenblum Collection. Stanford University.

Letter from Zhi Bingyi to Roy Hofheinz. September 27, 1979. Louis Rosenblum Collection. Stanford University.

Letter from Zhi Bingyi to William Garth IV and Louis Rosenblum. February 19, 1982. Rosenblum Collection. Stanford University.

Telex to Zhi Bingyi from Pres Low, Garth IV, and Rosenblum. June 23, 1980. Louis Rosenblum Collection. Stanford University.

CHARACTER GLOSSARY

10 lei zhongduan chengxu (INT 10) 10类终端程序

16 lei zhongduan chengxu (INT 16) 16类终端程序

24 zhen Hanzi dayinji (24-Pin Chinese Printer) 24针汉字打印机

Beijing hangkong xueyuan (Beijing Institute for Aeronautics and Astronautics) 北京航空学院

Beijing shi jisuanji yanjiusuo (Beijing Computer Research Institute) 北京市计算机研究所

bianma wuran ("Code Pollution") 编码污染

CCBIOS de CRT kongzhi chengxu (CCBIOS CRT Control Program) CCBIOS的CRT控制程序

CCBIOS de jianpan kongzhi chengxu (CCBIOS Keyboard Control Program) CCBIOS的键盘控制程序

chazifa gongzuozu (Character Retrieval Working Group) 查字法工作组

Chen Jianwen 陈建文

chongma huanchong qu (Chinese Character Duplicate Code Buffer) 重码缓冲区

Chung-Chin Kao 高仲芹

chushi daicanshu cunfangchu (CRT Initial Generation Parameter Storage) 初始代参数存放处

CRT shuaxin qu (CRT Refresh Zone) CRT刷新区

CRT zifu fashengqi (CRT Character Generator) 字符发生器

Cui Tienan 崔铁男

da moshi lianxiang ciku (Medium-Scale Associative Character Repository) 大模式联想词库

danxianti (terminal script) 单线体

daziji (typewriter) 打字机

di yi jixie gongyebu (First Ministry of Machine Building) 第一机械工业部

Dianbao xinshu (Chinese telegraph code) 電報新書

diannao (computer / "electronic brain") 电脑

diannao shu tongwen ("unification of writing for computers") 电脑书同文、

erjinzhi (binary) 二进制

fandong xueshu quanwei (Reactionary Academic Authority) 反动学术权威

fangzao (emulation / "copy-catting") 仿造

Haanzii Zhuaannbiaan Zimuhua (Chinese Transalphabet) 汉字转变字母化

Hanzi BASIC yuyan (Chinese BASIC) 汉字BASIC语言

Hanzi bianma (Chinese Character Encoding) 汉字编码

Hanzi COBOL yuyan (Chinese COBOL) 汉字COBOL语言

Hanzi shibie (Chinese character recognition / OCR) 汉字识别

Hanzi ziku xinpian (Chinese Character Library Chip) 汉字字库芯片

Hanzi shuru dasai (Chinese Input Competition) 汉字输入大赛

Hanzi xianshi (Chinese Character Display) 汉字显示

Hanzi xinxi chuli (Chinese Character Information Processing) 汉字信息处理

jineima (Internal Code) 机内码

jianzi shima (OSCO) 见字识码

jidianbu liusuo (Number 6 at the Ministry of Machine-Building Industries) 机电部六所

jingdian ouhe shi Hanzi jianpan (Electrostatic Chinese Keyboard) 静电耦合式汉字键盘

jisuanji (computer) 计算机

Jisuanji yanjiusuo (Computer Research Institute) 计算机研究所 (),

Kung Sheung Wan Pao (Kung Sheung Evening News) 工商晚報

Le shi jianpan (Loh's keyboard) 樂氏鍵盤

Le Xiuzhang (Shiu C. Loh) 樂秀章

Li Fan (Francis Fan Lee) 李凡

Li Jinkai 李金凯

lianxiang ciku weihu chengxu (Associative Character Repository Maintenance Program) 联想词库维护程序

Lin Yutang 林语堂

Lois Lew 劉淑蓮

mokuai (interrupt/INT) 模块

nanzhezi ("hard-to-cut characters") 难折字

Pinyin (*Pinyin*) 拼音

Pinyin-ization (*Pinyinhua*) 拼音

Qian Peide 钱培德

Riben chongdianqi gongye (OKI Corporation) 日本沖電器工業

Shanghai yinshua jishu yanjiusuo (Shanghai Printing Institute of Technology) 上海印刷技术研究所

shiliang Hanzi (vector font) 矢量汉字

shurufa 输入法

shuruma huanchong qu (Input Code Buffer Zone) 输入码缓冲区

sijibu (Fourth Machinery Ministry) 四机部

Sougou yunrufa (Sogou Cloud Input) 搜狗云输入法

tishihang (prompt line) 提示行

tuxing fangshi (graphics mode) 图形方式

Wang An 王安

Wang Jizhi 王辑志

Wang Xuan 王选

Wang Yongming 王永民

weixing jisuanji (microcomputer) 微型计算

xianshizi (display fonts) 显示字

xiao moshi lianxiang ciku (Small-Scale Associative Character Repository) 小模式联想词库

xinxihua (informationalization) 信息化

Xue Shiquan 薛士权

xuni shuaxin qu (Virtual Refresh Zone) 虚拟刷新区

yagan shi Hanzi jianpan (Pressure Sensitive Chinese Character Keyboard) 压感式汉字键盘

Yang Lien-Sheng (Lien-Sheng Yang) 楊聯陞

Yanshan jisuanji yingyong yanjiu zhongxin (Yanshan Computer Applications Research Center) 燕山计算机应用研究中心

zhenshi dayinji (Impact printer) 针式打印机

Zhi Bingyi 支秉彝

Zhima (Zhi Code, also known as OSCO) 支码

zhong moshi lianxiang ciku (Medium-Scale Associative Character Repository) 中模式联想词库

Zhongguo biaozhun jishu kaifa gongsi (Chinese Standards Technology Development Company) 中国标准技术开发公司

Zhongguo biaozhunhua yu xinxi fenlei bianma yanjiusuo (Chinese Standardization and Information Classification and Encoding Research Institute) 中国标准化与信息分类编码研究所

Zhongguo jisuanji yonghu xiehui (China Computer User Society) 中国计算机用户协会

Zhongguo Zhongwen xinxi xuehui (Chinese Character Encoding Working Group of the Chinese Information Society) 中国中文信息学会

Zhongguo Zhongwen xinxi yanjihui (Chinese Information Processing Society of China) 中国中文信息研究会

Zhongguo Zhongwen xinxi yanjiuhui (Chinese Information Processing Society of China) 中国中文信息研究会

Zhongguo Zhongwen xinxi yanjiuhui Hanzi bianma zhuanye weiyuanhui (Chinese Information Research Association Chinese Character Encoding Professionals Committee meeting) 中国中文信息研究会汉字编码专业委员会

Zhongwen jisuanji xiehui (Chinese Computing Society) 中文计算机协会

Zhongwen xinxi chuli guojia yanjiuhui (ICCIP) 中文信息处理国家研究会

Zhongwen xinxi yanjiuhui (Chinese Information Research Association) 中文信息研究会

zhongxing jianpan (medium-sized keyboard) 中型键盘

Zhou Youguang 周有光

zhuyin zimu (Chinese Phonetic Alphabet) 注音字母

zifu fangshi (symbolic mode) 字符方式

zifu fangshi xianshi (symbolic display mode/character display format) 字符方式显示

zifu zimo ku (character font library) 字符字模库

*zimo huanchongqu (*Character Bitmap Buffer Zone*)* 字模缓冲区

NOTES

INTRODUCTION

1. 提筆忘字.

2. Barbara Demick, "China Worries about Losing its Character(s)," *Los Angeles Times*, July 12, 2010, http://articles.latimes.com/2010/jul/12/world/la-fg-china-characters-20100712 (accessed September 2, 2014).

3. Victor Mair, "Character Amnesia," *Language Log*, July 22, 2010, https://languagelog.ldc.upenn.edu/nll/?p=2473; Jennifer 8. Lee, "In China, Computer Use Erodes Traditional Handwriting, Stirring a Cultural Debate," *New York Times*, February 1, 2001; Demick, "China Worries."

4. "'To Lift One's Brush and Forget the Character': What Have We Forgotten?" ('Tibi wangzi' Women jiujing wangdiao le shenme?) ['提笔忘字' 我们究竟忘掉了什么?], *New China Net (Xinhua wang)* [新华网]. November 4, 2013, http://edu.qq.com/a/20131104/002402.htm (accessed March 1, 2019).

5. *Hanzi weiji* 汉字危机; "'To Lift One's Brush and Forget the Character': Did Informationalization Cause the Chinese Character Crisis?" ('Tibi wangzi': shi xinxihua zaocheng le Hanzi weiji ma?) ['提笔忘字': 是信息化造成了汉字危机吗?], *China Youth Daily (Zhongguo qingnian bao)* [中国青年报], September 16, 2013, http://www.cernet.edu.cn/zhong_guo_jiao_yu/yiwujiaoyu/201309/t20130916_1017489.shtml.

6. Deborah Cameron, *Verbal Hygiene* (London: Routledge, 1995); Crispin Thurlow, "Generation Txt? The Sociolinguistics of Young People's Text-Messaging," *Discourse Analysis Online* 1, no. 1 (2003), https://extra.shu.ac.uk/daol/articles/v1/n1/a3/thurlow2002003-paper.html; Victoria Carrington. "Txting: The End of Civilization (Again)?" *Cambridge Journal of Education* 35 (2004): 161–175; Jannis Androutsopoulos, "Introduction: Sociolinguistics and Computer-Mediated Communication," *Journal of Sociolinguistics* 10, no. 4 (2006): 419–438; Tim Shortis, "'Gr8 Txtpectations': The Creativity of Text Spelling," *English Drama Media* 8 (June 2007): 21–26; David Crystal, *Txtng: The Gr8 Db8* (Oxford: Oxford University Press, 2008); Tamara Plakins Thornton, *Handwriting in America* (New Haven, CT: Yale University Press, 1996).

7. Cao Xuenan, "Bullet Screens (Danmu): Texting, Online Streaming, and the Spectacle of Social Inequality on Chinese Social Networks," *Theory, Culture and Society* 38, no. 3 (2019): 29–49.

8. In one sense, this book's focus on Input Method Editors brings it into dialogue with the subfields of software studies and the history of software. As "middleware," however, IMEs fit somewhat incongruously into these subfields, insofar as IMEs are a form of software that runs in between the user and other software applications—between a user and Microsoft Word, database programs, web browsers, or otherwise, that is. See Martin Campbell-Kelly, "The History of the History of Software," *IEEE Annals of the History of Computing* (December 18, 2007): 40–51. See also Matthew Fuller, *Behind the Blip: Essays on the Culture of Software* (Brooklyn, NY: Automedia, 2003).

9. There are also IMEs based upon fingerswipes, where one uses a smartphone, tablet, or computer track pad to retrieve Chinese characters by drawing them. In this case, the IME constantly updates its list of potential Chinese character matches based upon this data rather than keystrokes.

10. "Final Round of the 2013 'National Chinese Characters Typing Competition' Takes Place in the Pingqiao District of Xinyang City" (2013 "Quanguo Hanzi shuru dasai" zongjuesai zi Xinyang Pingqiaoqu juxing) [2013全国汉字输入大赛总决赛在信阳平桥区举行], *Henan xinwen* (broadcast December 12, 2013, 7:11 p.m.), https://zhuanlan.zhihu.com/p/37361698.

11. The full official English-language translation reads "Hold High the Great Banner of Socialism with Chinese Characteristics and Strive for New Victories in Building a Moderately Prosperous Society in All Respects." See Hu Jintao [胡锦涛], "Hold High the Great Banner of Socialism with Chinese Characteristics and Strive for New Victories in Building a Moderately Prosperous Society in All Respects," Report to the Seventeenth National Congress of the Communist Party of China, October 15, 2007, https://www.chinadaily.com.cn/china/2007-10/25/content_6204663.htm.

12. The competition, which was cosponsored by leading Chinese newspaper *Guangming Daily*, resembled hundreds just like it held across China and the Sinophone world since the 1980s. In contests like this, anywhere from dozens to hundreds of participants compete to transcribe a set of Chinese passages using a computer and a QWERTY keyboard, as quickly and accurately as possible. Just like Huang, they rattled off long alphanumeric strings that, as if by magic, transformed into on-screen Chinese-character passages. The other cosponsor was the Chinese Inventor's Association (Zhongguo faming xiehui).

13. To detect the keys that Huang depressed, particularly given his tremendous speed, it is necessary to pause the video and move frame-by-frame.

14. Translating between Chinese characters-per-minute and English words-per-minute requires careful accounting not only for the difference between "characters" and "words," but also for the idiosyncratic way that English-language typing competitions define the concept of "words" in "words-per-minute." Taking the second of these questions first, the number of "words" in a given passage is calculated differently by typing competition judges than by everyday individuals. First, judges tally up the total number of alphanumeric characters in the passage (including spaces and punctuation marks), then divide that number by 5 (considered the "average length"

of English words by typing competition leagues). This raises the admittedly thorny question of how to quantify and compare Chinese characters to English words. A reasonable technique would be to translate the original Chinese passage into English, and then compute a ratio. In accordance with the PRC state's officially endorsed English-language translation, the 31-character passage ". . . 高举中国特色社会 . . ." translates into the passage "Hold High the Great Banner of Socialism with Chinese Characteristics and Strive for New Victories in Building a Moderately Prosperous Society in All Respects." Although this passage contains 24 words by a layperson's count, when tallied in accordance with English-language typing competition rules, the actual count comes to 31.4 words in all (157 alphanumeric characters, spaces, and punctuation marks, divided by 5). For more on "rules of thumb" among Chinese translators, see the rich discussion on Quora with Kaiser Kuo: "On average, how many Chinese characters does it take to translate an English word?" https://www.quora.com/On-average-how-many-Chinese-characters-does-it-take-to-translate-an-English-word (accessed August 3, 2019).

15. Still other typing enthusiasts have posted astoundingly high speeds of 200, 242, and even 256 words-per-minute, albeit in the context of typing "sprints" lasting only a few seconds. Across the board, "sprint" competitions enable typists to achieve speeds unsustainable over any length of time.

16. Huang Zhenyu's "sprint" speeds also exceeded anything seen in the world of English-language typing. Sean Wrona, in a performance that has gone down in typing competition lore, once transcribed 21 words in just under 6 seconds (wherein each "word" was again defined as "each 5 characters, including all spaces and symbols," in accordance with North American standards for typing competitions). His speed for that performance was clocked at 256 words-per-minute, which would suggest that he completed the transcription in approximately 5.25 seconds overall. Lightning-fast as Wrona's performance was, Huang Zhenyu's initial "sprint" in the opening seconds of the 2013 competition was still faster. See https://www.youtube.com/watch?v=IozhMc6lPTU. Still others in this echelon include Jelani Nelson, Guilherme Sandrini, Yifei Chen, Nate Bowen, and a Youtuber named Kathy, https://www.youtube.com/watch?v=IFuMEnLthHs. See also https://seanwrona.com/typing.php. At SXSW, Wrona won the entire championship, at a speed of 124 words-per-minute https://www.youtube.com/watch?v=m9EXEpjSDEw.

17. In other cases, Chinese simply couldn't join the family, as in the case of hot metal composing—better known as Linotype and Monotype machines—which engineers tried and failed to accommodate to China's character-based script. For an in-depth investigation of the denigration of Chinese information technologies, see Thomas S. Mullaney, *The Chinese Typewriter: A History* (Cambridge, MA: MIT Press, 2017).

18. For a foundational work on the cultural logic and enduring appeal of "immediacy," with a focus on the Western world, see Jay David Bolter and Richard Grusin, *Remediation: Understanding New Media* (Cambridge, MA: MIT Press, 1998).

19. Michael H. Adler, *The Writing Machine: A History of the Typewriter* (London: Allen and Unwin, 1973); Wilfred A. Beeching, *Century of the Typewriter* (New York: St. Martin's Press, 1974).

20. Here I draw inspiration from, and join, scholars such as Nick Montfort, Noah Wardrip-Fruin, Matthew Kirschenbaum, Jacob Gaboury, and others who urge us to push beyond what Montfort refers to as "screen essentialism": a fixation on the visual output of computational devices. "The computer is not a visual medium," Jacob Gaboury reminds us. "We might argue it is primarily mathematical, or perhaps electrical, but it is not in the first instance concerned with questions of vision or image. Yet our engagement with computing technology is increasingly mediated through the interface of the screen. It is perhaps no surprise, then, that the vast majority of scholarship on computational media engages in an analysis of a computer's visual output—as text, image, and interaction—with little account given to the material processes that produce these images." This has special urgency for scholars working on Chinese computing and new media, where a fixation on output diverts attention from the myriad sites wherein Latin alphabetic dominance resides: encoding systems, the metallurgical properties of printer pins, the BIOS interrupts, bitmap grid sizes, and more. See Nick Montfort, "Continuous Paper: The Early Materiality and Workings of Electronic Literature," MLA Annual Conference, Philadelphia. December 2004. Text of talk available at https://nickm.com/writing/essays/continuous_paper_mla.html; Noah Wardrip-Fruin, *Expressive Processing: Digital Fictions, Computer Games, and Software Studies* (Cambridge, MA: MIT Press, 2009); Matthew Kirschenbaum, *Mechanisms: New Media and the Forensic Imagination* (Cambridge, MA: MIT Press, 2012); and Jacob Gaboury, "Hidden Surface Problems: On the Digital Image as Material Object," *Journal of Visual Culture* 14, no. 1 (2015): 40–60.

21. Chapter 5 joins the conversation on user-led innovation inaugurated in the classic work by Kline and Pinch. Ronald Kline and Trevor Pinch, "Users as Agents of Technological Change: The Social Construction of the Automobile in the Rural United States," *Technology and Culture* 37 (1996): 763–795.

22. Jack Goody, "Technologies of the Intellect: Writing and the Written Word," in *The Power of the Written Tradition* (Washington, DC: Smithsonian Institution Press, 2000), 133–138.

23. As with any neologism, other possibilities presented themselves along the way. "Subtext" and "subverse" both occurred to me but were ruled out because of their potential confusion with the notions of "underlying theme," "implicit meaning," or "iconoclastic." "Infragraph" was possible as well, playing upon terms such as "infra-sonic," signifying "below human audibility." This failed to capture the ephemeral nature of hypographs, however; hypographs are not "below" or "beneath" human perception, but rather transient and ultimately disposable steps en route to writing. "Underwriting" was another possibility, but it risked being confounded with unrelated terms from finance and banking. "Hypotext" was perhaps the closest contender, offering a potentially productive counterpoint with the well-known term "hypertext." If hypertext is text that links and leads to other, related text, "hypotext" could be thought of as text that links only to itself (i.e., that which it is meant to represent and retrieve). Ultimately, however, the choice of "-graph" does a better job, in my estimation, of capturing the potential expansiveness and diversity of the writing in question. After all, hypographs do not need to be text, in the

conventional sense of the word. Hypographs can be gestures, eye movements, all manners of symbols and markings, and much more. Anything that can be recorded can serve as the basis for hypography.

24. W. A. Martin, *The History of the Art of Writing* (New York: Macmillan, 1920).

25. The scale and ubiquity of hypography in Chinese computing and new media sets it apart from Bolter and Grusin's concept of "hypermediacy"—their term for immediacy's opposite. By and large, "hypermediacy" for Bolter and Grusin is the domain of the artist, designer, or critic. It is a purposeful, often expressly subversive, act by which one draws attention to the medium itself—the concrete poetry of the digital age, one might say. But in the Chinese context, hypography is quotidian. While it shares certain features with "hypermediation," it is decidedly not the arena of the artist or critic. Hypography is the domain of every single person who picks up every single digital device. Bolter and Grusin, *Remediation*.

26. My work has benefitted immensely through engagement with Brian Rotman and Lydia Liu, each having ventured into such questions well before me. Rotman argues that the emergence of phonetic script was a process by which humans "became besides themselves": a process of cognitive self-alienation that created new spaces of analytical abstraction, while foreclosing other cognitive possibilities. While the argument seems to echo earlier work by Alfred Bloom and others, who sought to find linkages between language and cognitive styles, Rotman does an admirable job of avoiding any stark or simplistic dichotomies between phonetic orthographies and Chinese. If the inscription of the spoken word was such an act, then hypography would seem to push such abstraction one step further. Hypography depends upon an ability to abstract abstraction itself: to have an out-of-body-out-of-body experience, so to speak, or to "become beside oneself, beside oneself." Meanwhile, Lydia Liu has written of printed English in the digital age as "ideographic English," marking a "postphonetic" turn in the Latin alphabet. "What is digital writing?," Liu reflects in an interview on the topic of her work. "It is a system of discrete ideographic symbols—numerals or letters that can be recognized and processed by the computing machine." These letters, once primarily in service of phonocentric writing, now take on an "ideographic" quality once thought to be possessed near-exclusively by Chinese script. See Brian Rotman, *Becoming Beside Ourselves: The Alphabet, Ghosts, and Distributed Human Being* (Durham, NC: Duke University Press, 2008); Lydia H. Liu, *The Freudian Robot: Digital Media and the Future of the Unconscious* (Chicago: University of Chicago Press, 2011); and "Interview with Lydia Liu," http://rorotoko.com/interview/20110615_liu_lydia_freudian_robot_digital_media_future_unconscious/?page=2. See also Alfred H. Bloom, "The Impact of Chinese Linguistic Structure on Cognitive Style," *Current Anthropology* 20, no. 3 (1979): 585–601; and Alfred H. Bloom, *The Linguistic Shaping of Thought: A Study in the Impact of Language on Thinking in China and the West* (Hillsdale, NJ: L. Erlbaum, 1981). For essential work on digital writing in the Euro-American context, see Richard A. Lanham, *The Electronic Word* (Chicago: University of Chicago Press, 1993); Matthew G. Kirschenbaum, *Track Changes: A Literary History of Word Processing* (Cambridge, MA: Harvard University Press, 2016); N. Katherine Hayles, *Writing Machines* (Cambridge, MA: MIT

Press, 2002); and N. Katherine Hayles, *How We Think: Digital Media and Contemporary Technogenesis* (Chicago: University of Chicago Press, 2012).

27. Examples of constructed language include Esperanto or Klingon, one of assistive script being Japanese Furigana (used to help readers discern the correct pronunciation of a kanji character). Examples of clipping and consonant reductions include "congrats" and "imedtly" (to replace "immediately"). Letter-number homophones include examples such as "r" (in lieu of "are"), "b" (in lieu of "be"), and in other languages, "6" (in lieu of the Italian "sei"). Braj B. Kachru, "The Bilinguals' Creativity," *Annual Review of Applied Linguistics* 6 (1985): 20–33; Braj B. Kachru, "The Bilingual's Creativity and Contact Literatures," in *The Alchemy of English: The Spread, Functions, and Models of Non-Native Englishes*, ed. Braj B. Kachru (Oxford: Pergamon Press, 1986), 159–170; Yamuna Kachru, "Code-Mixing, Style Repertoire and Language Variation: English in Hindu Poetic Creativity," *World Englishes* 8, no. 3 (1989): 311–319; Brenda Danet, *Cyberpl@y: Communicating Online* (Oxford: Berg Publishers, 2001); Hsi-Yao Su, "The Multilingual and Multi-Orthographic Taiwan-Based Internet: Creative Uses of Writing Systems on College-Affiliated BBSs," *Journal of Computer-Mediated Communication* 9, no. 1 (2003), JCMC912, https://doi.org/10.1111 /j.1083-6101.2003.tb00357.x; Larissa Hjorth, "Cute@keitai.com," in *Japanese Cybercultures*, ed. Nanette Gottlieb and Mark J. McLelland (London: Routledge, 2003); David Crystal, *A Glossary of Netspeak and Textspeak* (Edinburgh: Edinburgh University Press, 2004); Hirofumi Katsuno and Christine R. Yano, "Kaomoji and Expressivity in a Japanese Housewives' Chat Room," in *The Multilingual Internet: Language, Culture, and Communication Online*, ed. Brenda Danet and Susan C. Herring (New York: Oxford University Press, 2007), 278–302; Tim Shortis, "Revoicing Txt: Spelling, Vernacular Orthography and 'Unregimented Writing,'" in *The Texture of Internet: Netlinguistics in Progress*, ed. Santiago Posteguillo, María José Esteve, and Lluïsa Gea-Valor(Cambridge: Cambridge Scholar Press, 2007); Tim Shortis, "'Gr8 Txtpectations'"; Blake Sherblom-Woodward, "Hackers, Gamers and Lamers: The Use of l33t in the Computer Sub-Culture," unpublished paper, Swarthmore College, Swarthmore, Pennsylvania, 2008, https://www.swarthmore.edu/sites/default/files/assets /documents/linguistics/2003_sherblom-woodward_blake.pdf; Suresh Canagarajah, "Codemeshing in Academic Writing: Identifying Teachable Strategies of Translanguaging," *The Modern Language Journal* 95, no. 3 (2011): 401–417; Zhang Wei, "Multilingual Creativity on China's Internet," *World Englishes* (May 2015): 231–246; Li Wei, "New Chinglish and the Post-Multilingualism Challenge: Translanguaging ELF in China," *Journal of English as a Lingua Franca* 5, no. 1 (2016): 1–25; Li Wei and Zhu Hua, "Tranßcripting: Playful Subversion with Chinese Characters," *International Journal of Multilingualism* 16, no. 2 (2019): 145–161; Elena Giannoulis and Lukas R. A. Wilde, eds., *Emoticons, Kaomoji, and Emoji: The Transformation of Communication in the Digital Age* (New York: Routledge, 2020). A fascinating series of code-switching and stylization examples can be found in Hsi-Yao Su's work on Taiwan, as well as a riveting study of Hebrew and its representation with ASCII characters by Carmel Vaisman. See Carmel Vaisman, "Performing Girlhood through Typographic Play in Hebrew Blogs," in *Digital Discourse: Language in the New Media*, ed. Crispin Thurlow and Kristine Mroczek (Oxford: Oxford University Press, 2011), 177–196.

28. Greg Downey, "Constructing 'Computer-Compatible' Stenographers: The Transition to Real-Time Transcription in Courtroom Reporting," *Technology and Culture* 47, no. 1 (January 2006): 1–26; Miyako Inoue, "Word for Word: Verbatim as Political Technologies," *Annual Review of Anthropology* 47, no. 1 (2018): 217–232; Dongchen Hou, "Writing Sound: Stenography, Writing Technology, and National Modernity in China, 1890s," *Journal of Linguistic Anthropology* 30, no. 1 (2019): 103–122.

29. Sometimes, even letters that *do* appear on the keyboard do not stand for themselves. To record the initial consonant "F," for example (that is, an "F" that appears at the start of a word), the user does not in fact type the key which stands for "F" (even though there is one) but instead must depress two other keys, simultaneously: "T" and "P."

30. Phrased another way, if scholars had long misunderstood the "interface" between writing and orality, as Jack Goody famously argued, we now seem to misunderstand the relationship between various registers of writing. See Jack Goody, The Interface Between the Written and the Oral (Cambridge: Cambridge University Press, 1987). We must pay attention, to borrow from Benjamin Peters, to the "work that words do" in our attempts to understand and theorize digital technologies; in this case, words that fail to change in cadence with dramatic changes taking place in the world at large. See Benjamin Peters, "Introduction." In Digital Keywords: A Vocabulary of Information Society and Culture, ed. Benjamin Peters (Princeton, NJ: Princeton University Press, 2016), xv.

CHAPTER 1

1. This comment appeared on Thomas S. Mullaney (personal blog), August 21, 2010.

2. Thomas S. Mullaney, "Have you ever used a Chinese typewriter? Do you own one? 中文打字机你用过没有？办公室还有吗" (personal blog), June 22, 2010, https://thechinesetypewriter.wordpress.com/2010/06/22/have-you-ever-used-a-chinese-typewriter-do-you-own-one/. I donated to a nonprofit organization that digitizes historic films and makes them publicly available.

3. *Modern Business Machines for Writing, Duplicating, and Recording*, director not listed. Teaching Aids Exchange, 1947, film available online, http://www.archive.org/details/modern_business_machines_for_writing (segment begins at 12m35s).

4. I recently donated this collection to Stanford University, for use by the research community. See "Stanford Libraries Receives a Remarkable East Asian Information Technology Collection," Stanford Libraries, May 26, 2021, https://library.stanford.edu/node/172367 (accessed July 2, 2022).

5. "IBM Electric Chinese Typewriter." Museum of Business History and Technology. Wilmington, Delaware (hereafter MBHT).

6. Eugen Buhler and Christopher A. Berry, Machine adapted for typing Chinese ideographs," US Patent 2458339, filed May 3, 1946, and issued January 4, 1949; Burt to Mead, June 7, 1946; F. C. Frolander to R. R. Mead, June 11, 1946, Mergenthaler Linotype Company Records, 1905–1993, Archives Center, National Museum

of American History, Smithsonian Institution, Box 3628 (hereafter MLCR). See also Chung-Chin Kao, "Keyboard-controlled ideographic printer having permutation type selection," US Patent 2427214A, filed December 11, 1943, and issued September 9, 1947; and Chung-Chin Kao, "Chinese language typewriter and the like," US Patent 2412777A, filed June 28, 1944, and issued December 17, 1946; "Chinese Typewriter, Shown to Engineers, Prints 5,400 Characters with Only 36 Keys," *New York Times*, July 1, 1946, 26; J. C. Plasteras to R. R. Mead, June 13, 1946, MLCR.

7. My work (along with the title of this chapter) is heavily indebted to scholarship on Western computing by Jennifer Light, David Alan Grier, Janet Abbate, Corinna Schlombs, Mar Hicks, and Nathan Ensmenger. Jennifer S. Light, "When Computers Were Women," *Technology and Culture* 40, no. 3 (July 1999): 455–483; David Alan Grier, *When Computers Were Human* (Princeton, NJ: Princeton University Press, 2007); Janet Abbate, *Recoding Gender: Women's Changing Participation in Computing* (Cambridge, MA: MIT Press, 2017); Francesca Bray, "Gender and Technology." *Annual Review of Anthropology* 36 (2007): 37–53; Corinna Schlombs, "Women, Gender and Computing: The Social Shaping of a Technical Field from Ada Lovelace's Algorithm to Anita Borg's 'Systers,'" in *The Palgrave Handbook of Women and Science: History, Culture and Practice Since 1660* (London: Palgrave MacMillan, 2022), 307–332; Mar Hicks, *Programmed Inequality: How Britain Discarded Women Technologists and Lost Its Edge in Computing* (Cambridge, MA: MIT Press, 2018).

8. His name also appears in records as Kao Chung-Chin, C. C. Kao, and Gao Zhongqin.

9. Interview with Winston Kao, son of Chung-Chin Kao, April 23, 2014.

10. Erik Baark, *Lightning Wires: The Telegraph and China's Technological Modernization, 1860–1890* (Westport, CT: Greenwood Press, 1997), 82.

11. The code contained just under 7,000 characters, with additional empty spaces left in order for Chinese telegraphers to customize their code books as needed. Septime Auguste Viguier (Weijiye [威基謁]), *Dianbao xinshu* [電報新書] (Guangxu 18), in "Extension Selskabet—Kinesisk Telegrafordbog." 1871. Arkiv nr. 10.619, in "Love og vedtægter med anordninger," GN Store Nord A/S SN China and Japan Extension Telegraf. Rigsarkivet [Danish National Archives], Copenhagen, Denmark; Kurt Jacobsen, "A Danish Watchmaker Created the Chinese Morse System," *NIASnytt (Nordic Institute of Asian Studies) Nordic Newsletter* 2 (July 2001): 17–21. Viguier's original list of characters was later slightly adjusted by Deming Zaichu. In a signal contribution to the history of Chinese telegraphy, Deming Zaichu's full name was only recently reconstructed by Jing Tsu, a comparative literature scholar and Winter Olympics commentator. See Jing Tsu, *Kingdom of Characters: The Language Revolution That Made China Modern* (New York: Riverhead Books, 2022), 294.

12. Confronted by these and other challenges, some Chinese engineers and policymakers attempted to replace the four-digit telegraph code, while others sought simply to minimize its disadvantages. The first of these camps experimented with phonetic Chinese telegraph codes, in hopes of replacing the numerical system with

one that "spelled out" the sound of Chinese characters. The latter largely accepted the code as is but petitioned the international body governing telegraphic transmission to secure special, reduced rates for Chinese telegrams. While some of these efforts bore fruit—preferential tariffs were granted, for example—more radical calls for abandoning the cipher system failed. In part this was because phonetic systems were bedeviled by the problem of Chinese homophones: namely, the fact that sometimes dozens of distinct Chinese characters have the same phonetic pronunciations, thereby vastly increasing the possibility of misunderstanding when all one has is phonetic spellings to go on. For a detailed discussion of these efforts, see Mullaney, *The Chinese Typewriter: A History* (Cambridge, MA: MIT Press, 2017).

13. Although challenging to reconstruct the individual biographies of such Chinese telegraph operators and clerks, one small window we have is via the Thomas La Fargue Collection, 1872–1954, housed in Case 255 in the Washington State University Manuscripts, Archives, and Special Collections division. This collection contains biographical materials regarding young Chinese men who made their way through telegraph offices, as well as telegraph training programs. According to one contemporary source, there were 133 telegraph stations in Jiangsu province, for example, and 84 telegraph stations in Guangdong province. Even in more remote and sparsely populated parts of the country—such as Xinjiang, Gansu, Ningxia, Mongolia, and Qinghai—there were a combined 62 stations. The precise totals were 26 telegraph stations in Xinjiang, 24 in Gansu, 6 in Ningxia, 4 in Mongolia, and 2 in Qinghai. In Xinjiang, these were located in Tuhulu, Bachu, Gucheng, Yili, Yierkesitang, Tulufan, Tuokexun, Qincheng, Aertai, Hami, Dihua, Kukeshencang, Kuche, Wusu, Yanqi, Kashigeer, Wensu, Xingxingxia, Tacheng, Suilai, Ningyuan, Jinghe, Luntai, Huoerguosi, Zhenxi, and Emin. In Ningxia, these were located in Daba, Shizuishan, Wuzhongbao, Ningxia, Ninganbao, and Dengkou. In Mongolia, these were located in Daolin, Kulun, Wude, and Maimaicheng. In Qinghai, these were located in Xining and Guide. See *Dianma xinbian* (Shanghai: Shanghai Zhonghua shuju, n.d. ca. 1920, and Jiaotong bu, *Ming mi dianma xinbian*, 1935. Special thanks to Sijia Mao for assistance in collating this list.

14. James Purdon, "Teletype," in *Writing, Medium, Machine: Modern Technographies*, ed. S. Pryor and D. Trotter (London: Open Humanities Press, 2016), 120–136; Jay David Bolter and Richard Grusin, *Remediation: Understanding New Media* (Cambridge, MA: MIT Press, 1998).

15. R. H. Turner, "Chinese Type Casting Machine," November 19, 1943, 1. MLCR.

16. Chung-Chin Kao to A. P. Paine, February 7, 1944, MLCR; Chung-Chin Kao to C. H. Griffith, February 7, 1944, MLCR. Kao had a further meeting with J. T. Mackey on January 19, 1944. See Chung-Chin Kao to C. H. Griffith. February 7, 1944, MLCR.

17. F. C. Frolander to R. R. Mead, June 11, 1946, MLCR.

18. To be assessed against a royalty rate of 3 percent on net sales. "Suggested Outline X of License Agreement Between Chung-Chin Kao and Mergenthaler Linotype Company, NY," October 24, 1943, MLCR; Frolander to Mead, June 11, 1946.

19. C. H. Griffith to George A. Kennedy, December 1, 1943, MLCR.

20. C. H. Griffith to George A. Kennedy, February 11, 1944, MLCR; C. H. Griffith to J. T. Mackey, February 11, 1944, MLCR.

21. George A. Kennedy to C. H. Griffith, December 17, 1943, MLCR. For more on Kennedy's extensive work in the statistical analysis of Chinese, see in particular George A. Kennedy Papers. Manuscripts and Archives, Yale University Library, MS 308, Box 3, Folder 39. See also George A. Kennedy, ed. *Minimum Vocabularies of Written Chinese* (New Haven, CT: Far Eastern Publications, 1966).

22. George A. Kennedy to C. H. Griffith. February 17, 1944, MLCR.

23. Kennedy to Griffith, February 17, 1944.

24. Kennedy to Griffith, February 17, 1944.

25. This was not for lack of trying, either. In the 1920s, the company set out on an ultimately ill-fated attempt to break into the Chinese market, one that failed in spectacular fashion. For more on this failure, see Thomas S. Mullaney, "The Font That Never Was: Linotype and the 'Phonetic Chinese Alphabet' of 1921," *Philological Encounters/Brill* 3, no. 4 (November 2018): 550–566. See also R. Hoare, "Keyboard Diagram for Chinese Phonetic," Mergenthaler Linotype Collection, Museum of Printing, North Andover, MA (hereafter MOP), February 4, 1921; R. Hoare, "Keyboard Diagram for Chinese Phonetic Amended." Mergenthaler Linotype Collection, MOP, March 3, 1921; Mergenthaler Linotype Company, *China's Phonetic Script and the Linotype* (Brooklyn: Mergenthaler Linotype Co., April 1922), Smithsonian National Museum of American History Archives Center, Collection no. 666, box LIZ0589 ("History—Non-Roman Faces"), folder "Chinese," subfolder "Chinese Typewriter"; and "Chinese Romanized—Keyboard no. 141," Hagley Museum and Library, Accession no. 1825, Remington Rand Corporation, Records of the Advertising and Sales Promotion Department. Series I Typewriter Div. Subseries B, Remington Typewriter Company, box 3, vol. 1.

26. Silahis O. Peckley, No Title, *Asia Africa Intelligence Wire*, October 14, 2002; Obituary of T. Kevin Mallen, February 2, 2000, http://www.almanacnews.com/morgue /2000/2000_02_02.obit02.html (accessed December 12, 2012).

27. Frank Romano, *History of the Linotype Company* (Rochester, NY: RIT Press, 2014).

28. Chung-Chin Kao to A. P. Paine, February 7, 1944. MLCR.

29. A. P. Paine to J. T. Mackey, February 7, 1944, MLCR.

30. Griffith to Mackey, February 11, 1944.

31. Griffith to Mackey, February 11, 1944.

32. Griffith to Mackey, February 11, 1944.

33. Griffith to Mackey, February 11, 1944. Kao was becoming increasingly nervous. Mobilizing his own network of high-profile figures, Kao recounted meetings at Harvard with Dean George H. Chase, Harvard Printing Office director James W. MacFarlane, Harvard Yenching Institute director Serge Elisseeff, Harvard Chinese-Japanese Library director K. M. Jiu, and esteemed linguist Y. R. Chao, among others, all of whom reportedly expressed keen interest in Kao's autocasting machine, and assured

him that they would make contact with Mergenthaler soon—presumably to express their sentiments directly. See Chung-Chin Kao to J. T. Mackey, MLCR.

34. Eugene B. Mirovitch to J. T. Mackey, April 26, 1944, MLCR; C. H. Griffith to J. T. Mackey. April 27, 1944, MLCR.

35. Griffith to Mackey, April 27, 1944.

36. Chung-Chin Kao to C. H. Griffith. April 29, 1944, MLCR.

37. Kao to Griffith, April 29, 1944.

38. Kao to Griffith, April 29, 1944.

39. Kao to Griffith, April 29, 1944.

40. "IBM's Chinese Typewriter Demonstrated in New York," *Business Machines*, July 9, 1946; "Boon to China: Typewriter Has 5,400 Symbols," *Herald Tribune*, July 1, 1946; "Chinese Engineers Meet," *New York Times*, June 30, 1946, 9.

41. One of the surviving IBM electric typewriters is held at a semiprivate collection in Wilmington, Delaware, MBHT. This collection also houses Kao's original four-digit code book. A second IBM electric Chinese typewriter appeared a few years ago at auction, acquired by scholar and type designer Vaibhav Singh. In designing and prototyping the machine, Kao owed much to two IBM engineers in particular: Eugen Buhler and Christopher Berry. Both hailed from IBM's research laboratory located in Poughkeepsie, where the Kao's prototype was built. Berry joined IBM in 1923, at which time the company was known as the "Computing-Tabulating-Recording Company." He started as a service representative in a Philadelphia office with fewer than ten coworkers. He rose through the ranks briskly, transferring to Boston in 1927 to take up a position as service supervisor, and then in 1935, as a district manager. In 1943, Berry moved to the IBM Laboratory at East Orange, and the following year, relocated to Poughkeepsie at a time when IBM was producing carbines for the US military. Eugen Buhler was a Swiss-born engineer and inventor whose family emigrated to the United States in 1910 when Eugen was still a young boy. See "Lab's Chris Berry Marks 40th Anniversary with IBM," *IBM News* 1, no. 4 (February 25, 1964): 1, 4 (Personal archives of Richard Foss and John O'Farrell); "Chinese Typewriter, Shown to Engineers," 26; IBM 1947 press release. See also https://ethw.org/Oral-History:John_McPherson. For information on Eugen Buhler, see 1930 US Census information and his World War II Draft Registration Card (1942) via Ancestry.com.

42. IBM Brochure for Electric Chinese Typewriter. International Business Machines Corporate Archives, Somers, New York (hereafter IBM), 1. "Chinese Typewriter, Shown to Engineers," 26; Buhler and Berry, "Machine adapted for typing Chinese ideographs"; H. A. Burt to R. R. Mead, June 7, 1946, MLCR. Kao's designs evolved over time. In his earlier designs and patent materials, Kao envisioned a machine with a static, rather than revolving, drum, which except when triggered by the keyboarding process to move laterally and rotationally, would remain at rest. According to engineers at IBM, Kao was eventually convinced to adopt a continuously rotating drum, at their urging. See Frolander to Mead, June 11, 1946. See also Chung-Chin Kao, "Keyboard-controlled ideographic printer having permutation type selection,"

US Patent 2427214A; and Chung-Chin Kao, "Chinese language typewriter and the like," US Patent 2412777A; J. C. Plasteras to R. R. Mead, June 13, 1946, MLCR. The subsecond speed ensured that the character would not be blurred or smudged, since the character drum did not stop revolving during the registration process.

43. Chung-Chin Kao to Thomas S. Watson, February 2, 1946, IBM.

44. Chung-Chin Kao to Eugene B. Mirovitch, May 13, 1946, MLCR.

45. On the feminization of clerical labor in Asia see Janet Hunter, "Technology Transfer and the Gendering of Communications Work: Meiji Japan in Comparative Historical Perspective," *Social Science Japan Journal* 14, no. 1 (Winter 2011): 1–20. See also Kae Ishii, "The Gendering of Workplace Culture: An Example from Japanese Telegraph Operators," *Bulletin of Health Science University* 1, no. 1 (2005): 37–48; Mullaney, *The Chinese Typewriter*; Alisa Freedman, Laura Miller, and Christine R. Yano, eds., *Modern Girls on the Go: Gender, Mobility, and Labor in Japan* (Stanford, CA: Stanford University Press, 2013); and Raja Adal, "The Flower of the Office: The Social Life of the Japanese Typewriter in Its First Decade," presentation at the Association for Asian Studies Annual Meeting, March 31–April 3, 2011. On the feminization of clerical and other forms of white collar labor in the United States and Europe, see Sharon Hartman Strom, *Beyond the Typewriter: Gender, Class, and the Origins of Modern American Office Work, 1900–1930* (Chicago: University of Illinois Press, 1992); Margery W. Davies, *Woman's Place Is at the Typewriter: Office Work and Office Workers 1870–1930* (Philadelphia: Temple University Press, 1982); Brenda Maddox, "Women and the Switchboard," in *The Social History of the Telephone*, ed. Ithiel de Sola Pool (Cambridge, MA: MIT Press, 1977), 262–280; Susan Bachrach, *Dames Employées: The Feminization of Postal Work in Nineteenth-Century France* (London: Routledge, 1984); Michele Martin, *"Hello, Central?": Gender, Technology and Culture in the Formation of Telephone Systems* (Montreal: McGill-Queens University Press, 1991); and Ken Lipartito, "When Women Were Switches: Technology, Work, and Gender in the Telephone Industry, 1890–1920," *American Historical Review* 99, no. 4 (1994): 1074–1111.

46. IBM Electric Chinese Typewriter Four-Digit Code Tables. MLCR.

47. Kao to Griffith, April 29, 1944.

48. Born in Jinan, she also appears in the historical record as Huan Jung Kwoh and Grace Hjan-Jung Kwoh, Ancestry.com. Dutchess County, New York, Naturalization Records, 1932–1989; National Archives and Records Administration (hereafter NARA-MD), Passenger and Crew Lists of Vessels Arriving at Seattle, Washington, NAI Number: 4449160, "Records of the Immigration and Naturalization Service, 1787–2004," Record Group Number: 85, Series Number: M1383, Roll Number: 230, Ancestry.com.

49. Index to Marriages, New York City Clerk's Office, New York, New York, Volume Number, vol. no. 5, New York City Municipal Archives.

50. Park College 1939 Yearbook; Plasteras to Mead, June 13, 1946.

51. Eugene B. Mirovitch to J. T. Mackey, April 30, 1946, MLCR.

52. Mirovitch to Mackey, April 30, 1946.

53. Frolander to Mead, June 11, 1946.

54. Frolander to Mead, June 11, 1946.

55. Mirovitch to Mackey, June 19, 1946. MLCR.

56. Burt to Mead, June 7, 1946; and Mirovitch to Mackey, June 19, 1946.

57. IBM Brochure Chinese; "IBM's Chinese Typewriter Demonstrated in New York," July 9, 1946.

58. Photographic album from Chung-Chin Kao to J. T. Mackey. August 14, 1946. MLCR.

59. Chung-Chin Kao to Griffith, August 21, 1946, MLCR.

60. "Boon to China: Typewriter Has 5,400 Symbols," *Herald Tribune*, July 1, 1946; "Chinese Engineers Meet," *New York Times*, June 30, 1946, 9.

61. Chung-Chin Kao to Griffith, August 21, 1946.

62. Chung-Chin Kao to J. T. Mackey, October 1, 1946, MLCR.

63. *IBM Journal of Research and Development* 40 (1996): 34.

64. Chung-Chin Kao to R. H. Turner, October 1, 1946.

65. "Faster Chinese," *Time*, July 15, 1946, 86.

66. "Faster Chinese," *Time*, July 15, 1946, 86.

67. "Chinese Typewriter, Shown to Engineers," 26.

68. Grace and Yanghu Tong welcomed their first child in May 1944 when they lived in Lexington, Kentucky. By winter 1948, she was expecting their second child, Vivian, born on September 22, 1948. While this timing does not preclude Tong's continued involvement (the China trip took place in fall 1947), it is possible that pressures of family life and marital expectations began to intrude more heavily.

69. Rita K. Conley, "At Plant 3, Rochester," *Business Machines* 27, no. 20 (May 10, 1945): 5.

70. New York State, Birth Index, 1881–1942. According to a 1924 Albany city directory, one "Ying Eng" was a restaurant worker in Albany at this time. Given the small number of people of Chinese descent then living in Albany, the proximity of Albany and Troy, and the shared surname, it is almost certain that this is a relative of Lois Eng. The person's address is listed as "55 Green." Next to her name "Lois Eng," "Eng Ying Yum" is written in pencil (the record seems to be providing the young girl's Chinese name; in this case the surname would be "Eng" and her given name "Ying Yum."

71. Interview with Lois Lew, January 14, 2019.

72. Circa 1951, one Gay N. Lew, whose profession was listed in contemporary records as "typist," lived at 281 East Avenue in Rochester, NY. She is listed as living with Tony Lew. It is possible that Tony Lew (Anglicized name of Yuen Lew) was the husband of Lois Lew, and Gay N. Lew was her sister-in-law. City Directory, Rochester, NR, 1951, 542.

73. See https://www.ibm.com/ibm/history/exhibits/vintage/vintage_4506VV2045 .html.

74. Interview with Lois Lew, January 14, 2019.

75. For an in-depth analysis of the educational background of Chinese typists in the 1920s through 1950s, as well as for demographic information related to many hundreds of Chinese typists, see Mullaney, *The Chinese Typewriter*.

76. In the ship manifest, the 16-year-old Lois Eng is listed as being accompanied by "Will Moy," an uncle living at 41–51 Main St. in Flushing, NY. During our interview, Lois Lew made mention of a relative living in New York, which would seem to confirm this connection.

77. Materials related to IBM Electric Chinese Typewriter Demonstration in Shanghai. Shanghai Municipal Archives (hereafter SM), Q449-1-535.

78. Interview with Winston Kao, son of Chung-Chin Kao, April 23, 2014.

79. Receipt indicating the provision of one Model EH 1 (*Hengda shi keyi zhida*). Dated December 2, 1947. The paper is marked "Ziyuan weiyuanhui" 资源委员会.

80. "Revolutionary Chinese Typewriter Displayed," *The North-China Daily News*, October 21, 1947.

81. Chung-Chin Kao [高仲芹], "The Design and Applications of the Electric Chinese Typewriter" (Diandong Huawen daziji zhi sheji ji qi yingyong) [電動華文打字機之設計及其應用], *Kexue huabao* [科學畫報] 13, no. 12 (1947): 746–748, 747; "Education and Culture: Chung-Chin Kao Invents an Electric Chinese Typewriter" (Jiayu yu wenhua: Gao Zhongqin faming diandong Zhongwen daziji, yingchang qi chuang bian guoyin dianbao cidian) [教育與文化:高仲芹發明電動中文打字機、應昌期創編國音電報詞典], *Jiaoyu tongxun* [教育通訊] 4, no. 1 (1947): 29–30; "Electric Chinese Typewriter: First Electric Chinese Typewriter Displayed in New York at the National Trade Show, Invented by Chung-Chin Kao, Chinese" (Diandong Zhongwen daziji: di'yi jia diandong Zhongwen daizji ceng zai Niuyue quanguo shangye zhanlanhui zhong chenlie ke ji wei guoren Gao Zhongqin shi suo faming) [電動中文打字機:第一架電動中文打字機曾在紐約全國商業展覽會中陳列該機為國人高仲芹氏所發明], *Xinwen tiandi* [新聞天地] 18 (1946): cover; "News: Chung-Chin Kao Invents Electric Chinese Typewriter" (Xiao xiaoxi: Gao Zhongqin faming diandong Zhongwen daziji) [小消息:高仲芹氏發明電動中文打字機], *Tianjia banyuekan* [田家半月報] 13, no. 1/2/3/4 (1946): 3; "Chinese Science and the Requirements of Economic Development: Chung-Chin Kao Creates a Chinese Typewriter" (Zhongguo kexue yu jingji jianshe yaoxun: Gao Zhongqin chuangzhi Huawen daziji) [中國科學與經濟建設要訊:高仲芹創制華文打字機], *Minzhu yu kexue* [民主與科學] 1, no. 4 (1945): 64; "Gao Zhongqin Invented Automatic Chinese Typewriter" (Gao Zhongqin faming zidong shi Zhongwen daziji) [高仲芹發明自動式中文打字機], *Xinan shiye tongxun* [西南實業通訊] 12, no. 3/4 (1945): 59; "Science and Technology News: Domestic: Chung-Chin Kao Invents a Chinese Character Technology Application" (Kexue jishu xiaoxi: Guonei: Gao Zhongqin faming 'Zhongguo wenzi jishu yingyong') [科學技術消息:國內:高仲芹發明'中國文字技術應用'], *Kexue yu jishu* [科學與技術] 1, no. 4 (1944): 87; "Electric Chinese Typewriter"

(Diandong Huawen daziji) *Qingnian shiji* [青年世紀] 1, no. 1 (1946): 10; "Electric Chinese Typewriter" (Diandong Huawen daziji), *Tiandiren* [天地人] 2, no. 1 (1946): 36; "Electric Chinese Typewriter" (Diandong Huawen daziji), *Zhonghua shaonian* [中華少年] 3, no. 11 (1946): 1; "Invention: Electric Chinese Typewriter" (Faming: diandong Huawen daziji) [發明: 電動華文打字機], *Qingnian wenti* [青年問題] 3, no. 6 (1946): 21; "Electric Chinese Typewriter" (Diandong Zhongwen daziji), *Shenbao* (July 16, 1946): 3.

82. Promotion of the machine extended beyond those cities where physical demonstrations took place, moreover. Tianjin, which according to all available information was not one of the locations on the group's itinerary, was nevertheless the site for a small-scale advertising campaign. Promotional literature was sent to the Tianjin branch of the China Textile Industries Corporation. See "Electric Chinese Typewriter" (Dian Hua daziji) [電華打字機], January 17, 1948, 1–10. Tianjin Municipal Archives (hereafter TMA) J66-3-410. Addressed to the Tianjin branch of the China Textile Industries Corporation (Zhongguo fangzhi jianshe gongsi Tianjin fen gongsi) [中國紡織建設公司天津分公司].

83. "Electric Chinese Typewriter" (Diandong Huawen daziji), *Kexue* 29, no. 12 (1947): 378; "Electric Chinese Typewriter" (Diandong Huawen daziji), *Shizhao yuebao* [時兆月報] 42, no. 3 (1947): 33; "Electric Chinese Typewriter" (Diandong Huawen daziji), *Shizheng pinglun* [市政評論] 9, no. 12 (1947): 17; "Gao Zhongqin Invents an Electric Chinese Typewriter" (Gao Zhongqin faming diandong Zhongwen daziji), *Jiaoyu tongxun* 4, no. 1 (1947): 29–30; "New Invention" (Xin faming), *Guofang yuekan* 2, no. 1 (1947): 2; "Two Newly Invented Chinese Typewriters" (Zhongwen daziji liang qi xin faming) [中文打字機兩起新發明], *Kexue yuekan* 15 (1947): 23–24; Gao Zhongqin, "Diandong Huawen daziji zhi sheji ji qi yingyong," *Kexue huabao* 13, no. 12 (1947): 746–748; "Revolutionary Chinese Typewriter Displayed."

84. "Chinese Typewriter Inventor Chung-Chin Kao Brings His Invention to San Francisco Recently, Attends American Business Stationery Exhibition; Kao Accompanied by Demonstrators Chen Rujin and Liu Shulian to Demonstrate Typing on the Machine" (Huawen daziji famingren Gao Zhongqin jun, jin tui qi faming pin fu Sanfanshi canjia 'quan Mei shangye wenju zhanlanhui': dangzhong biaoyan dazi qingxing, sui Gao jun tongwang huichang biaoyanzhe you Chen Rujin, Liu Shulian liang nvshi) [華文打字機發明人高仲芹君,近攜其發明品赴三藩市參加全美商業文具展覽會:當眾表演打字情形,隨高君同往會場表演者有陳如金劉淑蓮兩女士], *Chinese-American Weekly* (*Zhong-Mei zhoubao*), [中美周報] 230 (1947): 1.

85. "Lab's Chris Berry Marks 40th Anniversary with IBM"; personal archives of Richard Foss and John O'Farrell.

86. Interview with Winston Kao, April 23, 2014.

87. Gao Yongzu [高永祖], "A Newly Invented Chinese Tele-Typewriter (Xin faming de Zhongwen dianbao daziji) [新發明的中文電報打字機]," *World Today*, March 16, 1962. Pardee Lowe Papers. Accession No. 98055–16.370/376. Hoover Institute Archives. Box No. 276 (hereafter HA). See also https://www.oki.com/en/130column/08.html (accessed July 2, 2018).

88. First, the keyboard was now manipulated automatically via a paper tape feed, and of course, the character drum was re-outfitted with Japanese kanji and kana characters. As before, the internal drum rotated continuously, shifting left and right to bring the correct symbols into the "type" position. "Japanese Language Telegraph Printer," US Patent 2728816, filed March 24, 1953, and issued December 27, 1955, assignor to Trasia Corporation, NY; see also Stacy V. Jones, "Telegraph Printer in Japanese with 2,300 Symbols Patented," *New York Times*, December 31, 1955, 19.

89. Jones, "Telegraph Printer in Japanese with 2,300 Symbols Patented"; "Japanese Language Telegraph Printer," US Patent 2728816, assigned to "Trasia Corporation," of which Kao was then vice president.

90. Interview with Winston Kao, son of Chung-Chin Kao, April 23, 2014.

91. Intriguingly, this experiment with, and debate over, Chung-Chin Kao's code came just a few years before the publication of a canonical study of the limits of human information processing by George A. Miller. See George A. Miller, "The Magical Number Seven, Plus or Minus Two: Some Limits on Our Capacity for Processing Information," *The Psychological Review* 63, no. 2 (March 1956): 81–97.

92. Alan Morrell, "Whatever Happened to . . . Cathay Pagoda?" *Democrat and Chronicle*, May 6, 2017, https://www.democratandchronicle.com/story/local/rocroots/2017/05/06/whatever-happened-cathay-pagoda/101345224/ (accessed January 2, 2019).

93. "Radical Machines Chinese in the Information Age," exhibition held October 18, 2018–March 24, 2019, at the Museum of Chinese in America. The author curated the exhibition.

CHAPTER 2

1. Paul N. Edwards, *The Closed World: Computers and the Politics of Discourse in Cold War America* (Cambridge, MA: MIT Press, 1997).

2. It would also undermine, they felt, any ambitions by Mao and the Communists to abolish Chinese character script and replace it with the Latin alphabet. "With the ultimate development of keyboard composing of Chinese," a US Army report later emphasized, "the need for language reform will no longer exist." See Quartermaster Research and Engineering Center, "Chinese Photocomposing Machine" (Natick, MA: Headquarters Quartermaster Research and Engineering Command, US Army, March 1960), 6. Louis Rosenblum Collection, Stanford University (hereafter LR).

3. The Graphic Arts Research Foundation was founded with the express goal of collaborating with American centers of phototypesetting research and manufacturing—above all, Lumitype and, later, Photon, Inc.—on the refinement and commercialization of the new technology. "Machine Heralds New Printing Era," *Boston Herald*, September 16, 1949, 1; "Printing Without Type," *Business Week*, October 1, 1949, 57. See also "The Press: Peace in Chicago," *Time*, September 26, 1949, https://content.time.com/time/subscriber/article/0,33009,800775,00.html (accessed February 19, 2022); Alan Marshall, *Du Plomb à la Lumière: La Lumitype-Photon et la Naissance des Industries Graphiques Modernes* (Paris: Maison des Sciences de L'Homme, 2003);

Frank J. Romano, *History of the Phototypesetting Era* (San Luis Obispo: Graphic Communication Institute, California Polytechnic State University, 2014); William W. Garth IV, *Entrepreneur: A Biography of William W. Garth, Jr. and the Early History of Photocomposition*, self-pub., 2002; "Visit of Caryl P. Haskins to Graphic Arts Foundation. Carnegie Corporation of New York Record of Interview," February 20, 1953, LR. Haskins was a former president of the Carnegie. N. Katherine Hayles, *Postprint: Books and Becoming Computational* (New York: Columbia University Press, 2021). As many as 140 different printing, publishing, and newspaper outfits "subscribed" to the Graphics Arts Research Foundation shortly following its formation, their $1,000 membership fees helping GARF further develop the technology—which subscribers would then be first in line to develop commercially. "Printing as it is now done is an obsolete art," the renowned analog-computer pioneer Vannevar Bush announced in a 1949 issue of *Business Week*. See "Printing Without Type," 57. See also "New Machine Sets Type on Film Instead of Metal," *Christian Science Monitor*, September 16, 1949, 1.

4. Samuel H. Caldwell, *Switching Circuits and Logical Design* (New York: Wiley, 1958); "S. H. Caldwell: 1904–1960," *MIT Technology Review* 63, no. 2 (December 1960): 4. As a core and longstanding member of the student and faculty body at MIT, Caldwell was also deeply shaped by currents in information theory and cybernetics. For more on the early era of cybernetic research, ranging from classic works to essential histories, see David A. Mindell, *Between Human and Machine: Feedback, Control, and Computing Before Cybernetics* (Baltimore, MD: Johns Hopkins University Press, 2004), 171, 278; Ronald Kline, *The Cybernetics Moment: Or Why We Call Our Age the Information Age* (Baltimore, MD: Johns Hopkins University Press, 2015), 20; Nathan L. Ensmenger, *The Computer Boys Take Over: Computers, Programmers, and the Politics of Technical Expertise* (Cambridge, MA: MIT Press, 2012), 125, 170; Bernard Dionysius Geoghegan, "From Information Theory to French Theory: Jakobson, Lévi-Strauss, and the Cybernetic Apparatus," *Critical Inquiry* 38, no. 1 (2011): 96–126; Claude E. Shannon and Warren Weaver, *The Mathematical Theory of Communication* (Urbana: University of Illinois Press, 1949); Norbert Wiener, *Cybernetics: Or, Control and Communication in the Animal and the Machine* (Cambridge, MA: MIT Press, 1961), 6; and Eden Medina, *Cybernetic Revolutionaries: Technology and Politics in Allende's Chile* (Cambridge, MA: MIT Press, 2014).

5. "S. H. Caldwell: 1904–1960"; Louis Rosenblum to Mr. Volta Torrey, editor, *MIT Technology Review*, regarding the passing of Prof. Samuel H. Caldwell," October 13, 1960, LR. Caldwell was also named a member of the American Academy of Arts and Sciences.

6. Interview with Ann Welch, granddaughter-in-law of Samuel Hawks Caldwell, September 5, 2013.

7. Interview with Ann Welch.

8. Joyce Chen, *Joyce Chen Cook Book* (Philadelphia: J. B. Lippincott Company, 1962), 223.

9. "Machine Seen as Possible 'Breakthrough' in Chinese Printing," File No. 147 (June 22, 1959), Pardee Lowe Papers. Hoover Institute Archives. Accession No. 98055–16.370/376. Box No. 276 (hereafter HI).

10. Samuel Caldwell, "Progress on the Chinese Studies," HI.

11. "Progress on the Chinese Studies," HI.

12. Samuel H. Caldwell, Ideographic type composing machine, US Patent 2950800, filed October 24, 1956, and issued August 30, 1960.

13. Caldwell, "Final Report on Studies Leading to Chinese and Devanagari," 14, LR.

14. Caldwell's Sinotype was not the first Chinese system to employ this mode of human-machine interaction. Rather, an experimental Chinese typewriter known as the MingKwai machine, invented in the 1940s by Lin Yutang, was the first. For an in-depth examination of MingKwai and Chinese retrieval-inscription machines, see Thomas S. Mullaney, *The Chinese Typewriter: A History* (Cambridge, MA: MIT Press, 2017).

15. Robert E. Harrist and Wen Fong. *The Embodied Image: Chinese Calligraphy from the John B. Elliott Collection* (Princeton, NJ: Princeton University Art Museum, in association with Harry N. Abrams, 1999), 4.

16. Yee Chiang. *Chinese Calligraphy: An Introduction to Its Aesthetics and Techniques* (Cambridge, MA: Harvard University Press, 1973 [1938]).

17. Harrist and Fong, *The Embodied Image*, 152.

18. Harrist and Fong, *The Embodied Image*, 152.

19. Caldwell, Ideographic type composing machine.

20. The index card also included a "serial number" at the top right to aid in record keeping. For this, Caldwell used the so-called Mathews Numbers from the well-known reference work by Robert Henry Mathews.

21. Samuel H. Caldwell, "Final Report on Studies Leading to Chinese and Devana-gari," 5, LR. Caldwell designed his binary system as a minimum-redundancy code, one of the pioneering developments of his MIT advisee David Huffman. In such a code, Caldwell summarized, "No code of any length forms the beginning of a longer code. Thus the code for stroke B is 11, and there is no other code that begins with the digits 11. Likewise, the code for stroke P is 1001, and no longer code begins with 1001." For the purposes of optimizing his system, this would have profound implications. First, it liberated Caldwell from assigning each stroke-letter a uniform-length code, which by necessity would have been equal to 5 binary digits (5 binary digits being the minimum needed to express the 21 total strokes). Using a minimum-redundancy code, Caldwell could use the insights he derived from his frequency analysis of strokes to assign much shorter binary codes to the most common strokes. For highly common strokes like the horizontal and vertical, these would be assigned two-digit codes. The next most frequent class of strokes—the downward-left stroke and the dot—would then be assigned binary codes of three digits, still far fewer than if he employed an evenly distributed 5-digit code throughout. At the other end of the spectrum, the highly infrequent stroke 乙, which appeared in a scant three one-hundredths of one percent of characters in Caldwell's sample, would be assigned a binary code of 11 digits. Even when accounting for much longer code sequences like this, stark differences in frequency assured Caldwell that the overall efficiency of the system would increase markedly.

22. 'CJK STROKE H' (U+31D0); Unicode U+31D2 (ˊ).

23. R = ∟ or Unicode U+31C4: CJK STROKE SW; T= U+31E0 (乙); H = U+31D6 (¬); M = U+31D9 (∟).

24. Graphic Arts Research Foundation, "Second Interim Report on Studies Leading to Specifications for Equipment for the Economical Composition of Chinese and Devanagari." December 1, 1953, HI. The report was addressed to the trustees and officers of the Carnegie Corporation of New York; Samuel Caldwell and W. W. Garth Jr., "Proposal for Studies Leading to Specifications for Equipment for the Economical Composition of Chinese and Devanagari." Marked "Confidential." Graphic Arts Research Foundation. Cambridge, MA. March 25, 1953, LR. Within this, $17,000 would be dedicated to the Chinese project; $10,000 to Devanagari; and $3,000 to administrative costs.

25. Caldwell and Garth Jr., "Proposal for Studies Leading to Specifications for Equipment for the Economical Composition of Chinese and Devanagari."

26. Caldwell, Ideographic type composing machine.

27. Caldwell, Ideographic type composing machine.

28. Caldwell and Garth Jr., "Proposal for Studies Leading to Specifications for Equipment for the Economical Composition of Chinese and Devanagari," 1.

29. Vannevar Bush to Garth Jr., August 19, 1954, LR.

30. Bush to Garth Jr., May 9, 1955, LR.

31. W.W. Garth Jr. to Vannevar Bush, December 14, 1953, LR.

32. W.W. Garth Jr. to Vannevar Bush, December 14, 1953, LR.

33. Caldwell's work during World War II is preserved in part in a small cache of materials held in the Institute Archives and Special Collections at MIT. (See Collection AC 0004, Box 44, Folder 66. Institute Archives and Special Collections at MIT). MIT, however, does not house the papers of the Graphic Arts Research Foundation per se (an understandable error made by at least one amateur historian). The most robust collections pertaining to Caldwell and GARF are held at the Museum of Printing in North Andover, Massachusetts (MOP), and at Stanford University in the Louis Rosenblum Papers (LR).

34. For more examples on the intimate relations between engineering, research, and the military during World War II, see also M. M. Irvine, "Early Digital Computers at Bell Telephone Laboratories," *IEEE Annals of the History of Computing* (July–September 2001): 22–42.

35. Florence Anderson (Associate Secretary, Carnegie Corporation) to W. W. Garth Jr. (President, Graphic Arts Research Foundation), Mergenthaler Linotype Company Records, 1905–1993, Archives Center, National Museum of American History. Smithsonian Institution. Box 3628 (hereafter MLCR); Edward P. Lilly, "Memorandum for the Executive Officer: Chinese Ideograph Type-setting Machine." (April 23, 1959). OCB Secretariat Series, Box 3: Ideographic Composing Machine. Dwight D. Eisenhower Presidential Library (hereafter DEPL); "Machine Seen as Possible 'Breakthrough' in Chinese Printing," HI.

36. Samuel H. Caldwell. "The Sinotype: A Machine for the Composition of Chinese from a Keyboard," *Journal of The Franklin Institute* 267, no. 6 (June 1959): 502.

37. "Memorandum for the Operations Coordinating Board: Interim Report of the Chinese Ideographic Composing Machine (CICM)," May 18, 1959. OCB Secretariat Series, Box 3: Ideographic Composing Machine. DEPL.

38. R. G. Crockett to T. S. Bonczyk, June 18, 1958.

39. "Memorandum of Meeting: OCB Ad Hoc Working Group on Exploitation of the Chinese Ideographic Composing Machine," May 7, 1959, DEPL.

40. "Memorandum for the Operations Coordinating Board: Interim Report of the Chinese Ideographic Composing Machine (CICM),"

41. "Memorandum for the Executive Office: Chinese Ideographic Composing Machine—Briefing Memo on Deferral of Board Consideration," May 20, 1959, DEPL.

42. R. G. Crockett to T. S. Bonczyk of the Quartermaster Research and Development Section, June 18, 1958. Subject: "Contract No. DA19–129-QM-458 Quarterly Progress Report of June 18, 1958, Report no. 12," LR.

43. Caldwell, "The Sinotype," 486–487.

44. Caldwell, "The Sinotype," 484.

45. Caldwell, "Progress on the Chinese Studies," 2, in "Second Interim Report on Studies Leading to Specifications for Equipment for the Economical Composition of Chinese and Devanagari." Report by the Graphic Arts Research Foundation, Inc. Addressed to the Trustees and Officers of the Carnegie Corporation of New York, HI.

46. Samuel Caldwell. "Progress on the Chinese Studies," 1–6, HI.

47. Caldwell, "The Sinotype," 478. Caldwell took this analysis further, in fact. "If the entire vocabulary is studied without weighting for frequency of character occurrence, the median length of the full spellings is 10.2 strokes, and the median length of minimum spellings is 6.7 strokes. For a relatively small sample of actual prose text the median length of full spelling was 8.3 strokes and that of minimum spelling was 6.1 strokes."

48. "Contract No. DA19–129-QM-458 Quarterly Progress Report of June 18, 1958, Report No. 12." Included in letter from R. G. Crockett to T. S. Bonczyk. June 18, 1958. Crockett listed as Project Administration. Addressee listed as Quartermaster Research and Development Section, Natick, MA.

49. "Chinese Language Photocomposition Machine," May 4, 1959, HI. Box No. 193.

50. "Memorandum of Meeting: OCB Ad Hoc Working Group on Exploitation of the Chinese Ideographic Composing Machine," May 19, 1959. Attendees included Colonel Weber (Defense), Edwin Kretzmann (State Department), Pardee Lowe (USIA), Albert L. Cox Jr. (OCB), Colonel Charles Welsh (Defense), Major Charles Frances (Defense), Richard See (NSF), and Kay Kitagawa (NSF).

51. "Memorandum for the Executive Office," May 20, 1959, DEPL.

52. "Memorandum for the Executive Office," May 20, 1959, DEPL.

53. "Memorandum for the Executive Officer," May 21, 1959, 2, DEPL.

54. K. E. Eckland to Pardee Lowe, n.d. (sometime August 10, 1959). HI, Box No. 193; Rosenblum to Torrey, *MIT Technology Review*, regarding the passing of Samuel H. Caldwell, LR.

55. Interview with Louis Rosenblum, April 21, 2014.

56. For insightful and cutting-edge work on the phenomenology of digital media, and the ways that existing models fail to understand it, see Mark B. N. Hansen. *Feed-Forward: On the Future of Twenty-First-Century Media* (Chicago: University of Chicago Press, 2014). For the classic account of the interface, and competing conceptualizations of mediation, see Alexander R. Galloway, *The Interface Effect* (Cambridge: Polity, 2012).

CHAPTER 3

1. Interview with Chan-hui Yeh, March 21, 2010.

2. "Toynbee Lectures." *VMI Cadet* (February 10, 1958): 2.

3. Interview with Chan-hui Yeh, March 21, 2010. This was far from an original idea, it's important to emphasize. Many before and after Toynbee have proposed as much.

4. Lu Xun, "Reply to an Interview from My Sickbed" (Bingzhong da jiumang qing-bao fangyuan) [病中答救亡情報訪員] (1938), in *Complete Works of Lu Xun* (*Lu Xun Quanji*) [魯迅全集], vol. 6 (Beijing: Renmin wenxue, 1981), 160.

5. John DeFrancis, *Nationalism and Language Reform in China* (Princeton, NJ: Princeton University Press, 1950); John DeFrancis, *The Chinese Language: Fact and Fantasy* (Honolulu: University of Hawai'i Press, 1984); Jeremy Norman, *Chinese* (Cambridge: Cambridge University Press, 1988); and John DeFrancis, *Visible Speech: The Diverse Oneness of Writing Systems* (Honolulu: University of Hawai'i Press, 1989).

6. Leslie Berlin, *Troublemakers: Silicon Valley's Coming of Age* (New York: Simon and Schuster, 2018).

7. Bernard Dionysius Geoghegan, "An Ecology of Operations: Vigilance, Radar, and the Birth of the Computer Screen," *Representations* 147, no. 1 (August 2019): 59–95; Jeremy Packer and Kathleen F. Oswald, "From Windscreen to Widescreen: Screening Technologies and Mobile Communication," *The Communication Review* 13, no. 4 (2010): 309–339; Jeremy Packer, "Screens in the Sky: SAGE, Surveillance, and the Automation of Perceptual, Mnemonic, and Epistemological Labor," *Social Semiotics* 23, no. 2 (2013): 173–195; Friedrich A. Kittler, "Computer Graphics: A Semi-Technical Introduction," trans. Sara Ogger, *Grey Room* 2 (Winter 2001): 30–45; Lev Manovich, "An Archeology of a Computer Screen," Moscow: Soros Center for Contemporary Art. http://www.manovich.net/TEXT/digital_nature.html; Jan Hurst et al., "Retrospectives: The Early Years in Computer Graphics at MIT, Lincoln Lab and Harvard," *SIGGRAPH '89 Panel Proceedings* (1989): 19–38; Jacob Gaboury, "Hidden Surface Problems: On the Digital Image as Material Object," *Journal of Visual Culture* 14, no. 1 (2015): 40–60; Jacob Gaboury, "Image Objects: An Archaeology

of Computer Graphics, 1965–1979," PhD dissertation, New York University, 2014; Charlie Gere, "Genealogy of the Computer Screen," *Visual Communication* 5, no. 2 (2006): 141–152.

8. Yeh enrolled in the Northrop Aeronautical Institute (now Northrop University) in Inglewood, California, receiving his pilot and mechanics licenses, and going on to work as a cargo aircraft mechanic at Slick Airways in Burbank. "Chan H. Yeh A Brief Biography" (personal collection of author, provided by Yeh).

9. "Chan H. Yeh A Brief Biography."

10. "IBM Goes West: A 73-Year-Long Saga, From Punch Cards to Watson," *Fast Company*, October 28, 2016, https://www.fastcompany.com/3064902/ibm-goes-west-a-73 -year-long-saga-from-punch-cards-to-watson.

11. Interview with Chan-hui Yeh, March 21, 2010.

12. Interview with Chan-hui Yeh, March 21, 2010.

13. Interview with Chan-hui Yeh, March 21, 2010.

14. Interview with Chan-hui Yeh, March 21, 2010.

15. Interview with Chan-hui Yeh, March 21, 2010; Interview with Chan-hui Yeh, April 18, 2010. Yeh took a one-year leave of absence from IBM in 1971, serving as a visiting professor of applied mathematics and nuclear engineering at National Tsing Hua University in Taiwan. The year was a transformative one, and the submission of his patent application immediately brought on increased attention of the Taiwanese military.

16. Interview with Chan-hui Yeh, March 21, 2010. Yeh's younger brother—Chan Jong ("CJ") Yeh— held a PhD in mechanical engineering as well.

17. Interview with Peter Samson, September 5, 2023. Historians of gaming will perhaps know Peter Samson best for his role in the development of *Spacewar!*, one of the first computer games produced. Historians of electronic music, meanwhile, will know Samson best as the inventor of the "Samson Box," one of the world's first digital synthesizers, commissioned by the Center for Computer Research in Musical and Acoustics (CCRMA) at Stanford University. See PDP-1 Restoration Project, "*Spacewar!*" Computer History Museum website, www.computerhistory.org/pdp-1 /spacewar/; D. Gareth Loy, "The Systems Concepts Digital Synthesizer: An Architectural Retrospective," *Computer Music Journal* 37, no. 3 (2013): 49–67. Samson features prominently in parts of Steven Levy's classic account. See Steven Levy, *Hackers: Heroes of the Computer Revolution*, 25th Anniversary Edition (Sebastopol, CA: O'Reilly Media), 2010.

18. The keyboard weighed 30 pounds. See "IPX Model 9600 Intelligent Keyboard User's Manual," included inside "IPX Model 9600 Intelligent K'Board," Computer History Museum. Mountain View, CA, Box 22 of 27, Folder 003047 (hereafter CHM).

19. "Technical Data: IPX 5486 Automatic Send-Receive (ASR) Telecommunications Terminal," Thomas S. Mullaney East Asian Information Technology History Collection, Stanford University (hereafter TSM).

20. "Technical Data: IPX 5486 Automatic Send-Receive (ASR) Telecommunications Terminal," TSM.

21. "IPX Model 9600 Intelligent Keyboard User's Manual," CHM.

22. Chan-hui Yeh, System for the electronic data processing of Chinese characters, US Patent 3820644A, filed May 7, 1973, and issued June 28, 1974. Even as Yeh drew inspiration from Chinese typewriters, however, he heaped criticism upon them as well. "The typist," Yeh wrote in his 1973 patent application, leveling criticism against Chinese typewriters, "must be trained to memorize the location of each character type positioned on the galley. This training takes at least four months or more." "A well-trained typist could only type about 20 to 30 characters per minute," he concluded. Yeh was actively thinking *with and against* mechanical Chinese typewriters, this is to say, his relationship with China's precomputing IT history a deeply ambivalent one. In many ways, this criticism is unsurprising, if not predictable. As we have seen in this book thus far, as well as in *The Chinese Typewriter*, the mechanical Chinese typewriter had by this point served as a focus of critique, if not scorn, for nearly a century by the time of Yeh's application.

23. Interview with Chan-hui Yeh, March 21, 2010; interview with Chan-hui Yeh, April 18, 2010.

24. See Chi Wang, *Building a Better Chinese Collection for the Library of Congress. Selected Writings* (Lanham, MD: The Scarecrow Press, 2012), 106.

25. Interview with Chan-hui Yeh, March 21, 2010.

26. Stanford University was a customer as well, in the production of its Stanford Library Chinese catalog. In 1979, the company worked with Julia Tung on the project, with Tung coming to Chan-hui Yeh's office in Sunnyvale to input the data. Specifically Stanford librarians used the IPX system, coupled with an IBM 370, to collate the *Bibliography of Chinese Government Serials, 1880–1949* index. Other institutions that expressed interest, but ultimately did not purchase the system, included the Tokyo University Computer Center in Japan and the *Dongya Daily* newspaper in Korea. Interview with Chan-hui Yeh, March 21, 2010.

27. Among these were the *Economic Daily News*, *Living Daily News*, and evening editions. Interview with Chan-hui Yeh, March 21, 2010.

28. See the September 16, 1982, issue of *United Daily News* for multiple articles and dozens of full-cover images of the new system in action.

29. The history of this 1972 trip is a fascinating one in its own right; see Severo Ornstein, *Computing in the Middle Ages: A View from the Trenches 1955–1983* (Bloomington, IN: AuthorHouse, 2002); Thomas E. Cheatham Jr., Wesley A. Clark, Anatoly W. Holt, Severo M. Ornstein, Alan J. Perlis, and Herbert A. Simon, "Computing in China: A Travel Report," *Science* (October 12, 1973): 134–140; and Thomas S. Mullaney, "The Origins of Chinese Supercomputing and an American Delegation's Mao-Era Visit," *Foreign Affairs*, August 4, 2016. The Wesley A. Clark Papers, housed at Washington University's Becker Library, contains archival material and correspondence related to the delegation, as do the papers of Alan J. Perlis (housed both at

Yale University and the University of Minnesota), the papers of Herbert A. Simon Papers (housed at Carnegie Mellon University), and other archival collections. Special thanks to Severo Ornstein for our interview, and for the extensive photographs and personal materials shared with the author.

30. Model numbers: 551型四阶非线性电子模拟计算机; 552型 六级非线性电子模拟计算机; 中国经济成就展览会. "Over the Past Two Years, Tsinghua University Prototypes 18 Reliable Electronic Computers of Varying Types" (Liangnian lai Qinghua daxue shizhi le 18 tai ji zhong butong leixing de gongzuo kekao de dianzi jisuanji) [两年来清华大学试制了18台几种不同类型的工作可靠的电子计算机], *Kexue jishu gongzuo jianbao* [科学技术工作简报] 30 (March 7, 1960): 64–70. Beijing Municipal Archives (hereafter BMA) 001-022-00494. See also John H. Maier, "Thirty Years of Computer Science Developments in the People's Republic of China: 1956–1985," *IEEE Annals of the History of Computing* 10, no. 1 (1988): 19–34.

31. Lorenz M. Lüthi, *The Sino-Soviet Split: Cold War in the Communist World* (Princeton, NJ: Princeton University Press, 2008).

32. "Tsinghua University Organizes 30-Plus Professors and Students to Test Electronic Computer Designed for Statistical Purposes" (Qinghua daxue zuzhi 30 duo ge shisheng wei shijiwei shi shituo tongji yong dianzi jisuanji) [清华大学组织30多个师生为市计委试试托统计用电子计算机], BMA 001-022-00494 (March 11, 1960).

33. Central Intelligence Agency Directorate of Intelligence. "China: Progress in Computers." Intelligence Memorandum. December 1972, 7.

34. The conference ran from December 17 to 23, 1964. See Beijing Institute of Electronics Electronic Computing Group (Beijing shi dianzi xuehui dianzi jisuanji zhuanye zu) [北京市电子学会电子计算机专业组]. "Announcement Regarding the Attendees Name List for the 1964 Beijing Institute of Electronics Electronic Computing Professionals Academic Conference" (Guanyu 1964 nian Beijing shi dianzi xuehui dianzi jisuanji zhuanye xueshuhui daibiao ming'e de tongzhi) [关于1964年北京市电子学会电子计算机专业学术会代表名额的通知], BMA 010-002-00431 (April 17, 1964).

35. Beijing Wireless Factory No. 3 (Beijing wuxiandian san chang) [北京無限電三廠], "Small-Scale General Purpose Transistorized Digital Computer Design Task Report" (Xiaoxing tongyong jingti guan shuzi jisuanji sheji renwu shu) [小型通用晶體管數字計算機設計任務書], BMA 165-001-00130 (ca. 1965): 11–14; "China: Progress in Computers," 7. The machine's average addition time was clocked at 150 microseconds; its average multiplication time, 960 microseconds; and its average division time, 1,080 microseconds. By comparison, the EDVAC machine (1944) clocked an average addition time of 864 microseconds and average multiplication time of 2,900 microseconds. RAYDAC (1953) measured at 38 and 240 microseconds, respectively. FUJIC (1956) performed addition or subtraction in 100 microseconds, and multiplication 1,600. The Chinese machine was still considerably behind the United States, this signifies, but still worthy of note.

36. The model 109C computer, designed at Beijing Institute of Computing Technology in 1965, already boasted a performance of 115 kiloflops. Although this speed

paled in comparison to the Control Data Corporation's 6600 computer—an American machine which, at three megaflops, was then the world's fastest computer—nevertheless it still testified to the strides that China had made without technical support from either Cold War superpower. "Chinese Progress in the Production of Integrated Circuits." Report by the CIA Directorate of Intelligence (March 12, 1985): 2.

37. Mary Allen Clark, "China Diary," *Washington University Magazine* (Fall 1972), 11. Bernard Becker Medical Library Archives, Washington University School of Medicine, St. Louis, Missouri, https://digitalcommons.wustl.edu/ad_wumag/48/; Cheatham et al., "Computing in China: A Travel Report."

38. As the US delegation's summary report explained, their Chinese counterparts "showed little interest in mini-computers, which have become prevalent in the United States in recent years and which, because of their simplicity and economy, have made possible many new applications." Instead, Chinese engineers peppered the American team with questions about "very big and very fast machines such as the CDC Star computer and the Burrough's B6700." In addition, "the Chinese we talked to indicated a strong interest in what they called the 'super computer.'" The delegation's inclusion of scare-quotes for "super computer" is a subtle, yet revealing artifact of their report, suggesting that term itself—coined some years earlier by US leaders in the field of computer science—remained novel enough to merit qualification. "One guesses they will continue the trend towards bigger and faster computers," the report continued presciently, "perhaps attempting a very large step next."

39. Wang Guanwei [王官伟], Chen Hongzhong [陈闳中] and Wang Xiaoyu [王晓宇], eds., *Natural Code Chinese Character Input Method Tutorial* (*Ziran ma Hanzi shurufa jiaocheng*) [自然码汉字输入法教程] (Shanghai: Tongji University Press, 1994) [*Tongji daxue chubanshe*], 1–10.

40. See Chinese Character Encoding Research Group of China (Zhongguo Hanzi bianma yanjiuhui), ed., *Compendium of Proposals for Chinese Character Encodings* (Shanghai: Kexue jishu wenxuan chubanshe), n.d. [ca. 1979], 19–24.

41. Peking University Chinese Character Information Processing Technology Research Office (Beijing daxue Hanzi xinxi chuli jishu yanjiushi) [北京大学汉字信息处理技术研究室], "Design Proposal for Medium-Sized Keyboard Chinese Character Information Processing and Input System," in *Chinese Information Processing* (*Hanzi xinxi chuli*) [汉字信息处理], ed. Chinese Language Journal Editorial Department (Zhongguo yuwen bianjibu) (Beijing: Zhongguo shehui kexue chubanshe, 1979), 1–18.

42. Peking University Chinese Character Information Processing Technology Research Office, "Design Proposal for Medium-Sized Keyboard Chinese Character Information Processing and Input System," 2.

43. See "Pressure Sensitive Chinese Character Keyboard" (*yagan shi Hanzi jianpan*) and "Electrostatic Chinese Keyboard" (*jingdian ouhe shi Hanzi jianpan*), in Chen Zengwu and Jin Lianfu, *Chinese Language Information Processing System* (Beijing: Zhongguo jisuanji yonghu xiehui zonghui [中国计算机用户协会总会], 1984), 78.

44. In Chen Zengwu and Jin Lianfu, *Chinese Language Information Processing System*.

45. In Chen Zengwu and Jin Lianfu, *Chinese Language Information Processing System*.

46. Peking University Chinese Character Information Processing Technology Research Office, "Design Proposal for Medium-Sized Keyboard Chinese Character Information Processing and Input System." For more on the joint project overseen jointly by Wuhan and Shenyang, as well as the work being done at the Yanshan Institute, see Chen Zengwu and Jin Lianfu, *Chinese Language Information Processing System*, 78.

47. Peking University . . . Research Office, "Design Proposal for Medium-Sized Keyboard Chinese Character Information Processing and Input System," 2.

48. Peking University . . . Research Office, "Design Proposal for Medium-Sized Keyboard Chinese Character Information Processing and Input System," 9.

49. Peking University . . . Research Office, "Design Proposal for Medium-Sized Keyboard Chinese Character Information Processing and Input System," 2–3.

50. Here, the choice of including full-body Chinese characters, which might first seem as a squandering of precious space, and an underutilization of the inherent modularity of Chinese characters, was defended as the most statistically sensible path to take. As both foreign and Chinese lexicographers had known for over a century by this point, the frequency distribution of Chinese characters is highly uneven. According to one study from the era, a set of just 8 Chinese characters—out of the more than 70,000 that make up the script—accounts for 10 percent of all Chinese usage. The 103 most commonly used Chinese characters account for a full 40 percent of all Chinese usage. What this meant, in the minds of the Peking University group, was that it would have been counterproductive to try and enforce strict subdivision and modularity upon such commonly used characters. The better solution was to leave these common characters intact, and to instead subdivide less frequent graphs. See Peking University . . . Research Office, "Design Proposal for Medium-Sized Keyboard Chinese Character Information Processing and Input System," 2.

51. The use of a "medium-sized" keyboard also afforded the designers a way to bypass the knotty question of structural ambiguity that bedeviled so many efforts to produce a fully transparent input system. As we recall from the discussion of Samuel Caldwell's Sinotype, Chinese characters do not always decompose into a tidy and straightforward set of elements. There can be ambiguities. Even when the elemental units of a character are widely accepted, moreover, the order by which they should be entered can be cause for disagreement. The classic example, and one cited by the Peking University group in their report, is the character *tu* (凸), which signifies "a protrusion." While everyone agrees that the pen or brush should begin at the top-left-most point, one set of Chinese writers is likely to move their pens to the right and down (⌐), while others are prone to proceeding down and to the left (⌞). By deciding in favor of the medium-sized keyboard, and against the "small-keyboard"/QWERTY route, the team was able to create a small bank of keys, in the lower-right of the board, dedicated to what they termed "hard to cut characters" (*nanzhezi*).

52. Samuel Dyer, *A Selection of Three Thousand Characters Being the Most Important in the Chinese Language for the Purpose of Facilitating the Cutting of Punches and Casting Metal Type in Chinese* (Malacca: Anglo-Chinese College, 1834); Marcellin Legrand, *Tableau des 214 clefs et de leurs variants* (Paris: Plon frères, 1845); Marcellin Legrand, *Spécimen de caractères chinois gravés sur acier et fondus en types mobiles par Marcellin Legrand* (Paris: n.p., 1859); K. T. Wu, "The Development of Typography in China During the Nineteenth Century," *The Library Quarterly* 22, no. 3 (1952): 288–301; and Ibrahim bin Ismail, "Samuel Dyer and His Contributions to Chinese Typography," *The Library Quarterly* 54, no. 2 (1984): 157–169. I also suggest that readers keep an eye out for Yun Xie [谢筠], a brilliant early-career scholar working on Chinese type history.

53. This approach to Chinese typewriting, referred to as "divisible type" printing, predates typewriting. Chinese divisible type printing was innovated in the mid-nineteenth century by printers in France, Germany, and the United States. For more, see Mullaney, *The Chinese Typewriter*. See also K. T. Wu, "The Development of Typography in China During the Nineteenth Century; "Chinese Divisible Type," *Chinese Repository* 14 (March 1845): 124–129; and Martin J. Heijdra, "The Development of Modern Typography in East Asia, 1850–2000," *East Asia Library Journal* 11, no. 2 (Autumn 2004): 100–168.

54. In certain respects, this strategy resembles the Chinese "meta-language" described by Allen. See Joseph R. Allen, "I Will Speak, Therefore, of a Graph: A Chinese Metalanguage," *Language in Society* 21, no. 2 (June 1992): 189–206.

55. Peking University . . . Research Office, "Design Proposal for Medium-Sized Keyboard Chinese Character Information Processing and Input System," 2.

56. See, for example, the online resource "Percentages of Letter Frequencies per 1000 Words," http://www.cs.trincoll.edu/~crypto/resources/LetFreq.html [accessed August 2, 2020].

57. Zhang Shoudong [张寿董], Xu Jianyi [徐建毅], and Zhang Jiansheng [张建生], *Fundamentals of Chinese Language Computing (Zhongwen xinxi de jisuanji chuli)* [中文信息的计算机处理] (Shanghai: Yuzhou chubanshe, 1984), 98–99. The English name of this publication is included in the original, by the authors.

58. Still other medium-sized keyboards were developed around this time in Japan, including the Toshiba TOSMEC keyboard, a 192-key system implemented at Japan's National Diet Library, and one designed by the Itochu corporation. See Shiu C. Loh [樂秀章], Ideographic character selection, US Patent 4270022A, filed June 18, 1979, and issued May 26, 1981; Lam Man-Wah, "Now . . . a Chinese Language Computer!" *South China Morning Post*, July 21, 1980, 22. For more on Loh, see also K. W. Ng, "An Intelligent CRT Terminal for Chinese Characters." *Microprocessing and Microprogramming* 8 (1981): 22–31. Called the "Ping" system, Loh's system was based upon the initial strokes by which characters are composed by hand. Loh identified five such "Ping" types: the horizontal stroke, the vertical stroke, the "bend stroke," the "kick stroke," and the dot or "point stroke." Loh, Ideographic character selection. Loh's estimation of English-language keystroke averages was slightly higher than that of

Peking University's team, but the point still holds: the average number of keystrokes required to produce a Chinese character was, on both systems, reportedly fewer (and significantly) than the number required to produce an English word. Although the system differed from Peking University's keyboard, nevertheless Loh's analysis arrived at strikingly similar results. Using his 256-key keyboard, Loh calculated, Chinese characters required an average of 2.7 keystrokes per character—making it, once again, competitive when compared to English text entry based on the complete spelling of words. While our discussion in this chapter is limited to the cases of IPX, Cable and Wireless, and the variety of medium-sized keyboards developed in mainland China, still other examples could be cited that further demonstrate this linkage between early experimental Chinese computer interfaces and the design principles of Chinese typewriting in the precomputing period. Consider, for example, the Kuno Tablet, developed at Harvard University by Susumo Kuno and his colleagues. Known also as the Harvard Graphics System, the project was funded by means of the National Science Foundation, ARPA, and the US Air Force. Running on a PDP-1 back-end, the Kuno Tablet was based upon the RAND tablet, a pen-based system in which operators using a stylus to select graphs by touching the surface of the screen. Atop the tablet were placed large-printed tables of Chinese characters—essentially the same as a "character tray bed guide" as used with Chinese typewriters—prepared using a Stromberg-Carlson 4020 recorder. The capacity of the tablet, as well, was nearly identical to the capacity of mechanical Chinese typewriters: 3,024 characters, as compared to 2,450 on mechanical typewriters. See Hideyuki Hayashi, Sheila Duncan, and Susumu Kuno, "Computational Linguistics: Graphical Input/Output of Nonstandard Characters," *Communications of the ACM* 11, no. 9 (1968): 613–618. See also S. Duncan, T. Mukaii, and S. Kuno, "A Computer Graphics System for Non-Alphabetic Orthographies," *Computer Studies in the Humanities and Verbal Behavior* (October 1969): 5.2–5.3, 5.7, 5.14; Zhang Shoudong et al., *Fundamentals of Chinese Language Computing*, 103; and Ichiko Morita, "Japanese Character Input: Its State and Problems," *Journal of Library Automation* 14, no. 1 (March 1981): 6–23, 9–10. In addition to these examples, Mathias and Kennedy refer to a 345-key keyboard they witnessed at the Institute of Scientific and Technical Information of China (ISTIC). The machine was, as the authors explained, a modified Japanese T4100. Each key on the machine was outfitted with twelve Chinese characters. See Jim Mathias and Thomas L. Kennedy, eds., *Computers, Language Reform, and Lexicography in China. A Report by the CETA Delegation* (Pullman: Washington State University Press, 1980).

59. Interview with Andrew Sloss, February 21, 2011.

60. "Difficult Oriental Languages Ready for Computer Technology," *The Telegraph-Herald*, June 4, 1978, 17.

61. Interview with Andrew Sloss; "Difficult Oriental Languages Ready for Computer Technology," 17.

62. Obituary of Robert Sloss, *The Darwinian: Newsletter of Darwin College* (Spring 2008): 14, https://www.darwin.cam.ac.uk/drupal7/sites/default/files/downloads/Alumni-Darwinian10-2008.pdf.

63. Obituary of Robert Sloss, 14.

64. Interview with Andrew Sloss.

65. R. W. Apple, "Two Britons Devise a Computer That Can Communicate in Chinese," *New York Times*, January 25, 1978.

66. Apple, "Two Britons Devise a Computer"; Philip Howard, "When Chinese is a String of Two-Letter Words," *The Times*, January 16, 1978.

67. Peter Nancarrow and Richard Kunst, "The Computer Generation of Character Indexes to Classical Chinese Texts," in *Sixth International Conference on Computers and the Humanities*. ed. Sarah K. Burton and Douglas D. Short (Rockville, MD: Computer Science Press, 1983), 772–780. For an introduction to early Chinese-English machine translation, see John W. Hutchins and Harold L. Somers, *An Introduction to Machine Translation* (Cambridge, MA: Academic Press, 1992). While beyond the scope of this book, readers are encouraged to learn more about Erwin Reifler, the Austrian-born Jewish linguist who, in his career at the University of Washington, served as a foundational figure in early Chinese-English machine translation research. His papers are housed at the University of Washington Special Collections.

68. Joseph Needham to Michael Loewe, September 26, 1979. Needham Research Institute (hereafter NI). SCC2/23/65.

69. During my visit to the East Asia Library at Cambridge, I consulted Sloss and Nancarrow's Chinese index cards firsthand, stored in upper shelves of the libraries high-density storage area. I tallied 54 card drawers in all, each one holding easily 1,000 or more Chinese vocabulary cards.

70. Apple, "Two Britons Devise a Computer That Can Communicate in Chinese"; Robert Sloss relied on his son's talent in computer programming as well. "We were the start of children doing computing," Andrew recalled to me. With the beginning of widespread and inexpensive microcomputers, such as the TRS-18 Level 2 (a Hong Kong copy of the TRS-18), Andrew became increasingly interested in computing, particularly around the age of 12 of 13. His father encouraged this interest, in part because he needed work to be done on the project. Interview with Andrew Sloss.

71. Robert P. Sloss and P. H. Nancarrow, "C.L.P. Ideo-Matic 66: A Pre-Production Prototype Encoder for Chinese Characters," Chinese Language Project, Cambridge, England, May 1976. FH.410.46. Cambridge University Library Special Collections (hereafter CSP), 6; interview with Andrew Sloss.

72. At the outset, the idea was to build a machine operated using an alphanumeric keyboard coded in Flexowriter code. They later abandoned this approach, however, and chose instead to have operators manipulate the character drum directly. To build their 66-by-66 grid, Sloss and Nancarrow employed a technology called the Cartographic Digitiser, manufactured by D-Mac LTD, based in Glasgow. Conventionally used for digitizing paper maps—but also used for aerial surveying, traffic analysis, forestry science, ship design, and the analysis of fat/lean content in bacon—the system involved placing a paper-bound original on the work surface, with the operator using the machine's "pencil" to trace key elements thereof. When the pencil came in contact with the surface of the original, it generated an electrical field that was tracked by sensors within the table itself—thereby creating a digitized

version. See James Dreaper, "Geared for Export: Three Cases Histories," *Design* (1968): 30–39; and interview with Andrew Sloss. Sloss and Nancarrow had to rely upon on support from Cambridge's Geology Department, and on the assistance of a colleague named Charles Aylmer, whose job it was to "chart out" the grid, converting it from its original paper-based form to a set of digitized, x-y coordinates. "Leaflet Advertising the 'Ideo-Matic' Chinese Character Encoder, Engineered by Robert Sloss and Peter Nancarrow of Cambridge University," NI, SCC2/24/7 (June 1981). See also "Simple Computer Conquers Chinese Translation Problem," *The Leader-Post*, January 28, 1978, 53; Robert P. Sloss and P. H. Nancarrow, "A Binary Signal Generator for Encoding Chinese Characters into Machine-Compatible Form," Chinese Language Project. Cambridge, England, 1976. FH.410.45, CSP.

73. Sloss and Nancarrow, "C.L.P. Ideo-Matic 66," 6.

74. Sloss and Nancarrow, "C.L.P. Ideo-Matic 66," 2.

75. One of the surviving prototypes of the Ideo-Matic 66 (also referred to as the Ideographic Encoder) can be found in the Porthcurno Telegraph Museum (hereafter PTM) and Cable and Wireless Archives (hereafter CW). These collections house other relevant materials as well, including "Ideographic Encoder Handbook (April 1978), 2.5.06, H1050B; "Summary of Cable & Wireless' History in China," DOC/CW/12/54; and "How Electronics Helped Solve a Chinese Puzzle" (1977), DOC/CW/12/262. Only two other machines are known to survive by the author: one housed at Cambridge University, and another at The National Museum of Computing (Bletchley Park).

76. "Leaflet Advertising the 'Ideo-Matic' Chinese Character Encoder."

77. Philip Howard, "When Chinese Is a String of Two-Letter Words," *Times*, January 16, 1978, 12; Apple, "Two Britons Devise a Computer."

78. Joseph Needham to Kung Sheung Wan Pao (September 25, 1978).

79. Joseph Needham to the *Kung Sheung Wan Pao* (*Kung Sheung Evening News*), September 25, 1978. In "Correspondence and other documents relating to an inaccurate report published in *Kung Sheung Evening News, Hong Kong*, about the computerisation of Chinese," reference SCC2/24/1, creator R. P. Sloss, Joseph Needham, Lee Tsung-Ying, September 25 to October 10, 1978, NI.

80. Apple, "Two Britons Devise a Computer."

81. "Chinese Computers: Dr. Yeh Chen-hui Won the Day in London," *Kung Sheung Evening News*, September 11, 1978. Clipping located in SCC2/24/1, NI.

82. Interview with Chan-hui Yeh, March 21, 2010.

83. Interview with Chan-hui Yeh, March 21, 2010.

84. IPX Materials. SCC2/24/11, NI.

85. The same is largely true of Japanese-built "medium-sized keyboards." Instead, pictures on Instagram and Pinterest show what appears to be the fate of at least one such machine, rusting somewhere in Japan.

86. Interview with Chan-hui Yeh, April 18, 2010.

87. One surviving IPX is housed in the off-site holdings of the Computer History Museum, and another at Stanford University. Additional IPX machines are likely still in the possession of Chan-hui Yeh's family.

88. Interview with Chan-hui Yeh, March 21, 2010.

89. Interview with Chan-hui Yeh, March 21, 2010. Ideographix's last employee left around 2005, Yeh told me, leaving behind this massive, empty space. Yeh continued to own the space, renting out of a portion of it, but keeping the rest to himself.

CHAPTER 4

1. Yang Jisheng, *The World Turned Upside Down: A History of the Chinese Cultural Revolution* (New York: Farrar, Straus and Giroux, 2021); Andrew Walder, *Agents of Disorder: Inside China's Cultural Revolution* (Cambridge, MA: Belknap Press, 2019); Roderick MacFarquhar and Michael Schoenhals, *Mao's Last Revolution* (Cambridge, MA: Belknap Press, 2008); Barbara Mittler, *A Continuous Revolution: Making Sense of Cultural Revolution Culture* (Cambridge, MA: Harvard University Asia Center, 2016); and Frank Dikötter, *The Cultural Revolution: A People's History, 1962–1976* (London: Bloomsbury Press, 2016).

2. Zhi Bingyi [支秉彝], "Recommendation for a New Method of Chinese Character Encoding" (Jianyi yi zhong Hanzi bianma xin fangfa) [建议一种汉字编码新方法], *Diangong yiqi* [电工仪器] (1975); Zhi Bingyi, "The On-Site Coding Chinese Encoding System and its Realization" (Jianzi shima Hanzi bianma fangfa jiqi zai yingyong-zhong shixian) [见字识码汉字编码方法及其在应用中实现], *Shanghai Wenhuibao* [上海文汇报], August 19, 1978; Zhi Bingyi, "A Cursory Discussion of On-Site Coding" (Qiantan jianzi shima) [浅谈见字识码], *Ziran zazhi* [自然杂志] (1978): 1; Zhi Bingyi, "The On-Site Coding Chinese Encoding Method and Its Implementation on Computers" (Jianzi shima Hanzi bianma fangfa ji qi jisuanji shixian) [见字识码汉字编码方法及其在计算机实现], *Zhongguo yuwen* [中国语文], 1979; Zhi Bingyi, *On-Site Code Chinese Character Code Book* (*Jianzi shima Hanzi bianma fangfa bianmaben*) [见字识码汉字编码方法编码本] (Shanghai: Shanghai yiqi yibiao yanjiusuo), 1982.

3. "Chinese Characters Have Entered the Computer" (Hanzi jinru le jisuanji) [汉字进入了计算机]," *Wenhuibao* (July 19, 1978), housed in Graphic Arts Research Foundation (October 1976), Box "Oct 94 Sinotype '81," Folder "Sinotype Vol VI Wang Pinyin Sequences," Graphic Arts Research Foundation Materials, Museum of Printing, North Andover, MA (hereafter MOP).

4. The history of script Romanization is a rich literature too extensive to treat here. For an introduction, see Mahvash Nickjoo, "A Century of Struggle for the Reform of the Persian Script," *The Reading Teacher* 32 (May 1979): 926–929; Shlomit Shraybom Shivtiel, "The Question of Romanisation of the Script and the Emergence of Nationalism in the Middle East," *Mediterranean Language Review* 10 (1998): 179–196; Mehmet Uzman, "Romanisation in Uzbekistan Past and Present," *Journal of the Royal Asiatic Society* 20 (January 2010): 61–74; and Dennis Kurzon, "Romanisation of Bengali and Other Indian Scripts," *Journal of the Royal Asiatic Society* 20 (January 2010): 61–74.

5. James B. Stepanek, "Microcomputers in China," *The Chinese Business Review*, May–June 1984, 26–29, 29. See also Huang Jinfu [黄金富], *The Weiwu Chinese Dictionary* (*Weiwu Zhongwen zidian*) [唯物中文字典] (Beijing: Jijie gongye chubanshe, 1988), 1.

6. Others included "capitalist roader," "ox devils," and "snake spirits." See Jie Li, *Shanghai Homes: Palimpsests of Private Life* (New York: Columbia University Press, 2014), 124.

7. "Chinese Characters Have Entered the Computer" (Hanzi jinru le jisuanji) [汉子进入了计算机]," MOP.

8. This forms part of the genealogy explored in my previous book with figures like Chen Lifu, Du Dingyou, Wang Yunwu, and others. See Thomas A. Mullaney, *The Chinese Typewriter: A History* (Cambridge, MA: MIT Press, 2017).

9. Wu Qidi [吴启迪], ed., *The History of Chinese Engineers, Volume III, Innovation and Transcendence: The Rise and Engineering Achievements of Engineers in the Contemporary Period* (*Zhongguo gongchengshi shi di'san juan chuangxin chaoyue: dangdai gongchengshi qunti de jueti yu gongcheng chengjiu*) [中国工程师史 第三卷 创新超越: 当代工程师群体的崛起与工程成就] (Shanghai: Tongji University Press, 2017).

10. According to Zhi's own telling, the remaining years of the Cultural Revolution—even after his formal release—was still one of incarceration, only this time confined to his home. Diary entry by Bruce Rosenblum, the morning after his January 1984 dinner with Dr. and Mrs. Zhi Bingyi, relayed to me via email from Bruce Rosenblum on March 26, 2017.

11. Zhi Bingyi, On-Site Coding Chinese Encoding Method (*Jianzi shima Hanzi bianma fangfa*), 14.

12. "China: Progress in Computers," 7.

13. "China: Progress in Computers," 7. Further advances in Chinese computer engineering came, along with opportunities for reverse engineering, as a result of ongoing relations with France, England, Canada, and even Japan. This included reported attempts by the PRC to acquire, by way of France, an IBM Stretch for use in the country's nascent nuclear program; along with either confirmed or rumored purchases of a Honeywell-Bull Gamma 3, a Data General Nova minicomputer, a Redifon R-2000, an Elliott Electric 803, and a Computer Ltd. 1903 and 1905, among others. See "Japan Sells China 'Strategic' Computer System," *New Scientist* (April 13, 1978): 69; Bohdan O. Szuprowicz, "CDC's China Sale Seen Focusing Western Attention," *Computerworld*, November 29, 1976, 51.

14. Otto W. Witzell and J. K. Lee Smith, *Closing the Gap: Computer Development in the People's Republic of China* (Boulder, CO: Westview Press, 1989), 12.

15. Zhi Bingyi, *On-Site Coding Chinese Encoding Method*, 14.

16. "Chinese Characters Enter the Computer, 1, 3; Zhi Bingyi, *On-Site Coding Chinese Encoding Method*, 14.

17. Zhi spoke at the National Academic Conference on Chinese Character Encoding (*Quanquo Hanzi bianma xueshu jiaoliuhui*) held in the city of Qingdao.

18. Chinese Character Encoding Research Group of China (Zhongguo Hanzi bianma yanjiuhui), ed. *Compendium of Proposals for Chinese Character Encodings (Hanzi bianma fang'an huibian)* [汉字编码方案汇编] (Shanghai: Kexue jishu wenxuan chubanshe, n.d. [ca. 1979]), iii, 30–34, 43–48, 79, 83–84. Diversity was the order of the day, not just in terms of the proposals offered, but also the backgrounds of the advocates. Niu Zhenhua [牛振华], the inventor of the one of the input systems included in the compilation, hailed all the way from the no. 1 middle school of the fifth division of the Aksu Bingtuan (XPCC) in Xinjiang. See Chinese Character Encoding Research Group of China, ed. *Compendium of Proposals for Chinese Character Encodings*, 79–82, which refers to the following input methods: *Hanzi cengci fenjie shurufa* [汉字层次分解输入法]; Yinxing siwei fudong bianma fa [音形四位浮动编码法]; *San jian bianmafa* [三键编码法]; *San zimu bianma fa* [三字母编码法]; *Hanzi xingmu bianmafa*) [汉字形母编码法]. See also Tianjin Chinese Character Encoding and Information Processing Research Society (Tianjin shi Hanzi bianma xinxi chuli yanjiuhui) [天津市汉字编码信息处理研究会], *Chinese Character Encoding Compilation (Hanzi bianma huibian)* [汉字编码汇编]. August 1979, Thomas S. Mullaney East Asian Information Technology History Collection. Stanford University (hereafter TSM).

19. *Proceedings of the First International Symposium on Computers and Chinese Input/Output Systems*, Taipei, Taiwan: Academia Sinica (hereafter AS), August 14–16, 1973, passim.

20. For the history of the early twentieth-century Chinese character retrieval crisis, see Mullaney, *The Chinese Typewriter*, as well as Uluğ Kuzuoğlu, "Codebooks for the Mind: Dictionary Index Reforms in Republican China, 1912–1937," *Information & Culture* 53, no. 3/4 (2018): 337–366.

21. This committee came out of the "Symposium on the Problem of a Unified Chinese Character Retrieval Method," held in April 1961, 关于统一汉字排检法问题座谈会. See also Ding Xilin [丁西林], *Chinese Stroke-Shape Look-Up System and Encoding System (Hanzi bixing chazifa bianmafa)* [汉字笔形查字法编码法] (Beijing: Beijing shifan daxue Zhongwen xinxi keti zu, 1979), 3, 8.

22. This is not to say that the specific protocols of Wubi input are the same as Wubi retrieval from the 1930s. This is to say, however, that a great many of the basic strategies seen in the era of input were already well developed during the Republican period, during the so-called Character Retrieval Problem period.

23. Zheng code (*Zhengma*); Taiji code (*Taiji ma*); Wubi input (*Wubi shurufa*); Yi input (*Yi shurufa*); Shape-Meaning Three Letter Code (*Xingyi sanma*); Stroke-Shape Code (*Bixing bianma*); Chinese Character Stroke-Shape Look-Up Encoding Method (*Hanzi bixing chazifa bianmafa*).

24. While a handful of input sequences outlined here begin with the letter "D," suggesting a phonetic dimension to the input method in question, all subsequent letters in those input sequences diverged completely from pinyin. "DJFZ," in the case of Beginning-and-End Sound-Shape Code, is not pinyin. Neither is "DJJM," in the case of the Double Stroke Sound-Shape input; "DQTK," in the case of the Double Pinyin Double Radical coding method; "DDDD," in the case of OSCO; or "DMLA,"

in the case of Natural Code, among other examples. Beginning-and-End Sound-Shape Code (*Shouwei yinxing shurufa*); Double Stroke Sound-Shape Input (*Shuangbi yinxing shurufa*); Double Pinyin Double Radical coding method (*Shuangpin shuangbu bianmafa*); Natural Code (*Ziran ma*). There is only one exception in the list—Hanyu pinyin input, which features the input sequence *d-i-a-n*. Again, I underscore here that, in examining various input methods in detail, our object is to illustrate these three broad domains, not to demarcate hard and fast limitations to the broader domain of hypography itself. As in the history of writing more broadly—by which is typically meant the history of *orthographic* writing—the study of any specific writing system (hiragana-style syllabaries, Arabic-style abugida, or otherwise) is a way of mapping out a broad terrain of writing, not to suggest that writing as a whole is exhausted by such examples. Hypography is an exceedingly diverse field of potentiality and, as in the case of orthographic writing, offers up more possibilities than human beings have, in fact, instantiated in the form of actually existing systems of writing. (Incidentally, two of the most insightful scholars on the question of Chinese script—scholars whose work urges us to explore the cultural techniques of Chinese script far beyond boundaries of nation and culture, and even beyond the limits of that which has ever existed, are in my estimation Zev Handel and Xu Bing.) See Zev Handel, *Sinography: The Borrowing and Adaptation of the Chinese Script* (Leiden: Brill, 2019). On "cultural techniques," see Bernard Siegert, *Cultural Techniques: Grids, Filters, Doors, and Other Articulations of the Real* (New York: Fordham University Press, 2015).

25. "A Guide to the OSCO-Method of Coding Chinese Characters for Entry into Olympia Chinese/English Memory Typewriters." Olympia Werke AG T1/04–10/84. Rolf Heinen Technology Collection. Drolshagen, Germany (hereafter RHTC); Zhi Bingyi, *On-Site Coding Chinese Encoding Method (Jianzi shima Hanzi bianma fangfa)* [见字识码汉字编码方法] (Shanghai: Shanghai Yiqi yibiao yanjiusuo, 1982).

26. Another question about hypographic semiotics, but one that we will not venture into here in depth, pertains to the question of error: What constitutes "error" and "ambiguity" within the context of hypography? Within the context of orthography, the concept of *ambiguity* brings certain possibilities to mind. First, there are homophones. *Is it principle or principal?* Then there are apostrophes. *Is it their's or theirs, it's or its?* Most fundamentally, of course, there is spelling itself. *Is it 'weird' or 'wierd'? Is it embarrassing, embarassing, or embarrasing? Liason or liaison? Artic or Arctic?* How does ambiguity work, however, when the function of letters is no longer to spell—but rather to *find*? For, were we to consider the examples above, and to imagine them in the context of a Chinese input-like process of criteria-candidates-confirmation, hardly any of them would fail to bring about the accurate, desired graph one had in mind. Even in the context of English language spell-checking, for example, to input "wierd" is sufficient to prompt most word processing systems, first, to flag the error, and second, to recommend a correction to the user. In spelling, therefore, "wierd" is just wrong. But in *input*, "wierd" is *as good as* "weird" in getting the job done. What this example demonstrates, is not that "anything goes" in the context of input, but rather that ambiguity itself functions in a different, perhaps unexpected way, when

we move from orthographic to hypographic writing. There is still confusion, that is, but this confusion does not behave in the same way.

27. "A Guide to the OSCO-Method," 3.

28. Cangjie was also referred to as Tsang-Chieh, Cheong Kit, and "Dragon Input." Chu Bong-Foo [朱邦复], *On Chinese Computing* (*Zhongwen diannao mantan*) [中文电脑漫谈] (Self-published: ca. 1982).

29. In some respects, we can draw an imperfect analogy between the "sampling" method and Optical Character Recognition (OCR), only one in which humans—rather than machines—were the entities being trained to identify particular, distinguishing features of Chinese characters for the purposes of identification. For more on the history of Chinese OCR, a topic not addressed in this work, see Wen Desheng [汶德胜], *Research on Computer Recognition of Printed Chinese Characters* (*Yinshua Hanzi jisuanji shibie de yanjiu*) [印刷汉字计算机识别的研究] (Zhongguo kexue jishu daxue [中国科学技术大学], 1988); Wu Youshou [吴佑寿] and Ding Xiaoqing [丁晓青], *The Principle Method and Realization of Chinese Character Recognition* (*Hanzi shibie yuanli fangfa yu shixian*) [汉字识别原理方法与实现] (Gaodeng jiaoyu chubanshe, 1992); and Ye Naibeng [叶乃犇], Zhang Xizhong [张炘中] and Xia Ying [夏莹], *Chinese Language Microcomputers and Chinese Character Recognition* (*Hanzi weixing jisuanji yu Hanzi shibie*) [汉字微型计算机与汉字识别] (Jixie gongye chubanshe [机械工业出版社], 1989); and Zhang Xizhong [张炘中], *Chinese Character Recognition Technology* (*Hanzi shibie jishu*) [汉字识别技术] (Beijing: Qinghua daxue chubanshe, 1992).

30. Huang was a professor of computer science at Ming Chuan College in Taiwan, and director of the computer science department. See Jack Kai-tung Huang to Louis Rosenblum, April 16, 1980, Louis Rosenblum Collection (hereafter LR). See also Jack Kai-tung Huang, "Principles of Chinese Keyboard Layout" (draft), April 25, 1981, LR.

31. As readers of *The Chinese Typewriter* may recall, one of the earliest and most successful character retrieval systems in modern Chinese history—"Four-Corner Look-Up," invented by Wang Yunwu—was premised upon the same logic, some five decades earlier.

32. Evgeniy Gabrilovich and Alex Gontmakher, "The Homograph Attack," December 2001, https://gabrilovich.com/publications/papers/Gabrilovich2002THA.pdf [accessed August 10, 2022].

33. Adi Stern, "Aleph = X: Hebrew Type on the Edge," presentation at the 2007 ATypI Annual Conference, Helsinki.

34. H. C. Tien, "On Learning Chinese," *World Journal of Psychosynthesis* 4, no. 7 (July 1972): 4–5.

35. H. C. Tien is a fascinating character in his own right, whose life and career story cannot be treated adequately here. Born in China in 1929, Tien was the son of Chinese diplomat Tian Fangcheng [田方城], and grandson of the Qing-dynasty official and Chinese logician Tian Wuzhao [田吳炤]. He spent a portion of his childhood in Portugal during World War II, where he studied English among a variety of other subjects. Tien went on to attend Adrian College in Michigan, supported by the Boxer Indemnity Fund, and later received two master's degrees: in neurology

from the University of Michigan, and in electrical engineering from Michigan State University. He later received a medical degree, attending the University of Michigan, and completed his psychiatric residency at Ypsilanti State Hospital in Michigan. As an impassioned, entrepreneurial, and many might say eccentric man, Tien published his own journal, the *World Journal of Psychosynthesis*, and was an active member of the East Lansing Human Rights Commission and various civil rights and anti–Vietnam War movements. For a more in-depth treatment of Tien, see Thomas S. Mullaney, "QWERTY in China: Chinese Computing and the Radical Alphabet," *Technology and Culture* 59, no. 4 Supplement (October 2018): S34–S65. For more on Tien's grandfather, see Joachim Kurtz, *The Discovery of Chinese Logic* (Leiden: Brill, 2011). See also Tian Wuzhao 田吳炤, trans., *Outline of Logic* (*Lunlixue gangyao*) [論理學綱要] (Shanghai: Shangwu yinshuguan, 1903; fourth ed., 1914); original publication: Totoki Wataru 十時弥, *Ronrigaku kōyō* (*Outline of logic*) [論理学綱要] (Tōkyō: Dai Nihon tosho, 1900). I am indebted as well to Aled Tien, the son of H. C. Tien, who I interviewed by phone on March 16, 2015.

36. Zhang Hongfan [张闳凡], "Chinese Character Shape-Letter Encoding Method (*Hanzi xingmu bianmafa*) [汉字形母编码法].

37. Zhang Hongfan, "Chinese Character Shape-Letter Encoding Method," 45, 83. Such techniques continued to be used well into the 1990s, moreover. The Shape-Meaning Three Letter Code (*Xingyi sanma*), for example, was invented in 1993 by Luo Yinghui during his time at South China Technology University in Guangzhou. See Luo Yinghui [罗英辉] and Zhang Yala [张亚拉], *Shape-Meaning Three Letter Code: A Revolution in Chinese Character Input Methods* (*Xingyi sanma: Hanzi shurufa de geming*) [形意三码: 汉字输入法的革命] (Guangzhou: Zhongshan daxue chubanshe, 1995), 39–40.

38. When Roy Hofheinz first met Zhi Bingyi, he misjudged Zhi's age by ten or twenty years, thinking that Zhi was in his late 70s or 80s, when he was in fact only 67. Reason suggests that Zhi's prematurely frail demeanor was due, at least in part, to the trying experiences he went through during China's Cultural Revolution period (1966–1976).

39. Hofheinz, "Memorandum of Conversation with Dr. Chih Ping-I," LR.

40. Weili Ye, *Seeking Modernity in China's Name: Chinese Students in the United States, 1900–1927* (Stanford, CA: Stanford University Press, 2001).

41. "Prof. Zhi Bingyi." College of Electrical Engineering, Zhejiang University, http://ee.zju.edu.cn/english/redir.php?catalog_id=18647&object_id=19246 [accessed May 5, 2012].

42. "Prof. Zhi Bingyi."

43. Zhi Aidi, born in 1908, was two years older than Zhi Bingyi. She was based at Leipzig University from 1939 to 1946, working in the university library for most of that period. In 1946, she joined Zhi Bingyi and returned to China, working as a translator in the German Embassy. Between 1947 and 1957, she held a variety of posts, including positions at Zhejiang University's College of Engineering, Tongji University's College of Engineering, Fudan University, Jiaotong University, and the

Shanghai office of the Chinese Academy of Sciences. In 1958, she went on to hold the position she would retire from: professor of German at the Shanghai Foreign Languages Institute. In addition to her native German, and her adoptive Chinese, she also spoke English and French. See *Dictionary of Famous Chinese Women* (*Huaxia funü mingren cidian*) [华夏妇女名人词典] (Beijing: Huaxia chubanshe, 1988), 110–111.

44. *Deceased Persons: Shanghai Cultural Yearbook 1994* (*Shishi renwu: Shanghai wenhua nianjian 1994*) [逝世人物上海文化年鉴 1994], 297–298.

45. Photograph of Chinese typist with "Eputima" label behind her, TSM.

46. The joint venture comprised Zhi Bingyi, Olympia Werke (a subsidiary of AEG-Telefunken, based in Wilhelmshaven, Germany), and a small group of investors from the China Technical Import and Export Corporation, the Shanghai Instruments Research Institute, the China Instruments Industry Administration, and the Hong Kong Sun Hung Kai Consortium. See "Olympia 1011 Chinese Word Processor System," FBIS reprint from *China Computer World* (*Beijing jisuanji shijie*) no. 20 (October 20, 1982): 4. Translation in *China Report: Science and Technology* no. 189 FBIS (March 1, 1983): 12–13; see also "Olympia 1011," *Electronics* 54, no. 9–13 (1981): 71.

47. The Olympia 1011 ran on 3 Intel 8085 processors, one serving as the CPU, one dedicated to controlling the printer, and one dedicated exclusively to Chinese character generation. The machine was equipped with 10K of RAM, along with 60K of ROM, also apportioned for the generation of Chinese character bitmaps. The device featured a standard, QWERTY-style keyboard, a 3-inch floppy disk drive, and a 22-inch display capable of displaying on a small number of Chinese characters at a time. Overall, the machine could handle a maximum of 4096 Chinese characters. Timing-wise, the consortium had a concrete deadline in mind: China's upcoming national census, set to take place in 1981, and in service of which a set of 1,000 Olympia 1011 machines, valued at more than US$5 million, were waiting to be used by census officials and data entry specialists; "Olympia 1011," 71; "West Germany—Co-Operation in Typewriters," *BBC Summary of World Broadcasts*, July 30, 1980; "Olympia 1011 Chinese Word Processor System," 12, 14–15.

48. The Harvard University Business School Special Collections (hereafter HUBSSC) houses perhaps the largest collection on Wang Laboratories, including materials on their "Ideographic" team. Within this HUBSSC collection, see "Chinese Ideographic VS Programmers Manual," Box 136, Folder 4; "Ideographic Professional Computer Database Reference Guide," Box 137, Folder 2. See also the following materials, located in Subseries IB2 ("Horace Tsiang Records, 1974–1992"): "Chinese Computer, 1981–1982," Box 97, Folder 7; and "Chinese Machine, 1981–1982," Box 97, Folder 8; "Jenny Chuang, 1986," Box 97, Folder 1. For documentation on Wang Labs Taiwan, see Subseries IA1, particularly boxes 28 through 32.

49. Autumn Stanley, *Mothers and Daughters of Invention: Notes for a Revised History of Technology* (New Brunswick, NJ: Rutgers University Press, 1995), 482; "An Wang Oral History," interviewed by Richard R. Mertz, October 29, 1970. Computer Oral History Collection, 1969–1973, 1977. Archives Center, National Museum of American History.

50. The world of Chinese computing, it bears pointing out, was still a rather small and intimate group. For example, while studying physics at National Taiwan University, Jenny Chuang became close with the wife of Chung-Chin Kao, the same inventor we met in chapter 1 who had developed the experimental Chinese typewriter for IBM. From 1965 to 1967, Mrs. Lew tutored Chuang twice each week in preparation for her US Embassy interview (in which she would need to write an English-language essay in 15 minutes). In 1967, Chuang received her BS in physics, and in the same year, sat for the embassy exam. Evidently, Mrs. Kao was an apt tutor, and Chuang an apt student, for she passed her exam and made her way to the United States that same year. Enrolling in Brown University, Chuang pursued her MS with a focus on high energy particle physics. While working on her thesis, titled "The Pion-Electron Scattering at 4.1 BeV/c in Hydrogen Bubble Chamber," she also served as a physics instructor at the College of Mt. Saint Vincent in Yonkers, New York. Chuang's earliest aspiration, she recalls, was to be the "second Marie Curie," and to design missile systems for China. Never could she have known that her career would take her instead to Wang Laboratories, let alone that she would help build one of the first Chinese- and Japanese-language word processors. Chuang was working at Harvard as a key punch operator, earning $90 per week, when a unique opportunity emerged, thanks in part to a letter of recommendation written by P. M. Yen of Summit Industry in New York. Addressed to An Wang, it alerted Wang to a young woman named Jenny Chen, no doubt outlining her stellar educational background. In 1971, Chen (now Chuang) was hired by Wang Laboratories, where she went on to write the floating-point math package for the Wang 2200 operating system, accurate to 16 digits. There were no female programmers within its labor force—Chuang became the first. See interview with Jenny Chuang, and follow-up email exchanges on September 3, 2013, and September 6, 2013; CV of Jenny Chuang, included in materials housed at Iowa State University; Stanley, *Mothers and Daughters*, 482. Chuang met Mrs. Kao by way of Vicky Ching, the famous restauranteur and owner of Ming's in Palo Alto.

51. Interview with Jenny Chuang, and follow-up email exchanges on September 3, 2013, and September 6, 2013. Other members of the team, as recalled by Jenny Chuang, included Ken Mei, Virginia Chung, and a woman by the name of Michelle whose surname Jenny Chuang was unable to recall, as well as junior members of the team working on quality assurance.

52. Circa 1974 and 1975, just prior to the creation of the Ideographic Word Processing group, libraries in Taiwan had started to experiment with Three-Corner Coding as a way of organizing library card catalog entries. See *The Collected Works of Margaret C. Fung, PhD* (Taipei: Showwe Information Corporation, 2004), 143.

53. This is not to suggest that these shared Taiwanese links made Three-Corner the inevitable choice. Taiwan was home to a number of other library cataloging systems, as well, many of them also being experimented with by libraries and other institutions. On top of this, of course, the vast majority across the broader Sinophone world were using other systems, with Three-Corner Input being largely unknown—or

simply not used—by most institutions responsible for sorting Chinese language material. In addition to the shared Taiwanese heritage of the two teams—the team behind Three-Corner and the team behind Wang's Ideographic project—there was yet further serendipity that we cannot reconstruct from the historical record.

54. *Wang Code Computer: Special Issue on Five-Stroke Technology* (*Wangma diannao: Wubi zixing jishu zhuankan*) 6 (October 1991). The October 1991 issue of the journal was listed as the sixth installment, making it likely that the *Wangma diannao* was launched either in early 1991 or late 1990.

55. *Chinese Computer* (*Zhongwen diannao*), January 1, 1986, inaugural issue. As noted, the fourth system to break into the market was Chu Bong-Foo's Cangjie input, taken up by the IBM Corporation for use on their Multistation 5550. At the urging of W. Michael Clark, then manager for IBM China, the company adopted an aggressive marketing posture for the machine circa 1985. Donating 100 workstations to four Chinese universities on the mainland—Tsinghua, Peking, Fudan, and Shanghai Jiaotong—the goal was to develop a beachhead in the mainland Chinese market. Ultimately, the Multistation 5550 was a failure. Zimin Wu and J. D. White, "Computer Processing of Chinese Characters," *Information Processing and Management* 26, no. 5 (1990): 686–687.

56. Zimin Wu and White, "Computer Processing of Chinese Characters," 685.

57. Zimin Wu and White, "Computer Processing of Chinese Characters," 685; Stepanek, "Microcomputers in China," 27.

58. Zimin Wu and White, "Computer Processing of Chinese Characters," 685. It is important to note that, at this time, computer ownership and usage was still limited.

59. The 5550 was an expansion of IBM's earlier, Japanese system. The system featured a 61-key, QWERTY-style keyboard outfitted with Cangjie symbols. Francis Chin, "IBM Shows Chinese PC at Information Week." *Asia Computer Weekly*, December 23, 1983, 24; "Chinese Computer," *South China Morning Post*, December 7, 1983, 33; "Chinese Computer Launched," *South China Morning Post*, July 17, 1984, 27; Bob King, "Enter the Dragon: IBM Launches Chinese Style Computer," *Financial Times*, February 8, 1984, 13; *IBM Multistation 5550* (*Hanzi bianma shouce*) [汉字编码手册], October 1984, IBM Corporation (note that the English language title is included in the original), TSM; Stepanek, "Microcomputers in China," 28. IBM seems to have rebranded the input system, calling it the "IBM zuijia jianpan shuru xunxu." The machine featured a US$10,000 price tag, 256 kilobytes of main memory, two disk drives, and a dot matrix printer. The Japanese predecessor was launched in 1983. See also Chin, "IBM Shows Chinese PC," 24 .

60. Pan Defu [潘德孚] and Zhan Zhenquan [詹振权], *Chinese Character Encoding Design* (*Hanzi bianma shejixue*) [汉字编码设计学], ca. 1997, 26, TCM.

61. Louis Rosenblum to Zhi Bingyi, November 1, 1978, LR.

62. "Electromechanics Multilingual Typesetters," *Electronic Age: RCA in 1962 Year-End Report by RCA* (Winter 1962–63): 27–29. F. E. Shashoua, "Photocomposition Machine for the Chinese Language." *RCA, Camden, NJ, RCA Tech. Paper* 101 (1964).

63. "Electromechanics Multilingual Typesetters"; Fred E. Shashoua, Warren R. Isom, and Harold E. Haynes, "Machine for Composing Ideographs," US Patent 3325786, filed June 2, 1964, issued June 13, 1967. Assigned to the Radio Corporation of America (Delaware).

64. "Electronic Type-Setting Machine for Chinese Invented" (November 3, 1962), Pardee Lowe Papers. Accession No. 98055–16.370/376. Box No. 276. Hoover Institute Archives (hereafter HI).

65. Walter N. Mott, "Service Test of Ideographic Composing Machine Final Report," Fort Bragg, NC: US Army Airborne, Electronics and Special Warfare Board (March 1968).

66. Mott, "Service Test," 1–3.

67. Mott, "Service Test," 9.

68. Mott, "Service Test," 6, 22. The machine weighed precisely 2,288 pounds.

69. Mott, "Service Test," 3–6.

70. For one of the last reports on the ICM produced by RCA, see RCA Advanced Technology Labs. "Language Manual for Use with Ideographic Composing Machines." Camden, NJ: RCA, February 1970.

71. "Museum of Printing Long-Time Board Member Louis Rosenblum Passes," https://museumofprinting.org/news-and-events/long-time-board-member-louis -rosenblum-passes/; email from Bruce Rosenblum to the author, March 26, 2017.

72. Circa 1980, the GARF board of directors comprised Prescott Low (chairman/ treasurer), a 34-year-old William Garth IV (elected GARF president in January 1980), James Hagler, Susumo Kuno, Alfred Moran, Ithiel deSola Pool, Annemarie Schimmel, and Richard Solomon. See Graphic Arts Research Foundation, "GARF 5-Year Plan 1980," LR, 12–18.

73. By this point, Hofheinz Jr. had authored many studies, including *Rural Administration in Communist China* (1962), *Chinese Communist Politics in Action* (1969), and *The Origins of Chinese Communist Concept of Rural Revolution* (1974), among others.

74. The method by which the various generations of Sinotype were counted is confusing at certain points in our history. Undoubtedly, the system that would come to be known as Sinotype I was the one built by Samuel Caldwell and his team at GARF. In theory, Sinotype II should refer to the machines then developed by the Radio Corporation of American (RCA) and the Quartermaster MC. It was during the later stages of this generation that holographic storage was incorporated. By this logic, the system developed by Roy Hofheinz and his associates should count as Sinotype III. This in turn would make the subsequent system developed by Louis Rosenblum, Bruce Rosenblum, and their team at GARF, Sinotype IV. But in most literature on the subject, including by the Rosenblums themselves, the Apple II–based Sinotype system is consistently referred to as Sinotype III. See "Summary of the Discussions and Samples on the Sinotype Project" (December 18, 1975): Appendix I, 2. Graphic Arts Research Foundation, LR.

75. John Forster to Hugh, April 15, 1980, 1, LR.

76. Graphic Arts Research Foundation, "Sinotype II Demonstration." August 1, 1979, 1–2, LR.

77. See, for instance, Wei Tang, "A Computational Chinese Character System with Multiple Input Methods" (*Keyong duozhong shuru fangfa de dianzi jisuanji Hanzi xitong*), *Wenzi gaige* [*Writing Reform*], 4 (1985): 47–48.

78. The original article appeared in the October 1979 issues of the journal *Nature*. Zhi Bingyi, "An Introduction to 'On Sight Encoding of Character,'" *Nature Magazine* 1, no. 6 (October 1979): 350–353, 367. Hofheinz's translation was marked as "Internal Circulation to GARF," and is housed in LR.

79. Hofheinz outlined his misgivings with Zhi's system as well in his memo to Rosenblum. With regards to the encoding system itself, the OSCO system struck him as one that "ordinary people will not master easily." What is more," Hofheinz opined, "reverse recognition"—that is, the ability, not just to see the character and know the code, but to see the code and discern the character—"will be virtually impossible," especially for speakers of Cantonese, Shanghainese, or other nonstandard forms of Chinese. Nevertheless, the same could easily be said of the Caldwell system, in which few speakers of Chinese could, simply when provided the stroketypes, know unambiguously which Chinese character was being referenced. Rosenblum and his colleagues at GARF decided that Zhi's OSCO system might prove to be superior to Caldwell's input, and thus decided to explore the idea of adding OSCO as a third input method on the Sinotype II (running alongside the Standard Telegraph Code and Caldwell methods). See Roy Hofheinz to Louis Rosenblum, December 7, 1978, LR; and Louis Rosenblum, "Photocomposition of Chinese: Input Systems and Output Results," n.p., n.d., 305–344, 310.

80. Louis Rosenblum. "Photocomposition of Chinese."

81. Note that GARF was not provided Zhi Bingyi's full OSCO code, the inventor being reticent to do so without securing a formal license. For that reason, the GARF team outfitted Sinotype with only a subset of the OSCO code, suitable for demonstration purposes. Email from Bruce Rosenblum to the author, January 8, 2023.

82. Li Liuping to Roy Hofheinz, September 13, 1979, LR. Li Liuping was listed in the document as secretary of the General Bureau of Instruments and Meters; Graphic Arts Research Foundation, "Technical Report," 1. Hofheinz formally accepted the invitation in a letter dated September 27, 1979. Roy Hofheinz to Li Liuping, September 27, 1979, LR. See also Zhi Bingyi to Roy Hofheinz., September 27, 1979, LR.

83. Hofheinz to Li Liuping, September 27, 1979; Rosenblum. "Photocomposition of Chinese: Input Systems and Output Results," 332; See also "Richard Solomon, Former Diplomat Who Helped Nixon Open Relations with China, Dies," *Wall Street Journal*, March 14, 2017, https://www.wsj.com/articles/richard-solomon-former -diplomat-who-helped-nixon-open-relations-with-china-dies-1489532159. In preparation for the visit, the company Optronics helped GARF create a high-resolution Chinese font. Interview with Bruce Rosenblum on March 3, 2017; follow-up Email from Bruce Rosenblum to the author. March 26, 2017.

84. Zhi Bingyi to Hofheinz, September 27, 1979, LR.

85. Email from Bruce Rosenblum to the author, April 8, 2017.

86. Graphic Arts Research Foundation, "Technical Report," December 6, 1979, 3.

87. Graphic Arts Research Foundation, "Technical Report," 4–5; "Bill Garth's Narrative of Demonstrations of Chinese Text Processing Computers in Beijing and Shanghai in 1979 and Negotiations in Beijing in 1980," prepared by William Garth IV for the author and sent via email on April 17, 2014.

88. Graphic Arts Research Foundation, "Technical Report," 5.

89. Rosenblum, "Photocomposition of Chinese: Input Systems and Output Results," 314–315.

90. Graphic Arts Research Foundation, Press release announcing news of joint protocol signed in Shanghai, November 1979, LR.

91. William Garth IV and Louis Rosenblum to Roy Hofheinz, January 14, 1980, LR; Graphic Arts Research Foundation, "Technical Report," 4.

92. Graphic Arts Research Foundation, "Trip Report. Peking, China October 29–November 22, 1980," December 1, 1980, LR; "Bill Garth's Narrative of Demonstrations of Chinese Text Processing Computers in Beijing and Shanghai in 1979 and Negotiations in Beijing in 1980," prepared by Bill Garth for the author and sent via email on April 17, 2014.

93. For discussion of Sinotype as an "intelligent terminal," see Graphic Arts Research Foundation, "Sinotype II Demonstration," 1–2. For Zhi's letter, see Zhi Bingyi to Rosenblum, March 3, 1980, LR. Hofheinz to William Garth IV, January 16, 1980, LR; Telex to Zhi Bingyi from Pres Low, Garth IV, and Rosenblum, June 23, 1980.

94. Paul E. Ceruzzi, *A History of Modern Computing* (Cambridge, MA: MIT Press, 2003 [1998]), chap. 7.

95. *CROMEMCO Microcomputer Software Data Compilation 1–6* (*CROMEMCO weixing jisuanji ruanjian ziliao huibian 1–6*) [CROMEMCO 微型计算机软件资料汇编1–6]. Beijing: Tsinghua University Computing Center, 1980. Housed in TCM; Stepanek, "Microcomputers in China," 26.

96. Photograph of Harry Garland and Roger Melen (1980), Wikimedia Commons, https://commons.wikimedia.org/wiki/File:Harry_Garland_and_Roger_Melen_-_with_Cromemco_shipment_to_China_(1980).jpg [accessed June 17, 2020].

97. "Prof. Zhi Bingyi."

98. Zhi Bingyi to William Garth IV and Louis Rosenblum, February 19, 1982. LR.

99. GARF began to court other input designers over the course of late 1979 and into the early 1980s to learn about their systems, compare them to that of Zhi Bingyi, and to assess the possibility of implementing their input techniques on Sinotype. On the one hand, it was a good thing that Sinotype II was capable of handling more than one input system at a time, yet at the same time, limitations of memory still applied—at least for now. While it was clear that computing hardware was on the verge of a sea change—one in which, with the steady, geometric growth in memory capacity, the day was not far off when one could imagine a personal

computer capable of handling dozens of Chinese input systems with ease—GARF was caught in something of a transitional stage where the most they could hope for was perhaps three, four, or five: Caldwell input, Standard Telegraph Code, OSCO, and perhaps one or two others. With such a limited array of possibilities, confronted with the dizzying number of new systems being debuted with each passing day, how could they be sure that they had backed the right horse in Zhi Bingyi and OSCO? One of the first inventors contacted by GARF during this time was Li Jinkai, a professor working at Beijing Normal University. Li had developed an input system that differed considerably from that of Zhi Bingyi, and it was making headway in technology circles in mainland China. Richard Solomon had an extensive meeting with Li Jinkai, after which he provided GARF an 82-page report. Li put forth a claim that his machine required an average of 3.28 strokes per Chinese character, with an average speed of 80 characters per minute or between 4,600 and 5,000 per hour. Li's system was called the 6-digit Three Corner Code Method (TCCM) and relied entirely on numerals—no letters. Because of this, Louis Rosenblum's thoughts turned naturally to the Chinese Telegraph Code with which he was familiar, despite the fact that the Three Corner System shared no protocological connection with the telegraph code. The system was based on 6 basic strokes, a cross, and a box, containing around 4,000 characters in all. The larger code sequence was 9 numerals long, while the average code was around 5. When this average was weighted with attention to the frequency of the Chinese characters themselves, the average code length was approximately 4. See "Bill Garth's Narrative of Demonstrations of Chinese Text Processing Computers in Beijing and Shanghai in 1979 and Negotiations in Beijing in 1980."

100. See Louis Rosenblum to Richard Solomon, June 21, 1982, LR; Richard Solomon to William Garth IV and Louis Rosenblum. August 2, 1982, LR.

101. Richard Solomon to William Garth IV and Louis Rosenblum, August 2, 1982, LR.

102. Yet another complicating factor pertained to the ownership of the Sinotype II prototype, and a series of sideline conversations between Zhi Bingyi and Roy Hofheinz that GARF later took issue with. In either December 1979 or January 1980, Chinese representatives requested a quote directly from Hofheinz as to the cost for purchasing or leasing the equipment which the delegation had used during the November 1979 demonstration in China. As indicated by this letter, and by subsequent exchanges between Hofheinz, Zhi, and GARF, the hardware itself was owned by Roy Hofheinz (and perhaps John Forster), but it was being used by GARF. As a result, ambiguity reigned regarding the capacity in which Hofheinz was conducting himself in exchanges with Zhi and others on the Chinese side. Was he operating as a representative of GARF (which was GARF's assumption and desire, of course), or as an independent party (who, in theory, had the right to offer for purchase the hardware in his possession)? In a January 14, 1980, letter to Zhi Bingyi, Hofheinz wrote: "If you are interested in acquiring the demonstrator, please let me know the exact configuration and we will quote you our best price." Were the specifications to remain exactly the same as the machine used during the November 1979

demonstrations, Hofheinz explained, the total estimated cost would be US$79,350 (a figure that included an estimated $10,000 in installation fees, along with the 25 percent surcharge added to US manufactures). While never disputing Hofheinz's ownership of the 1979 prototype, by mid-1980s, GARF leadership was protesting Hofheinz's individual dealings with Zhi, presumably because they undercut GARF's goals of establishing a longer-standing relationship with mainland China, rather than a one-off sale. Roy Hofheinz to Zhi Bingyi, January 14, 1980, LR; Hofheinz to Zhi Bingyi, May 21, 1980. It is important to emphasize that GARF never disputed ownership. See, for example, a letter from Hofheinz to Garth IV in which Hofheinz also provided (at GARF's request) a quote to GARF for the prototype, at the price of $58,600. Roy Hofheinz to Garth IV, January 16, 1980, LR.

103. Zhi Bingyi to Garth, IV and Rosenblum, February 19, 1982, LR.

104. In retrospect, leadership at GARF perhaps realized that signs of this transformation were present even during the successful visit in 1979—and, furthermore, that Hofheinz had tried to warn them. One afternoon, four Chinese representatives—hailing from the First Ministry of Machine Building and the *People's Daily*—visited Hofheinz in his hotel room for what they called a "casual conversation." "It soon became apparent," Hofheinz noted in a detailed memo to GARF, "that they had a serious proposal." The vision they had was far more rudimentary and bare bones, but also more flexible and better attuned to Chinese realities than the grander scheme that had been discussed in the open meetings. They were not interested in developing a complex network of time-shared editing terminals. Rather than online preparation of text, they wanted a system in which text was prepared offline, with one page of the *People's Daily* being prepared by one station. Rather than relying on networked transmission, they wanted the output of each station loaded onto a floppy disk "for transmission to the central make-up station." As Hofheinz summarized it, the system that his counterparts wanted was "something much closer to the pagitron," referring to the system developed by Optronics. Even GARF's estimation of typefaces was an overshoot. They needed five typefaces, the Chinese representatives explained, not ten. Finally, they wanted the input system to rely upon the Z80 microprocessor, since there were plans to manufacture such microprocessors in China in the near future. They referred to their idea as a "middle-sized system." "During the course of discussions," GARF's own report noted, "we were asked many times if it would be possible to sell the equipment we had with us." At the end of the day, however, GARF either failed to pick up on these signals, or to dismiss them. Instead, the vision which GARF ultimately proposed to their Chinese counterparts—a vision of multiyear planning, joint training programs, and the like—ultimately spelled doom for the relationship. See Roy Hofheinz, "Conversation from 2 p.m. to 5 p.m., November 14 in the Chinchiang Hotel with Representatives of the *People's Daily* and the First Ministry of Machine Building." Memorandum, November 14, 1979, LR; and Graphic Arts Research Foundation, "Technical Report," 3.

105. Zhang Shoudong et al., *Fundamentals of Chinese Language Computing*, 106.

CHAPTER 5

1. Bohdan Szuprowicz, "Expanding Chinese Micro Market Triggering Frenzy," *Computerworld*, May 21, 1984, 10; Otto W. Witzell and J. K. Lee Smith, *Closing the Gap: Computer Development in the People's Republic of China* (Boulder, CO: Westview Press, 1989), 35. See also Richard Baum, "DOS ex Machina: The Microelectric Ghost in China's Modernization Machine," in Denis Fred Simon and Merle Goldman, eds. *Science and Technology in Post-Mao China* (Cambridge, MA: Harvard University Asia Center, 1988), 347–374.

2. John F. Burns, "China's Passion for the Computer," *New York Times*, January 6, 1985. In November 1984, the Xintong Company was founded in the neighborhood, the first joint-stock enterprise in the neighborhood focused on science and technology.

3. Burns, "China's Passion for the Computer," 3. To gain a sense of the dynamism of Chinese computing circa 1985, a remarkable German-language source from the era provides an extensive (but likely still noncomprehensive) list of many dozens of international and local Chinese companies and outfits, both large and small, venturing into the new domain. The list includes such household names as Brother, Fujitsu, Hitachi, IBM, Logitec, and NEC, but also myriad others like Action Computer Enterprise, Bookmark Tele-Communication, Bright Forward (*Mingjian youxian gongsi*), and more who have largely disappeared from the annals of computing history (but which nevertheless had committed, in some cases, substantial investments toward the development of Chinese-character information-processing ventures). See Richard Suchenwirth, *Systeme zur Datenverarbeitung in chinesischer Schrift. Eine Marktübersicht*, Beijing, April 15, 1985. Original housed in the private collection of the Rolf Heinen in Drolshagen, Germany. Photocopy in author's collection.

4. For a fascinating introduction to the rise of "information culture" beginning around this time in China, see Xiao Liu, *Information Fantasies: Precarious Mediation in Postsocialist China* (Minneapolis: University of Minnesota Press, 2019).

5. Helen M. Wood, Donald J. Reifer, and Martha Sloan, "A Tour of Computing Facilities in China," *Computer*, January 1985, 80–87, 85.

6. For an analysis of the deeper history of *fangzao* in modern China, see Eugenia Lean, *Vernacular Industrialism in China: Local Innovation and Translated Technologies in the Making of a Cosmetics Empire, 1900–1940* (New York: Columbia University Press, 2020). See also Silvia M. Lindtner, *Prototype Nation: China and the Contested Promise of Innovation* (Princeton, NJ: Princeton University Press, 2020).

7. "Kickin' the Bucket: 12 Outrageous Fake KFC Restaurants," https://weburbanist.com/2014/05/04/kickin-the-bucket-12-outrageous-fake-kfc-restaurants/.

8. For a vivid account of more conventional forms of piracy in the realm of computing and software, see Sang Ye, "Computer Insects," in *The China Reader: The Reform Era*, ed. Orville Schell and David Shambaugh, 291–296. New York: Vintage Books, 1999. On the subversive potential of the "pirate" figure in late twentieth- and early twenty-first-century global economics, see Kavita Philip, "What is a Technological

Author? The Pirate Function and Intellectual Property," *Postcolonial Studies* 8, no. 2 (2005): 199–218.

9. Depending on the system and software, Chinese bitmap fonts intended for the screen were referred to alternatively as *danxianti* ("terminal script" 单线体) and *xianshizi* ("display fonts" 显示字), among other terms. See *24-Pin Chinese Printer User Manual* (*24 zhen Hanzi dayinji yonghu shouce*) [24针汉字打印机用户手册], n.p., n.d., 37; Luo Yinghui [罗英辉] and Zhang Yala [张亚拉], *Shape-Meaning Three Letter Code: A Revolution in Chinese Character Input Methods* (*Xingyi sanma: Hanzi shurufa de geming*) [形意三码: 汉字输入法的革命] (Guangzhou: Zhongshan daxue chubanshe, 1995), 40.

10. This figure accounts for either the simplified or traditional character form, but not both, and with no accompanying metadata.

11. To outfit a machine with both simplified and traditional forms of these characters, these memory requirements would need to be doubled.

12. During the many presentations I've given on Chinese information technology, a question that often emerges pertains to simplified and traditional Chinese characters, and the challenges (if any) posed by these two different kinds of character-forms on various types of Chinese text technologies. In most cases, as I emphasize, the question of simplified versus traditional characters has little to no impact on the technology in question. For example, the transmission of Chinese characters by way of the four-digit Chinese telegraph is entirely indifferent to the question of what those Chinese characters forms are. As we move into the era of personal computing, however, especially the early period in which memory was so limited, and at such a premium, the question of simplified versus traditional characters does become a salient factor. With most commercially available printers offering printer heads with 9 pins—translating into maximum bit map sizes of either 14 or 18 dots in height—this meant that many common "traditional" form Chinese characters could quite literally not be printed using these printers, except with a full three passes of the printer head. While it is difficult to determine with any certainty, the early history of computer memory and bitmap Chinese fonts may have contributed to the popularity of, or turn towards, simplified Chinese characters.

13. This was a "Song-style" (*songti*) font.

14. These included a 24-by-24 Heiti font, a 24-by-24 Kaiti font, and a 24-by-24 Fangsong font. See *UCDOS High-Level Chinese Character System User's Manual* v. 2.0 (*UCDOS gaoji Hanzi xitong yonghu shouce v. 2.0*) [UCDOS高级汉字系统用户手册 v. 2.0] (Beijing: *Zhongguo kexueyuan xiwang gaoji diannao jishu gongsi* [中国科学院希望高级电脑技术公司], 1990), 4–5. For the curious reader, these were 5.25-inch, 360 KB floppy disks. For customers of CCDOS 4.2, meanwhile, a full 35 floppy disks were dedicated to a set of just 7 fonts, referred to as Printer Character Databases (*dayin ziku*). These included Fangsong (24-by-24) (2 disks: FSLIB24), Kaiti (24-by-24) (2 disks: KTLIB24), Kaiti (32-by-32) (3 disks: KTLIB32), Heiti (48-by-48) (7 disks: HTLIB48), Songti (48-by-48) (7 disks: STLIB48), Kaiti (48-by-48) (7 disks: KTLIB48), and Fangsong (48-by-48) (7 disks: FSLIB48). See Pei Jie, *CC-DOS V4.2 Chinese Character Operating System User Guide* [CC-DOS V4.2 汉字操作系统使用指南] (Shanghai: Shanghai jiaotong daxue chubanshe, 1991), 142.

15. The group's expertise went far beyond China, moreover, once again indicating GARF's global ambitions and scope (other members included, for example Annemarie Schimmel, a professor of Indo-Muslim culture at Harvard). Letter addressed to Judy Ling from the offices of the Graphics Arts Research Foundation, June 9, 1981; written on GARF letterhead, which included a list of advisory member members. "Richard Solomon, Former Diplomat Who Helped Nixon Open Relations with China, Dies," *Wall Street Journal*, March 14, 2017: https://www.wsj.com/articles /richard-solomon-former-diplomat-who-helped-nixon-open-relations-with-china -dies-1489532159.

16. Email from Bruce Rosenblum to the author, April 1, 2017; Email from Bruce Rosenblum to the author, April 8, 2017.

17. Email from Bruce Rosenblum to the author, April 8, 2017.

18. Email from Bruce Rosenblum to the author, April 8, 2017.

19. Emails from Bruce Rosenblum to the author, March 26, 2017, and April 8, 2017.

20. Louis Rosenblum, B. D. Rosenblum, G. P. Low, and W. W. Garth, *Computerized Typesetting of Chinese: Text Processing on the Sinotype III* (Research Triangle Park, NC: Instrument Society of America, 1983).

21. Email from Bruce Rosenblum to the author, April 8, 2017.

22. Email from Bruce Rosenblum to the author, April 8, 2017.

23. Email from Bruce Rosenblum to the author, April 8, 2017.

24. Louis Rosenblum to Zhi Bingyi, June 30, 1981, LR.

25. By Bruce's calculation, in order to store the bitmap for each Chinese character, along with descriptors for two Chinese input methods per character, meant that each 2K PROM chip would be able to hold a maximum of 51 Chinese characters—a number which declined to a mere 28 characters per PROM were he to outfit each character with four Chinese input method descriptors, rather than just two.

26. Email from Bruce Rosenblum to the author, April 8, 2017.

27. Rosenblum to Zhi Bingyi, June 30, 1981, LR.

28. Rosenblum to Zhi Bingyi, June 30, 1981, LR.

29. Rosenblum to Zhi Bingyi, June 30, 1981, LR. When shut down, Sinotype III would presumably then erase these temporary graphs, and resort to its original, hardwired set of high-frequency characters. This, however, is not specified in any of Louis Rosenblum's correspondences or notes.

30. GARF was not alone in exploring this technique, moreover. Chen Shu described a similar approach circa 1982, wherein a Chinese character set of between seven and ten thousand characters were stored on floppy discs. When booting up on the machine, one of the first operations performed by the computer would be to load the most common of these characters into onboard RAM, after which point memory retrieval could then be performed at significantly faster rates. The lower-frequency characters, however, would remain in external floppy storage only—resulting in far slower retrieval times for these graphs. See Chen Shu, "Symposium on Chinese

Character Processing Systems Reviewed," FBIS report based on *Jisuanji shijie* 20 (October 1982): 13, in *FBIS* 83, 20–22.

31. James B. Stepanek, "Microcomputers in China," *The Chinese Business Review*, May–June 1984, 31.

32. As noted above, Bruce Rosenblum briefly explored the possibility of using PROMs, which in some respects would have been a comparable technique as "Chinese-on-a-chip" cards. At that moment, commercially available PROMs made this approach unviable.

33. Stepanek, "Microcomputers in China," 31.

34. See www.pcang.com, accessed March 26, 2019; Wu Xiaojun [吴晓军], *2.13 Series Chinese Character System User Manual* (*2.13 xilie Hanzi xitong yonghu shouce*) [2.13 系列汉字系统用户手册] (Beijing: Jixie gongye chubanshe, 1993); Chen Xiangwen [陈相文], ed., *The Five-Stroke and Natural Code Chinese Character Input Methods on a Microcomputer* (*Weixing jisuanji wubi zixing ji ziranma Hanzi shurufa*) [微型计算机五笔字型及自然码汉字输入法]. Tianjin: Nankai daxue chubanshe, 1994, 1; Zheng Yi [郑邑], ed., "Foreword," in *The Apple II Microcomputer and Chinese System* (*Apple II weiji ji Hanzi xitong*) [Apple II 微机及汉字系统] (Shanghai: Tongji daxue chubanshe, 1985); Liu Shuji et al., *Impact Printers: Principles of Usage and Repair* (*Zhenshi dayinji yuanli shiyong yu weixiu*) [针式打印机原理使用与维修] (Beijing: Guoji gongye chubanshe, 1988), 134; Sun Qiang [孙强]. "Chinese Character Matrix Generator That Can Generate Multiple Fonts" (*Yi zhong neng shengcheng duozhong ziti de Hanzi zimo fashengqi*) [一种能生成多种字体的汉字字模发生器]." CN1031140A. Assignee: Beijing Sitong Group. Date of Application April 10, 1987. Patent Date February 15, 1989; Zimin Wu and J. D. White, "Computer Processing of Chinese Characters: An Overview of Two Decades' Research and Development," *Information Processing and Management* 26, no. 5 (1990): 681–692, 687.

35. One further attempt to address the challenges of storing Chinese bitmap fonts involved early and longstanding explorations into vector fonts, known in Chinese as *shiliang Hanzi* (矢量汉字). In vector fonts, it is not a bitmap itself which is stored in memory, but a series of coordinates with which the character is dynamically produced using the computer's native graphics processing capacities. As early as 1969, two researchers at the University of Wisconsin, Madison—Shou-chuan Yang and Charlotte Yang—demonstrated that it was possible to decompose Chinese characters into a series of x-y coordinates, rather than into a conventional bitmap raster. Using their technique, Chinese characters could be drawn on screen, almost like a children's "connect-the-dots" puzzle. Using as their example the Chinese character for "brave" (勇 *yong*, in its traditional form), they explained that this graph could be represented using 23 pairs of spatial coordinates, with each pair describing the origin and terminus of a single line segment. When compared to a conventional 16-by-16 bitmap raster, the differences in memory were sizable: 256 bits (or 32 bytes) to store this character in bitmap, versus just 192 bits (24 bytes) using the "connect-the-dot" system. What is more, a conventional 16-by-16 grid would likely be incapable of rendering this particular character, since it is in traditional, rather than simplified format. To achieve bare-minimum legibility, a 24-by-24 bitmap raster would likely

be necessary, which in turn would require 576 bits (or 72 bytes). "Taking this as a basis of estimation," the researchers concluded, "the overwhelmingly numerous 10,000 Chinese characters can be decomposed and packed in 40K of memory." The researchers at the University of Wisconsin were not alone. Chinese vector fonts—or "compressed Chinese characters" (*yasuo Hanzi*), as they were sometimes referred to in Chinese—were also being explored at Harvard, the CIA, the Graphic Arts Research Foundation (GARF), Apple, Stanford University, and elsewhere. While more economical from the standpoint of memory, these skeleton fonts were more computationally taxing to produce. Rather than simply printing a bitmap raster to the monitor, each individual line segment had to be calculated and drawn, thus depending on the systems CPU. More than this, skeleton fonts were even less aesthetically acceptable than even the admittedly rough low-resolution bitmaps. While crude, bitmap fonts nevertheless strove to preserve at least some of the orthographic features of hand-drawn Chinese characters, including the curvature of strokes, stroke terminals, and the like. In Chinese vector fonts, by comparison, curvature was abolished altogether. "A curve of a stroke," Yang and Yang explained bluntly in their 1969 report, "is treated as many short straight line segments." As such, skeleton fonts were not, by and large, implemented on any significant commercial scale, and remained more a theoretical exercise than a practical solution. See Shou-chuan Yang and Charlotte W. Yang, "A Universal Graphic Character Writer. In International Conference on Computational Linguistics COLING 1969: Preprint No. 42, Sånga Säby, Sweden. See this link for info: https://aclanthology.org/C69-4201/. At Stanford University, visiting researchers attempted to develop a vector-based Chinese font using the METAFont program, developed by Donald Knuth. Referred to as LCCD ("Language for Chinese Character Design"), the program never yielded any commercial success. Extensive documentation on Metafont and its permutations are housed in the Donald Knuth papers at Stanford University.

36. Chan H. Yeh, "System for the Electronic Data Processing of Chinese Characters," US Patent 3820644, filed May 7, 1973, and issued June 28, 1974; interview with Chan-hui Yeh, March 21, 2010.

37. To achieve this effect, it was essential for Chinese engineers to use Western-built printers that offered this kind of precision. Two such machines identified in the literature were the Epson FX and Epson MX. See Chen Zengwu and Jin Lianfu, *Chinese Language Information Processing System* (Beijing: Zhongguo jisuanji yonghu xiehui zong-hui, 1984), 84–85. On account of this technique, many dot-matrix Chinese printouts from this era exhibit Chinese characters with an almost "bumpy" texture to them—a quality derived from the fact that, when using this "zipper" technique, the dots in the Chinese character bitmap actually overlapped one another, ever so slightly. Unlike low-resolution Latin alphabetic bitmap printing, then, where one often finds small yet still discernible gaps between all of the dots, the dots in Chinese bitmap printouts often touch one another (but only in systems that employed this mod).

38. Wang Jizhi [王辑志], "My Autobiography (25): Making a Chinese Character Card (*Wo de zizhuan (25): zhizuo Hanka*)," http://blog.sina.com.cn/wangjizhi (accessed January 2, 2018).

39. The quoted English phrase, translated here in Chinese: *jingguo yi dian jishu kaifa gongzuo hou.*

40. Because there are many more Chinese characters needed in Chinese than in Japanese, a core part of Wang's copycatting process was to increase the number of Chinese character library chips (*Hanzi ziku xinpian*).

41. Zheng Yi, ed., *The Apple II Microcomputer and Chinese System* (*Apple II weiji ji Hanzi xitong*) (Shanghai: Tongji daxue chubanshe, 1985), 10. The size of the screen monitors, at 200-by-640, enabled an estimated 10 rows, with 40 characters per line—an improvement, to be sure, but still far below English-language display capacity, which boasted 25 rows of 80 characters each. See Chen Zengwu and Jin Lianfu, *Chinese Language Information Processing System*, 221. By 1990, increased monitor sizes steadily enabled more rows of Chinese text—from 10 rows of text to 25—and yet each improvement in screen real estate changed fundamentally little about the Chinese deficit when compared to English. See *UCDOS High-Level Chinese Character System User's Manual*, 23.

42. Although the pop-up menu was a ubiquitous feature of Chinese computing from the 1980s onward, this feedback technique dates back to the 1940s. Earlier in this book, as well as in *The Chinese Typewriter*, we have already seen pop-up menus on multiple occasions, although not in regards to computer monitors. The "Magic Eye" component of the 1947 experimental Chinese typewriter designed by Lin Yutang was, in effect, the first "pop-up menu" in history, albeit a mechanical one. Its purpose, as readers will recall, was to present the operator with as many as 8 character candidates based on the user's key sequence. See Thomas S. Mullaney, *The Chinese Typewriter: A History* (Cambridge, MA: MIT Press, 2017).

43. Pei Jie, *CC-DOS V4.2 Chinese Character Operating System User Guide*, passim.

44. *UCDOS High-Level Chinese Character System User's Manual*, 6, 31.

45. Gu Jingwen [顾景文], *Driver Design for the Display of Chinese Characters and Images on a Microcomputer* (*Weiji Hanzi yu tuxing de xianshi jiqi jiekou chengxu sheji*) [微机汉字与图形的显示及其接口程序设计]. Shanghai: Tongji daxue chubanshe, 1995, 10–13.

46. Qian Peide [钱培得], ed., *CC-DOS Chinese Character Operating System for Microcomputers* (*Weixing jisuanji Hanzi caozuo xitong CC-DOS*) [微型计算机汉字操作系统CC-DOS] (Xi'an: Shaanxi dianzi bianjibu, 1988), 1; Qian Peide. "An Analysis of CC-DOS" (CC-DOS fenxi) [CC-DOS 分析], Microcomputer Applications (Weijisuanji yingyong) [微计算机应用] 5 (1985): 1–10.

47. My thanks to Jim Beveridge for his insightful edits and corrections in my discussion of Interrupts and the BIOS.

48. Chen Zengwu and Jin Lianfu, *Chinese Language Information Processing System*, 147–150.

49. Chen Zengwu and Jin Lianfu, *Chinese Language Information Processing System*, 220.

50. The screen buffer also had to be expanded, with the design of three new code zones: the Character Bitmap Buffer Zone, the CRT Refresh Zone, and the Virtual

Refresh Zone. In the first of these zones, which was stored in assembly language from code point CS0078 to CS0099, the Chinese character bitmap codes were stored—divided into "left-side area" (*zuobian qu*) and "right-side area" (*youbian qu*). The CRT Refresh Zone was stored in two zones: from code point 8000 to 953F, and from code point A000 to BF3F. The Virtual Refresh Zone was stored in three zones: 000B to 087F, 0880 to 104F, and 1050 to 181F. See Qian Peide, "An Analysis of CC-DOS," 3; Chen Zengwu and Jin Lianfu, *Chinese Language Information Processing System*, 223–224; and Gu Jingwei [顾景文], "Driver Design," 20–21. See also Gu Jingwen, *Driver Design for the Display of Chinese Characters and Images on a Microcomputer*, 18–19; Qian Peide, "An Analysis of CC-DOS," 3.

51. Chris Hutton, "Writing and Speech in Western Views of the Chinese Language," in *Critical Zone 2: A Forum of Chinese and Western Knowledge*, ed. Q. S. Tong and D. Kerr (Hong Kong: Hong Kong University Press, 2006), 83–105; Q. S. Tong. "Inventing China: The Use of Orientalist Views on the Chinese Language," *Interventions* 2 (2000): 11–14.

52. Chen Zengwu and Jin Lianfu, *Chinese Language Information Processing System*, 214.

53. Qian Peide, "An Analysis of CC-DOS," 5; Chen Zengwu and Jin Lianfu, *Chinese Language Information Processing System*, 218.

54. Qian Peide, "An Analysis of CC-DOS," 5.

55. Qian Peide, "An Analysis of CC-DOS," 6.

56. Within these look-up tables, known as "Chinese Character Input Code Scan Tables," each entry comprised 4 bytes. To economize memory, as many input system codes as possible were stored in the same look-up table. In one such look-up table, for example, the first segment of each entry was dedicated to an input system known as Shouwei code, which occupied the first byte of memory and then a portion of the second. The second segment of each entry was dedicated to Pinyin input code, which was by far the longest of the bunch, occupying most of the second byte of memory, all of the third, and a substantial portion of the fourth.

57. Chen Zengwu and Jin Lianfu, *Chinese Language Information Processing System*, 208.

58. Qian Peide, "CC-DOS," 2. "CCBIOS Keyboard Control Program" (*CCBIOS de jianpan kongzhi chengxu*).

59. In the span between CS 1848 and CS 2797 lived the "CCBIOS INT 10 program" (*10 lei zhongduan chengxu*). From CS 2BD5 to 9595 lived the scanning tables that enabled conversion between Chinese character input codes (*Hanzi shurufa*) and Chinese character internal codes (*Hanzi jineima*). In the span of CS 98B3 to AA76 lived the "CCBIOS INT 16 program" (*16 lei zhongduan chengxu*). Finally, in the span between CS AAC0 and CS AB43 lived the main execute code (*zhixing daima*) itself. It was this address—AAC0, the entry address for cccc.exe—where existing Interrupts in ROMBIOS would be redirecting, effectively capturing all keyboard, monitor, and printing controls, and rerouting them through cccc.exe. See Qian, "CC-DOS," 2

60. Qian Peide, "CC-DOS," 2.

61. Qian Peide, ed., *CC-DOS Chinese Character Operating System for Microcomputers*, 1.

62. Qian, "CC-DOS," 1. "CC-DOS 是对PC-DOS的扩充 (*CC-DOS shi dui PC-DOS de kuochong*)."

63. Chen Zengwu and Jin Lianfu, *Chinese Language Information Processing System*, 161.

64. Others included China Star (*Zhongguo zhi xing*), 2.13, GWBIOS (*Changcheng langchao lei*) [长城浪潮类], Super-CCDOS (SPDOS) (*Jinshan chaoxiang lei*) [金山超想累], LXBIOS (*Lianxiang shi Hanka lei BIOS*) [联想式汉卡类], UCDOS (*Xiwang Hanzi xitong*) [希望汉字系统], HDCCDOS, MECCDOS, WMDOS, New Era Chinese Operating System (*Xin shidai Hanzi caozuo xitong*), HZDOS, WMDOS (Wang Code Operating System) (*Wangma caozuo xitong*), XSD-1, Nanking DOS, ET (Taiwan, based on MS-DOS), KC (Taiwan, based on MS-DOS), 01 (Taiwan, based on MS-DOS), CDPS-2 (Chinese Data Processing System), Shengshu Hanyu xitong [声数汉语系统], and CW 中文语词处理系统, among others. See Chen Xiangwen [陈相文], ed., *The Five-Stroke and Natural Code Input Methods for Microcomputers* (*Weixing jisuanji wubi zixing ji ziranm Hanzi shurufa*) [微型计算机五笔字型及自然码输入法] (Tianjin: Nankai daxue chubanshe, 1994), 1–4; Zimin Wu and J. D. White, "Computer Processing of Chinese Characters," 687; Super-CCDOS (SPDOS) (*Jinshan chaoxiang lei*) [金山超想累] was developed by the Hong Kong Jinsan Company circa 1988. Chen Qixiu [陈启秀], *Chinese Character Input and Word Processing: 100 Questions* (*Hanzi shuru yu wenzi chuli 100 wen*) [汉字输入与文字处理100问]. Nanjing: Jiangsu kexue jishu chubanshe, 1996, 19. CDPS-2 (Chinese Data Processing System) was designed and manufactured by the Electronics Research and Service Organization of the Industrial Technology Research Institute. See *The Collected Works of Margaret C. Fung, PhD* (Taipei: Showwe Information Corporation, 2004), 147. UCDOS (*Xiwang Hanzi xitong*) [希望汉字系统] was developed beginning in 1986 by the Chinese Academy of Sciences Xiwang Gaoji Diannao Jishu Gongsi. Bao Yueqiao [鲍岳桥], Gan Dengdai [甘登岱] and Liu Qinghua [刘庆华], *Hope Soft UCDOS 3.1 Training Manual* (*Hope Soft UCDOS 3.1 peixun jiaocheng*) [Hope Soft UCDOS 3.1 培训教程学] (Beijing: Xuefan chubanshe, 1994), preface; *UCDOS High-Level Chinese Character System User's Manual*, 1. See also Luo and Zhang, *Shape-Meaning Three Letter Code*, 40. In August 1983, the first version of WMDOS—the Wang Code Operating System (*Wangma caozuo xitong*)—followed by version 2.0 in August 1984. In March 1986, the third version of WMDOS was released. By 1990, WMDOS had reached version 5.0. See Chen Xiangwen, *The Five-Stroke and Natural Code Input Methods for Microcomputers*, 1–4; Wu Xiaojun, *2.13 xilie Hanzi xitong yonghu shouce*, 242–243; CW中文语词处理系统 was developed by Peking University and CASS 语言文字应用研究所. Wu Xiaojun, *2.13 xilie Hanzi xitong yonghu shouce*, 231.

65. Work on CP/M, for instance, was led by researchers at the Chinese Academy of Science Computing Research Center (*Zhongguo kexueyuan jisuansuo*). Chen Zengwu and Jin Lianfu, *Chinese Language Information Processing System*, 150–155; "Maojiang Wang, "On the Interface Between the High-Level Languages and Chinese Character Information," *Computer Standards and Interfaces* 6, no. 2 (1987): 181–186, 181. See

also Wang Qihong, "PASCAL Compiler with Chinese Identifier" (Ke yong Hanzi biaoshifu de PASCAL bianyi chengxu) [可用汉字标识符的PASCAL编译程序], *Journal of Changchun Post and Telecommunication Institute (Changchun youdian xueyuan xuebao)* [长春邮电学院学报] 2 (1987): 61–71. On COBOL, see Yang Huimin [杨惠民] and Jiang Zifang [蒋子放], *Sinicized COBOL Language for Microcomputers (Programming Methods and Skills)* (*Weixing jisuanji Hanzihua COBOL yuyan*) [微型计算机汉字化COBOL语言] (Beijing: Dianzi gongye chubanshe, 1987). Regarding FoxBASE, see Liu Fuying [刘甫迎] and He Xiqiong [何希琼], *Chinese Language FoxBASE+ Novel Relational Database* (*Hanzi FoxBASE+ Xinying guanxi shujuku*) [汉字FoxBASE+新颖关系数据库] (Zhongguo kexueyuan changdu jisuanji yingyong yanjiusuo qingbao shi, November 1987).

66. High-level programming languages are another area in need of further research. Whether one is working in COBOL, Basic, C, or otherwise, all industry-standard programming languages in history have been premised upon the English language. Whether in terms of operators ("set," "display," "move," and others that rely on English-language words or abbreviations) or more subtle biases (such as the assumption of left-to-right text and, of course, the use of ASCII alphanumerics), every programmer must be in some sense "conversant" in both English and the Latin alphabet in order to program seriously (let alone professionally). Just as Chinese engineers sought to mod the BIOS and disk operating systems, so too did they target programming languages such as COBOL and Basic. "Chinese COBOL" (*Hanzi COBOL yuyan*) was created, as were Chinese versions of PASCAL, among others. Chinese versions of SuperCalc, VisiCalc, Multiplan, Lotus 1-2-3, Fortran, Wordstar, dBaseII, dBaseIII, and more were also developed. For a fascinating early treatise on the program of Chinese-language high-level programming languages, see Yaohan Chu, "Structure of a Direct-Execution High-Level Chinese Programming Language Processor." *ACM '74: Proceedings of the 1974 Annual Conference* 1 (January 1974): 19–27. See also Zhu Shili [朱世立], *Common Methods in Chinese TRUE BASIC Language and System Engineering* (*Hanzi TRUE BASIC yuyan he xitong gongcheng changyong fangfa*) [汉字 TRUE BASIC 语言和系统工程常用方法] (Beijing: Dianzi gongye chubanshe, 1988).

67. As Gu Jingwen noted, each engineering team undertook BIOS-hacking differently. The China Computer Technical Service Corporation (*Zhongguo jisuanji jishu fuwu gongsi*, now known as China National Software) Sinicized the RSX-11M-PLUS operating system, running it as a Chinese character operating system (*Hanzi caozuo xiting*) on the DJS-180—a Chinese-built clone of the PDP-11. See Gu Jingwen, *Driver Design for the Display of Chinese Characters*, 19.

68. Gu Jingwen, *Driver Design for the Display of Chinese Characters*, 241–242. Gu does not specify which version of Wordstar he is referring to in this context. It goes without saying, but the reverse engineering of software, especially by developers eager to add new functionalities to existing products, was not a practice exclusive to China. What distinguishes the Chinese case from those in the United States, for example, is that Chinese modding was essential for baseline usability, not merely the expansion of features.

69. David Chen, "The Race is On to Design New Chinese System," *South China Morning Post*, September 1, 1987, 25.

70. Prime examples include the Toshiba P321 Printer, Panasonic KX-P2123, Star Micronics NX-2420 Rainbow and NX-2420 Multi-Font, Star AR-2463, NEC NK 3826, NEC Pinwriter P6200, M2040, M1724, T3070, P1350, P1351, Epson LQ2500, Epson LQ1500, Epson LQ 1000K, Brother M-1724, M1570, OKI OK1, OKI 8320, and OKI 5320, AR3240, AR2463, TH-2100, and KC-3070. See https://www.atarimagazines .com/compute/issue75/Toshiba_P321_Printer.html; https://www.atarimagazines.com /compute/issue144/G10_Panasonic_KXP2123.php; https://www.atarimagazines.com /compute/issue123/28_2_NEW_PRODUCTS_4-STAR_PRINTERS.php; https://www.atari magazines.com/compute/issue130/42_NEC_Pinwriter_P6200.php; *UCDOS High-Level Chinese Character System User's Manual*, 24–25; While the majority of these printers were Japanese-built, the TH-2100 and KC-3070 were both identified as Chinese printers in a number of sources. See, for example, Liu Qi [刘奇], *Chinese Language dBASE-II Relational Database Management System* (*Hanzi xing dBASE-II guanxi shujuku guanli xitong*) [汉字型dBASE-II关系数据库管理系统] (Beijing: Beijing Yanshan shiyou huagongsi qiye guanli xiehui, 1984). See also Liu Shuji et al., *Impact Printers: Principles of Usage and Repair* (*Zhenshi dayinji yuanli shiyong yu weixiu*) (Beijing: Guoji gongye chubanshe, 1988), 134–138.

71. Laser printing is a particularly important area for further research. For an introduction to Wang Xuan, a foundational figure within high-resolution Chinese graphics and commercial printing, see Wang Xuan [王选], *Beijing University Founder Group Book Typesetting Technology and Application* (*Beida fangzheng shuban paiban jishu he yingyong*) [北大方正书版排版技术和应用] (Beijing: Founder Group, 1993).

72. Chinese character bitmaps and metadata still occupy orders of magnitude more memory than their alphanumeric counterparts, for example, only now both are sloshing about in such an immense ocean of low-cost storage, and bulleting through immeasurably faster computational processes, that de facto the differential matters little if at all. A Chinese font might be big, that is, but it's now nothing compared to the space occupied by even a single video file, let alone immense programming applications.

CHAPTER 6

1. Pei Jie, *CC-DOS V4.2 Chinese Character Operating System User Guide* (*CC-DOS V4.2 Hanzi caozuo xitong shiyong zhinan*) (Shanghai: Shanghai jiaotong daxue chubanshe, 1991), 22, 141–142.

2. By depressing ALT and the F1 function key, for example, the user could select Quwei input. ALT and F2 toggled to Shouwei input, ALT and F3 to Pinyin input, and so on. Pei Jie, *CC-DOS V4.2 Chinese Character Operating System User Guide*, 140.

3. Pei Jie, *CC-DOS V4.2 Chinese Character Operating System User Guide*, 75.

4. Ling Zhijun, *The Lenovo Affair: The Growth of China's Computer Giant and Its Takeover of IBM-PC*, trans. Martha Avery (Malden, MA: Wiley, 2006); James Cortada, *IBM: The Rise and Fall and Reinvention of a Global Icon* (Cambridge, MA: MIT Press, 2019).

5. *How to Study Hanyu Pinyin* (*Zenyang xuexi Hanyu Pinyin fang'an*) [怎样学习汉语拼音方案] (Beijing: Wenzi gaige chubanshe, 1958); Zhou Youguang [周有光], *General*

Theory of Chinese Character Reform (*Hanzi gaige gailun*) [汉字改革概论] (Beijing: Wenzi gaige chubanshe, 1961); Zhou Youguang [周有光], *Basics of Pinyin Letters* (*Pinyin zimu jichu zhishi*) [拼音字母基础知识] (Beijing: Wenzi gaige chubanshe, 1962); Zhou Youguang [周有光], *Questions Regarding Pinyin-ization* (*Pinyinhua wenti*) [拼音化问题] (Zhongguo wenzi gaige weiyuanhui yanjiuzu [中国文字改革委员会研究组], 1978).

6. David Moser, *A Billion Voices: China's Search for a Common Language* (New York: Penguin, 2016).

7. Zhou Youguang [周有光], *The Pinyin-ization of Telegrams* (*Dianbao pinyinhua*) [電報拼音化] (Beijing: Wenzi gaige chubanshe, 1965).

8. To use technical terminology, pinyin input strings cannot serve as the basis for "prefix-free codes" in which no code segment serves as the beginning of any other code segment in the overall system.

9. In the case of Sinotype, for example, Samuel Caldwell and his colleagues made frequent use of so-called Mathews Numbers when trying to identify Chinese characters in memory.

10. See Chen Jianwen, *Research Report on the Double Pinyin Code Chinese Character Information Processing System*, 1; and Chi Wang, *Building a Better Chinese Collection for the Library of Congress. Selected Writings.* Lanham: The Scarecrow Press, 2012, 99. Chen Jianwen and Xue Shiquan were based in the library of Nanjing University, and Qian Peide in the computing department at Suzhou University. Chen Jianwen [陈建文], The Double Pinyin Chinese Character Encoding Method: Purpose and Usage (*Hanyu shuangpin Hanzi Bianma de yongtu he yongfa*) [汉语双拼编码汉字的用途和用法] (Nanjing: Nanjing daxue tushuguan, February 1986); Xue Shiquan and Qian Peide, *Guanyu shuangpin bianma Hanzi chuli xitong de shiyan baogao* [关于双拼编码汉字处理系统的实验报告] (n.p.: February 1986); Chen Jianwen [陈建文], Xue Shiquan, and Qian Peide, *Double Pinyin Encoding Chinese Character Information Processing System Development Report* (*Shuangpin bianma Hanzi xinxi chuli xitong yanzhi baogao*) [双拼编码汉字信息处理系统研制报告]. Nanjing and Suzhou: Nanjing daxue tushuguan and Suzhou daxue jisuanji jiaoyanshi, n.d. (ca. May 1986); *Opinions on the Use of Double Pinyin Encoding Chinese Character Information Processing System* (*Shuangpin bianma Hanzi xinxi chuli xitong shiyong yijian*) [双拼编码汉字信息处理系统使用意见] (Nanjing: Nanjing daxue jisuanjichang, May 28, 1986); *Double Pinyin Encoding Chinese Character Information Processing System User Report* (*Shuangpin bianma Hanzi xinxi chuli xitong yonghu baogao*) [双拼编码汉字信息处理系统用户报告] (Nanjing: Nanjing daxue jisuanjichang, May 26, 1986).

11. Cai Rongbo [蔡荣波], ed., *Computational Chinese Input Methods: Methods, Skills and Training* (*Diannao Hanzi shuru de fangfa jiqiao yu xunlian*) [电脑汉字输入的方法技巧与训练] (Guangzhou: Guangdong keji chubanshe, 1993), 128.

12. Chen Jianwen, *Research Report on the Double Pinyin Code Chinese Character Information Processing System* (*Shuangpin bianma Hanzi xinxi chuli xitong yanzhi baogao*) [双拼编码汉字信息处理系统研制报告] (Nanjing: Nanjing University Library and the Suzhou University Mathematics Department, ca. 1986), 4.

13. Chen Jianwen, *Research Report on the Double Pinyin Code*, 1–6. By February 1986, Double Pinyin input had also been installed on a few IBM PC/XT computers.

14. Zhang Tinghua [张廷华], *Double Pinyin Double Radical Encoding System Table of 4000 Commonly Used Chinese Characters* (*Shuangpin shuangbu bianmafa siqian changyong Hanzi bianma biao*) [双拼双部编码法四千常用汉字编码表] (Shanxi: Shanxi Linyi zhongxue, 1979), 2; Cai Rongbo, ed. *Computational Chinese Input Methods*, 128.

15. Zhang Guofang [张国防], *Fifty Character Element Computational Chinese Input Usage Manual* (*50 ziyuan jisuan Hanzi shurufa shiyong shouce*) [五十字元计算汉字输入法使用手册] (Beijing: Zhongguo jiliang chubanshe, 1989).

16. *Shanxi Province Chinese Character Encoding Research Special Issue* (*Shanxi sheng Hanzi bianma yanjiu zhuanji*) [山西省汉字编码研究专辑]. Shanxi sheng kexue jishu qingbao yanjiusuo [山西省科学技术情报研究所]. March 1979. Thomas S. Mullaney East Asian Information Technology History Collection, Stanford University (hereafter TSM).

17. China submitted the code for official recognition by the International Organization of Standardization (ISO). "China Will Apply for International Registration of Chinese Graphic Character Set for Computers," *Xinhua General Overseas News Service*, March 27, 1981; Chen Yaoxing, "Introduction to Chinese Codes for Information Interchange" (Xinxi jiaohuan yong Hanzi bianma zifu ji jianjie), *Wenzi gaige* [*Writing Reform*] 4, 1983, 5–7. Within the broader Sinographic, Chinese-Japanese-Korean (CJK) world, a number of other countries enacted standards as well. In 1978, Japan promulgated JIS C6226, designed to function as the Japanese equivalent of ASCII. Based on a 2-byte structure, the code comprised a total of 6,349 commonly used Japanese kanji. In 1981 and 1982, Korea also enacted a national standard for Korean Hancha as well, with a code comprising 1,692 characters. Two years later, in 1984, Taiwan's Chinese Character Analysis Group of the Council for Cultural Planning and Development promulgated the CCCII (Chinese Character Code for Information Interchange). Steadily, the CJK information environment was stabilizing, providing a foundation upon Pinyin input could begin to rely, even as profound challenges of interoperability existed between these competing standards. Encoding systems were but one part of this massive deluge of new standards. Chinese dot matrix bitmaps were formalized as well, in an attempt to confront the wide array of competing grid sizes and layouts one finds throughout the late 1970s and early 1980s. As readers recall from the preceding chapters, the IPX system developed by Chan-hui Yeh in the 1970s featured a 20-by-24 bitmap grid, whereas Wang Laboratories, GARF, H. C. Tien, and others employed 15-by-16, 16-by-16, and 18-by-18, among others. The goal here was less a question of aesthetics than of interoperability, insofar as the regulation of Chinese bitmaps would enable font designers and Chinese character generator card manufacturers to rely upon agreed-upon standards, thereby making their products compatible across systems, at least within the PRC. In June 1983, the Ministry of Electronics Standardization Research Institute convened a conference dedicated just to Chinese character bitmaps titled the "Work Meeting on a Draft National Standard for Chinese Character Matrices Used in Information Exchange" (Xinxi jiaohuan yong Hanzi zhenzixing de guojia biaozhun qicao gongzuo huiyi).

The meeting focused on three matrices in particular: 16-by-16, 24-by-24, and 32-by-32. This conference, as well as others that followed, culminated in a sweeping series of official GB standards. GB 5007.1, GB 5007.2, and related guidelines established an initial national standard for 24-by-24 Chinese bitmaps. GB 11458.1-1989 established initial national standard for 15-by-16 Chinese bitmaps; as did GB 11459.1-1989 for 24-by-24 Chinese bitmaps. Individual font styles were standardized as well, including "Black" (or "Gothic") Chinese (*heiti*), Song-style (*Songti*), "Fangsong-style" (*fangsong*), Kai-style, and others. GB 12036-1989 established the standard for the 32-by-32 Chinese gothic, for example, as did GB 12040-1989 for 36-by-36 Chinese gothic and GB12044-1989 for 48-by-48 Chinese gothic, among other dimensions Song-style, Fang Song-style, and Kai-style bitmaps were encoded in multiple standards, including GB 12037-1989 (Song, 36-by-36 format), GB 12034-1989 (Fangsong, 32-by-32), GB 12038-1989 (Fangsong, 36-by-36), GB 12042-1989 (Fangsong, 48-by-48), GB 12039-1989 (Kai-style, 36-by-36), and GB 12043-1989 (Kai-style, 48-by-48), among others. Vector fonts were standardized as well, as encoded in GB/T 13845–1992, and GB/T 13848–1992, among others. In the 1990s and beyond, a new era in CJK standardization was born, thanks to Unicode. A global history of Unicode, which falls beyond our chronological scope, and the ambitions of the present work, has recently become all the more possible (and necessary) to write with Stanford University's recent acquisition of the Unicode papers. See Zimin Wu and J. D. White, "Computer Processing of Chinese Characters: An Overview of Two Decades' Research and Development," *Information Processing and Management* 26, no. 5 (1990): 681–692; Ching-chun Hsieh, Kai-tung Huang, Chung-tao Chang, and Chen-chau Yang, "The Design and Application of the Chinese Character Code for Information Interchange (CCCII)," Louis Rosenblum Collection. Stanford University (hereafter LR). See also Jack Kai-tung Huang and Timothy D. Huang, *An Introduction to Chinese, Japanese and Korean Computing* (Singapore: World Scientific, 1989); Ken Lunde, *CJKV Information Processing: Chinese, Japanese, Korean & Vietnamese Computing* (Sebastopol, CA: O'Reilly Media, 2009); and the Unicode Collection, M2864, Stanford University Special Collections. Chen Zengwu and Jin Lianfu, *Chinese Language Information Processing System*, 82; GB 5007.1-2010: Information technology—Chinese ideogram coded character set (basic set)—24 dot matrix font. Note that kanji bitmaps were also standardized in Japan around this time, for the same reasons. See also Haruhisa Ishida, "Chinese Character Input/Output and Transmission in Japanese Personal Computing," *IEEE* (1979): 402–409, 405.

18. There were limits to this pinyin bias, it should be noted. Although the first layer of characters in the code was sequenced according to pinyin phonetic value, the second layer of the GB Code—which contained 3,008 more characters and character components—was organized according to radical-stroke organization (the very same way that Chinese typewriter tray beds were arranged in the pre-1949 period).

19. See, for example, Ming-Yao Lin and Wen-Hsiang Tsai, "Removing the Ambiguity of Chinese Input by the Relaxation Technique," *Computer Processing of Chinese and Oriental Languages* 3, no. 1 (May 1987): 1–24; and Gee-Swee Poo, Beng-Cheng Lim, and Edward Tan, "An Efficient Data Structure for Hanyu Pinyin Input System,"

Computer Processing of Chinese and Oriental Languages 4, no. 1 (November 1988): 1–17.

20. This is not to say that structure-based input systems disappeared, or that these philologically minded enthusiasts simply gave up. To the contrary, every year they have continued to submit new patent applications and to advocate on behalf of their systems. At the same time, however, the tide has turned away from structure-based to phonetic input, foreclosing for these individuals the kinds of possibilities that had existed from the 1950s to the early 1990s. Nor is this to suggest that there weren't inventors of structure-based IMEs who boasted deep expertise in any number of fields (as we saw in the example of Zhi Bingyi). My point, quite simply, is that one did not need to be an electrical engineer to invent a structure-based input system.

21. N. B. Dale, "An Overview of Computer Science in China: Research Interests and Educational Directions," *ACM SIGCSE Bulletin* 12, no. 1 (February 1980): 186–190; Zhang Shoudong et al., *Fundamentals of Chinese Language Computing* (*Zhongwen xinxi de jisuanji chuli*) [中文信息的计算机处理] (Shanghai: Yuzhou chubanshe, 1984), 1; Jim Mathias and Thomas L. Kennedy, eds., *Computers, Language Reform, and Lexicography in China. A Report by the CETA Delegation* (Pullman: Washington State University Press, 1980, 34–35; "First Domestically Produced Microcomputer to Enter the International Market" (*Shouci jinru guoji shichang guochan weixing jisuanji*) [首次进入国际市场国产微型计算机], *Jiefang ribao*, October 26, 1982, 3; "A Brief History of the Development of Chinese Computer (1956–2006)," http://www.nobelkepu.org.cn/art/2008/11/13/art_1025_66163.html [accessed May 9, 2015]; Zhang Shoudong et al., *Fundamentals of Chinese Language Computing*, 2.

22. Zhang Shoudong et al., *Fundamentals of Chinese Language Computing*, 4; Chen Shu, "Symposium on Chinese Character Processing Systems Reviewed," FBIS report based on *Jisuanji shijie* 20 (October 1982): 13, in *FBIS* 83, 20; "Brief Note on 1986 International Conference on Chinese Computing" (1986 guoji Zhongguo jisuanji huiyi jiankuang) [1986国际中国计算机会议简况], *Journal of Chinese Information Processing* 1, no. 1 (1986): 44, 62; "Invitation Card to the Computational Chinese Information Processing Systems Exhibition (Jisuanji Zhongwen xinxi chuli xitong zhanlanhui qingjian) [计算机中文信息处理系统展览会请柬], TSM. Many of China's earliest computing journals—periodicals dedicated to Chinese computing—were launched during this time as well. In 1980, for example, *Computer World* was launched, followed over the next few years by *Chinese Information*, the *Journal of Chinese Information Processing*, the *Journal of Computer Science and Technology*, and more. See "A Brief History of the Development of Chinese Computer (1956–2006)"; and Chen Liwei, "Develop Chinese Information Technology, Promote Computer Use in More Situations" [发展中文信息技术, 促进计算机推广应用], *Journal of Chinese Information Processing* 1, no.1 (1986): 5–7.

23. For a fascinating study on the rise of China's technocracy in the post–Cultural Revolution period, see Joel Andreas, *Rise of the Red Engineers: The Cultural Revolution and the Origins of China's New Class* (Stanford, CA: Stanford University Press, 2009).

24. Examples of 1970s-era conferences included such as the First International Symposium on Computers and Chinese I/O Systems, which took place in 1973 in Taiwan; a follow-up meeting in Taiwan in 1975; the Chinese Character Encoding Academic Exchange Meeting (*Hanzi bianma xueshu jiaoliuhui*) in Qingdao in 1978; and the ACLS Conference on East Asia Character Processing in Automated Bibliographic Systems, held at Stanford University in 1979. See *Proceedings of the First International Symposium on Computers and Chinese Input/Output Systems*, Taipei, Taiwan: Academia Sinica, August 14–16, 1973; Zimin Wu and J. D. White, "Computer Processing of Chinese Characters," 681; "Chinese Computer on Way," *South China Morning Post*, August 21, 1975, 32.

25. Examples include the International Computer Conference, held in Hong Kong in 1980; the Chengde conference in September 1982; a pair of conferences in Beijing dedicated to Chinese character encoding (in July and December 1982), a conference in December 1982 in Nanjing dedicated specifically to Chinese-character peripherals (including printers, monitors, and keyboards), the Chinese Information Research Association Chinese Character Encoding Professionals Committee meeting in Chengdu in March 1983, the International Conference on Chinese Information Processing in Beijing during October 1983, the First International Symposium on Computers and Chinese Input/Output Systems; Miconext 83 in April 1983; the Chinese Information Processing Society of China in 1983, the ICCIP in 1983; and the International Conference on Chinese Computing, held in 1986 by the Chinese Computing Society. See Jim Mathias and Thomas L. Kennedy, eds., *Computers, Language Reform, and Lexicography in China*, 36; Zhang Shoudong et al., *Fundamentals of Chinese Language Computing*, 4; Chen Shu, "Symposium on Chinese Character Processing Systems Reviewed," 20; "Brief Note on 1986 International Conference on Chinese Computing," 44, 62. The Chengde conference was held September 12–19, 1982; Zimin Wu and J. D. White, "Computer Processing of Chinese Characters," 681; Chen Shukai [陈树楷] and Jiang Decun [姜德存]. "Overview of the Activities of the Chinese Information Processing Society of China in 1982" (Zhongguo Zhongwen xinxi yanjiuhui 1982 nian huodong zongshu) [中国中文信息研究会1982年活动综述], in *1983 Spring Festival Symposium Proceedings* (*1983 chunjie zuotanhui wenji*), ed. Zhongguo Zhongwen xinxi yanjiuhui, 105–110, 109 (n.p.: Beijing, February 1983).

26. According to one estimate, by Chen Shukai, six computer-related events took place in China in 1979, with a total of 798 essays produced, and with some 1,169 people in attendance. By 1982, that number grew to twenty events, 1,358 essays, and 3,233 attendees. See Chen Shukai [陈树楷], "Review of Academic Activities in 1982" (*1982 niandu xueshu huodong shuping*) [一九八二年度学术活动述评]," in *1983 Spring Festival Symposium Proceedings*, ed. Zhongguo Zhongwen xinxi yanjiuhui, 142. Examples of major trade shows include the Data Show (1981), the Chinese Computer Exhibition (1986), Software Exhibition '86, Software Exhibition '87, China Software Show (1987), the Hong Kong Productivity Council Software Exhibition (1988), and Software Exhibition '89, held at Hong Kong Exhibition Centre, among many other possible examples. See "Asia Computer Plaza," *South China Morning Post*,

February 22, 1986, 21; "Software Exhibition '87," *South China Morning Post.* May 25, 1987, 47; "Software Exhibition '88," *South China Morning Post*, August 22, 1988, 7; and "Software Exhibition '89," *South China Morning Post*, October 10, 1989, 28.

27. Sino-Japanese exchange has been critical to the history of modern Chinese information technology from the 1800s onward. Any analysis which fails to consider it necessarily falls short of understanding Chinese in the era of modern IT. For an analysis of Sino-Japanese exchange in the precomputing era, see Thomas S. Mullaney, *The Chinese Typewriter: A History* (Cambridge, MA, 2017); and Thomas S. Mullaney, "Controlling the Kanjisphere: The Rise of the Sino-Japanese Typewriter and the Birth of CJK," *Journal of Asian Studies* 75, no. 3 (August 2016): 725–753.

28. Ichiko Morita, "Japanese Character Input: Its State and Problems," *Journal of Library Automation* 14, no. 1 (March 1981): 6–23; Ken Lunde, *CJKV Information Processing*, 228–229; Nanette Gottlieb, *Word-Processing Technology in Japan: Kanji and the Keyboard* (Richmond, VA: Curzon, 2000). For an introduction to critical early work on Japanese word processing, see Joseph D. Becker, "Multilingual Word Processing," *Scientific American* 251, no. 1 (July 1984): 96–107.

29. Duncan B. Hunter, "Chinese Computer that's Got Character," *South China Morning Post*, January 17, 1986, 30.

30. A rudimentary but powerful form of predictive text became fundamental to mechanical Chinese typewriting in the 1950s, even before the advent of computing (a subject examined at length in Mullaney, *The Chinese Typewriter*).

31. Liu Yongquan [刘涌泉], "A Discussion of Problems Related to Character Compound Repositories," 8–11.

32. Qin Dulie, "Principles of Developing Expert Systems of Traditional Chinese Medicine," in *Expert Systems and Decision Support in Medicine*, ed. Otto Rienhoff, Ursula Piccolo, and Berthold Schneider (Berlin: Springer, 1988), 85–93, 92.

33. Zhang Guofang [张国防], *Fifty Character Element Computational Chinese Input Usage Manual (50 ziyuan jisuan Hanzi shurufa shiyong shouce)* [五十字元计算汉字输入法使用手册] (Beijing: Zhongguo jiliang chubanshe, 1989), 238.

34. Qi Yuan, "Chinese Information Technology and Processing Natural Languages" (Zhongwen xinxi jishu he ziran yuyan chuli) [中文信息技术和自然语言处理], *Journal of Chinese Information Processing* 1, no. 1 (1986): 33–36.

35. Zhang Guofang, *Fifty Character Element Computational Chinese Input Usage Manual*, 238. Translation from original (目前，现代汉语的实际应用输入速度，已经明显地超过了西文的速度。将来，字、词处理再经过深一度地研究，建立起更高水平或更理想的词语库，绝对有希望在从容击键的状态下，达到或大大超过口述的语率速度。).

36. John DeFrancis, *Visible Speech: The Diverse Oneness of Writing Systems* (Honolulu: University of Hawai'i Press, 1989), 266–267. William Hannas expressed a far sharper view years later. See William C. Hannas, *The Writing on the Wall: How Asian Orthography Curbs Creativity* (Philadelphia: University of Pennsylvania Press, 2003).

37. Zhonghua renmin gongheguo guojia biaozhun, "General Word Set for Chinese Character Keyboard Input" (Hanzi jianpan shuru yong tongyong ciyu ji) [汉字键盘

输入用通用词语集], GB/T 15732–1995. Circulated August 1, 1995, adopted April 1, 1996. For earlier efforts on standardized digital character sets, see also Liu Yongquan, "A Discussion of Problems Related to Character Compound Repositories" (Tantan ciku wenti) [谈谈词库问题], *Journal of Chinese Information Processing* 1, no. 1 (1986): 8–11.

38. The three groups included the Chinese Character Encoding Working Group of the Chinese Information Society (*Zhongguo Zhongwen xinxi xuehui*), the Chinese Standardization and Information Classification and Encoding Research Institute (*Zhongguo biaozhunhua yu xinxi fenlei bianma yanjiusuo*), and the Chinese Standards Technology Development Company (*Zhongguo biaozhun jishu kaifa gongsi*).

39. Chen Xiangwen, *The Five-Stroke and Natural Code Input Methods for Microcomputers*, 110–112.

40. Pei Jie, *CC-DOS V4.2 Chinese Character Operating System User Guide*, 140–141. Users also had access to a program LXCK.exe, called the "Associative Character Repository Maintenance Program" [联想词库维护程序]. See also *UCDOS High-Level Chinese Character System User Manual v. 2.0* (*UCDOS gaoji Hanzi xitong yonghu shouce v. 2.0*) [UCDOS高级汉字系统用户手册 v 2.0]. Beijing: Zhongguo kexueyuan xiwang gaoji dianna jishu gongsi, 1990, 6, 38.

41. *UCDOS High-Level Chinese Character System User Manual*, 37.

42. China Great Wall Computer Group (Zhongguo changcheng jisuanji jituan gongsi) [中国长城计算机集团公司], *Great Wall 95 DOS Chinese System Great Wall Tianhui Standard Chinese Character System User's Manual* (*Changchang 95 DOS Zhongwen xitong changcheng tianhui biaozhun Hanzi xitong yonghu shouce*) [长城95DOS中文系统长城天汇标准汉字系统用户手册], n.d., 9.4; *Tianhui ABC Chinese Input Method* (*Tianhui ABC Hanzi shurufa*) [天汇ABC汉字输入法], TSM.

43. See Matt Scott [Mate Sikete 马特·斯科特], "The Background Technology and Story of Microsoft's Engkoo Pinyin Input Method" (Weiruan yingku pinyin shurufa beihou de jishu he gushi) [微软英库拼音输入法背后的技术和故事], http://tech.sina .com.cn/it/csj/2013-01-25/08418014659.shtml; and "Sogou Cloud Input: An Introduction to Cloud Computing" (Sougou yunrufa: yunjisuan jieshao) [搜狗输入法云计算介绍], https://pinyin.sogou.com/features/cloud/.

44. For an introduction to the darker potentials of "cloud input" see Thomas S. Mullaney. "How to Spy on 600 Million People: The Hidden Vulnerabilities in Chinese Information Technology," *Foreign Affairs* (June 5, 2016), https://www.foreignaffairs .com/articles/china/2016-06-05/how-spy-600-million-people.

45. See David Moser, *A Billion Voices*; Gina Anne Tam, *Dialect and Nationalism*; and Elisabeth Kaske, *The Politics of Language in Chinese Education, 1895–1919* (Leiden: Brill, 2008).

CONCLUSION

1. As of December 2022 this search engine can be found here: https://pss-system .cponline.cnipa.gov.cn/conventionalSearch.

2. Email from Jiang Wei [蒋薇], the granddaughter of inventor Jiang Kun [蒋琨], to the author, September 15, 2013.

3. "Sound Code Input System" (Yinma shurufa) [音码输入法], PRC Patent Document CN 1277379A, December 20, 2000; meeting with Jiang Kun [蒋琨] and Jiang Wei [蒋薇], Shanghai, September 17, 2013.

4. "A Girl from Shanghai and Her Father Create a 'New Character Retrieval Method': To Look Up Characters One Need Only Rely on Character Radicals" (Shanghai guniang yu qi fu chuang 'jianzi xinfa' chazi zhi xu yiju bushoujian) [上海姑娘与其父创检字新法查字只需依据部首件], *Dongfanwang*, October 20, 2013.

5. Michael Adas, *Machines as the Measure of Men: Science, Technology, and Ideologies of Western Dominance* (Ithaca, NY: Cornell University Press, 1989); Jennifer Karns Alexander, *The Mantra of Efficiency From Waterwheel to Social Control* (Baltimore, MD: Johns Hopkins University Press, 2008).

6. Cao Xuenan, "Bullet Screens (Danmu): Texting, Online Streaming, and the Spectacle of Social Inequality on Chinese Social Networks," *Theory, Culture and Society* 38, no. 3 (2019): 29–49; Yongmin Zhou, *Historicizing Online Politics: Telegraphy, the Internet, and Political Participation in China* (Stanford, CA: Stanford University Press, 2005); Weiyu Zhang, *The Internet and New Social Formation in China: Fan Publics in the Making* (New York: Routledge, 2016); Guobin Yang, *The Power of the Internet in China: Citizen Activism Online* (New York: Colombia University Press, 2011); Yizhou Xu, "The Postmodern Aesthetic of Chinese Online Comment Cultures," *Communication and the Public* 1, no. 4 (2016): 436–451; Xinyuan Wang, *Social Media in Industrial China* (London: UCL Press, 2016); Li Jinying, "The Interface Affect of a Contact Zone: Danmaku on Video Streaming Platforms," *Asiascape: Digital Asia* 4, no. 3 (2017): 233–256. Having tackled many of the challenges of Chinese character information processing, indeed, some Chinese engineers have turned their attention to information processing in the writing systems of Chinese ethnic minorities. See Dai Qingxia [戴庆夏], Xu Shouchun [许寿椿], and Gao Xikui [高喜奎], *Chinese Ethnic Minority Languages and Computer Information Processing* (*Zhongguo ge minzu wenzi yu diannao xinxi chuli*) [中国各民族文字与电脑信息处理] (Zhongyan minxue xueyuan chubanshe, 1991).

7. Michel Hockx, *Internet Literature in China* (New York: Columbia University Press, 2015).

8. One potential we can rule out immediately is the idea that Chinese engineers and computer users in some way value or pride themselves in this wide assortment of input methods. From the 1980s onward, the proliferation of input methods has never been a point of pride for Chinese engineers, entrepreneurs, government officials, or everyday users. A "unification of writing for computers" (*diannao shu tongwen*) is what one writer called for in the 1980s, harkening back to the very beginnings of Chinese imperial history, when Legalist thinkers in the Qin dynasty called for the standardization of Chinese writing amidst a jumble of variant forms in different kingdoms and regions. Another observer referred to the situation circa 1986 as "Code Pollution" (*bianma wuran*). Nothing stemmed the tide of new input

systems, however. Between 1985 and 1990, a survey revealed more than 500 different Chinese input methods, no fewer than one hundred of which were in actual use. "Code pollution" showed no signs of abating. See Qian Yuzhi [钱玉趾], "Prospects for the Development of a 'Unified Script for Computers'" (*'Diannao shu tongwen' de fazhan qianying*) [电脑书同文的发展前景], *Zhongwen xinxi* 1 (1992): 6–9; Pan Defu and Zhan Zhenquan, *Chinese Character Encoding Design* (*Hanzi bianma shejixue*), ca. 1997, TCM, 26; Zimin Wu and J. D. White, "Computer Processing of Chinese Characters: An Overview of Two Decades' Research and Development," *Information Processing and Management* 26, no. 5 (1990): 681–692, 684.

9. Theodore Huters, *Bringing the World Home: Appropriating the West in Late Qing and Early Republican China* (Honolulu: University of Hawai'i Press, 2005).

10. For the history of natural-language Chinese typewriter tray beds and the development of predictive text technologies in Mao-era China, see Thomas S. Mullaney, *The Chinese Typewriter: A History* (Cambridge, MA: MIT Press, 2007); and Thomas S. Mullaney, "The Movable Typewriter: How Chinese Typists Developed Predictive Text during the Height of Maoism," *Technology and Culture* 53, no. 4 (October 2012): 777–814.

11. For two critical works on feed-forward and generative anticipatory technologies, see Mark B. N. Hansen, *Feed-Forward: On the Future of Twenty-First-Century Media* (Chicago: University of Chicago Press, 2014); and Shane Denson, *Discorrelated Images* (Durham, NC: Duke University Press, 2020).

12. "Robot Writes *LA Times* Earthquake Breaking News Article," *BBC.com*, March 18, 2014, https://www.bbc.com/news/technology-26614051; Jaclyn Peiser, "The Rise of the Robot Reporter," *New York Times*, February 5, 2019; "The First Article Written by an Artificial Intelligence in the Guardian," *Web24.news*, September 15, 2020. Google recently unveiled functionality that enables you, instead of physically typing a response to your interlocutor, to select from a menu of options that have presumably been custom-designed for you by Google's machine reading algorithm.

13. Elizabeth Eisenstein, *The Printing Press as an Agent of Change: Communications and Cultural Transformations in Early-Modern Europe* (Cambridge: Cambridge University Press, 1980).

14. On "butt signatures," see Brigid O'Connell, "New 'Smart Chair' Designed to Improve Posture and Protect Back Health," *The West Australian*, perthnow.com, June 16, 2015, https://www.perthnow.com.au/news/new-smart-chair-designed-to -improve-posture-and-protect-back-health-ng-1233bbbe13e3f6536fca4e13445f4d6a. On urine stream signatures, see Cecily Mauran, "Pee and Me: The Gadget That Turns Toilets into Urine Labs. Why 2023 Could Be the Year of Personal Pee-Testing," *Medium.com*, January 3, 2023, https://mashable.com/article/withings-u-scan-urine -analysis-ces-2023. For the classic account on the "Gutenberg Galaxy," see Marshall McLuhan, *The Gutenberg Galaxy: The Making of Typographic Man* (Toronto: The University of Toronto Press, 1962).

15. I say this, not as a critique of the systems mentioned, but as a broader observation regarding the overall marginality of nonconventional text input systems for

English. It is precisely the brilliance of many of these systems—such as ShapeWriter by Shumin Zhai—that make their nonusage by everyday Anglophone computer users so relevant for our discussion. This is not to suggest, furthermore, that there has been a paucity of research into alternate interfaces, haptics, and more. The question here is not one of experimentation, prototyping, and research, but of a deeply embedded, normative assumption. For the definitive study on haptics, see David Parisi, *Archaeologies of Touch: Interfacing with Haptics from Electricity to Computing* (Minneapolis: University of Minnesota Press, 2018). For an earlier work on text entry systems on mobile devices, see I. Scott MacKenzie and Kumiko Tanaka-Ishii, *Text Entry Systems Mobility, Accessibility, Universality* (Amsterdam: Morgan Kaufman, 2007).

16. Special thanks to my former undergraduate advisee Chuan Xu who produced an exquisite term paper on Engelbart's Chorded Keyboard based on firsthand archival work in the Douglas C. Engelbart Papers housed in Stanford's Special Collections and University Archives. Although unpublished, his paper inspired and shaped this passage.

17. Consider this: If English-speaking computer users truly wanted to engage in hypographic input, they could do so right now, without any need to install dedicated IMEs. "Spell check"—a now ubiquitous feature preinstalled on all commercially available word processing applications—is all that's necessary to get started. Whether by underlining erroneous spellings in red, highlighting them, generating pop-up menus that contain suggested corrections, or automatically changing a text string into what the system has determined to be the most likely "correct" form, spell checking is a widely accepted and used part of word processing, and has been for over a decade. Why, then, don't Anglophone computer users harness this technology by spelling *incorrectly* or *efficiently*? If you want to type the word "better," why insist on typing B-E-T-T-E-R, when B-T-T-R or B-T-R would achieve the same result (i.e., by letting the program do its job, and then selecting the correct spelling)? If you are trying to type the word "medium," why not enter the string M-D-I-M, M-D-U-M, M-E-D-M, or even M-D-M? Some of these input strings would result in more than one potential graph, of course (M-D-M might elicit recommendations, not just for "medium," but also "madam"), but in the long run the added step of interacting with the spell checker pop-up menu would be compensated for by all the keystrokes you didn't have to depress. For most English speakers, such a proposition will seem like an absurd waste of time at best, and pathological at worst.

18. Paul A. David, "Clio and the Economics of QWERTY," *American Economic Review* 75, no. 2 (1985): 332–337.

19. Jared Diamond, "The Curse of QWERTY: O Typewriter? Quit Your Torture!" *Discover Magazine*, April 1997.

20. A question that surfaces regularly when speaking about my research is: *What about voice recognition?* Might this constitute the long-desired moment of emancipation? *Isn't this the answer to overcoming QWERTY?* No. Although voice-to-text technology is undoubtedly an advancement, especially in the context of mobile and "hands-free" computing, the underlying premise of voice-to-text reconstitutes the

existing technolinguistic order, simply by other means. Instead of the QWERTY-based expectation of what-you-type-is-what-you-get—of the assumption that the act of writing is fundamentally an act of "spelling" things out, piece by piece—we would simply have a speech-based one: namely, *what-you-say-is-what-you-get*. Instead of text composition being an act of "spelling" words out in full on a keyboard-based interface, it would involve "sounding them out" in full through speech. However you cut it, it would be the same technolinguistic prison, only with stronger bars (stronger, because we would be tempted to believe falsely that we are "free"). What is more, as vividly illustrated in work by the late and beloved Halcyon Lawrence, voice-to-text programs exert novel forms of discipline on their users along lines of dialect. See Halcyon M. Lawrence, "Siri Disciplines," in *Your Computer Is on Fire*, ed. Thomas S. Mullaney, Benjamin Peters, Mar Hicks, and Kavita Philip (Cambridge, MA: MIT Press, 2021), 179–198.

21. Eleanor Smith called QWERTY the "culprit" that prevents typists from exceeding the average speed of human speech (120 words per minute) with most of us "crawling" at 14 to 31 wpm. Robert Winder, meanwhile, called the QWERTY keyboard a "conspiracy." Robert Winder, "The Qwerty Conspiracy." *The Independent*, August 12, 1995; Eleanor Smith, "Life after QWERTY," *The Atlantic*, November 2013.

22. To suggest that the escape from QWERTY has happened, or that it is underway, is decidedly *not* the same as suggesting that the non-Western world has in some way become "liberated" from the profoundly unequal global information infrastructure we discussed at the outset. Even today, connected Arabic text cannot be achieved on some Adobe programs, for example, a remarkable fact when considering the kinds of advances made by this company in the arena of personal computing, as well as the fact that more than 1 billion people use one form or another of Arabic/Arabic-derived writing system. See, for example, Andrea Stanton, "Broken is Word," in *Your Computer Is on Fire*, ed. Thomas S. Mullaney, Benjamin Peters, Mar Hicks, and Kavita Philip (Cambridge, MA: MIT Press, 2021), 213–230. The argument I am trying to make is that contemporary IT has entered into a new phase in which Western Latin alphabet hegemony continues to exert material, and in many ways conceptual control over the way we understand and practice information, but that *within* this condition, new spaces of accommodation and even resistance have been fashioned.

BIBLIOGRAPHY

24-Pin Chinese Printer User Manual (*24 zhen Hanzi dayinji yonghu shouce*) [24针汉字打印机用户手册]. n.d, n.p. Thomas S. Mullaney East Asian Information Technology History Collection. Stanford University.

Abbate, Janet. *Recoding Gender: Women's Changing Participation in Computing.* Cambridge, MA: MIT Press, 2017.

Adal, Raja. "The Flower of the Office: The Social Life of the Japanese Typewriter in Its First Decade." Presentation at the Association for Asian Studies Annual Meeting, March 31–April 3, 2011.

Adas, Michael. *Machines as the Measure of Men: Science, Technology, and Ideologies of Western Dominance.* Ithaca: Cornell University Press, 1989.

Adler, Michael H. *The Writing Machine: A History of the Typewriter.* London: Allen and Unwin, 1973.

Alexander, Jennifer Karns. *The Mantra of Efficiency From Waterwheel to Social Control* (Baltimore, MD: Johns Hopkins University Press, 2008).

Allen, Joseph R. "I Will Speak, Therefore, of a Graph: A Chinese Metalanguage." *Language in Society* 21, no. 2 (June 1992): 189–206.

"An Wang Oral History." Interviewed by Richard R. Mertz. October 29, 1970. Computer Oral History Collection, 1969–1973, 1977. Archives Center, National Museum of American History.

Andreas, Joel. *Rise of the Red Engineers: The Cultural Revolution and the Origins of China's New Class.* Stanford, CA: Stanford University Press, 2009.

Androutsopoulos, Jannis. "Introduction: Sociolinguistics and Computer-Mediated Communication." *Journal of Sociolinguistics* 10, no. 4 (2006): 419–438.

Ann, T. K. [安子介]. *Chinese Character List A: Ann's System of Coding Chinese Characters.* Hong Kong: Stockflows Co., Inc., 1985.

Apple, R. W. "Two Britons Devise a Computer That Can Communicate in Chinese." *New York Times*, January 25, 1978.

"Asia Computer Plaza." *South China Morning Post*, February 22, 1986, 21.

Baark, Erik. *Lightning Wires: The Telegraph and China's Technological Modernization, 1860–1890*. Westport, CT: Greenwood Press, 1997.

Bachrach, Susan. *Dames Employées: The Feminization of Postal Work in Nineteenth-Century France*. London: Routledge, 1984.

Bao Yueqiao [鲍岳桥], Gan Dengdai [甘登岱], and Liu Qinghua [刘庆华]. *Hope Soft UCDOS 3.1 Training Manual* (*Hope Soft UCDOS 3.1 peixun jiaocheng*) [Hope Soft UCDOS 3.1 培训教程学]. Beijing: Xuefan chubanshe, 1994.

Baum, Richard. "DOS ex Machina: The Microelectric Ghost in China's Moderniza-tion Machine." In *Science and Technology in Post-Mao China*. Ed. Denis Fred Simon and Merle Goldman, 347–374. Cambridge, MA: Harvard University Asia Center, 1988.

Becker, Joseph D. "Multilingual Word Processing." *Scientific American* 251, no. 1 (July 1984): 96–107.

Beeching, Wilfred A. *Century of the Typewriter*. New York: St. Martin's Press, 1974.

Beijing Institute of Electronics Electronic Computing Group (Beijing shi dianzi xuehui dianzi jisuanji zhuanye zu) [北京市电子学会电子计算机专业组]. "Announce-ment Regarding the Attendees Name List for the 1964 Beijing Institute of Electronics Electronic Computing Professionals Academic Conference" (Guanyu 1964 nian Bei-jing shi dianzi xuehui dianzi jisuanji zhuanye xueshuhui daibiao ming'e de tongzhi) [关于1964年北京市电子学会电子计算机专业学术会代表名额的通知]. Beijing Municipal Archives (BMA) 010-002-00431 (April 17, 1964).

Beijing Wireless Factory No. 3 (Beijing wuxiandian san chang) [北京無限電三廠]. "Small-Scale General Purpose Transistorized Digital Computer Design Task Report" (Xiaoxing tongyong jingti guan shuzi jisuanji sheji renwu shu) [小型通用晶體管數字計算機設計任務書]." Beijing Municipal Archives (BMA) 165-001-00130 (c. 1965): 11–14.

Berlin, Leslie. *Troublemakers: Silicon Valley's Coming of Age*. New York: Simon and Schuster, 2018.

"Bill Garth's Narrative of Demonstrations of Chinese Text Processing Computers in Beijing and Shanghai in 1979 and Negotiations in Beijing in 1980." Prepared by William Garth IV for the author and sent via email on April 17, 2014.

Bloom, Alfred H. "The Impact of Chinese Linguistic Structure on Cognitive Style." *Current Anthropology* 20, no. 3 (1979): 585–601.

Bloom, Alfred H. *The Linguistic Shaping of Thought: A Study in the Impact of Language on Thinking in China and the West*. Hillsdale, NJ: L. Erlbaum, 1981.

Bolter, Jay David, and Richard Grusin. *Remediation: Understanding New Media*. Cam-bridge, MA: MIT Press, 1998.

"Boon to China: Typewriter Has 5,400 Symbols." *Herald Tribune*, July 1, 1946.

Bray, Francesca. "Gender and Technology." *Annual Review of Anthropology* 36 (2007): 37–53.

"A Brief History of the Development of Chinese Computer (1956–2006)." http://www.nobelkepu.org.cn/art/2008/11/13/art_1025_66163.html. Accessed May 9, 2015.

"Brief Note on 1986 International Conference on Chinese Computing" (1986 *guoji Zhongguo jisuanji huiyi jiankuang*) [1986国际中国计算机会议简况]. *Journal of Chinese Information Processing* 1, no. 1 (1986): 44, 62.

Buhler, Eugen, and Christopher A. Berry. "Machine Adapted for Typing Chinese Ideographs." US Patent 2458339. Filed May 3, 1946, and issued January 4, 1949.

Burns, John F. "China's Passion for the Computer." *New York Times*, January 6, 1985, 3.

Cai Rongbo [蔡荣波], ed. *Computational Chinese Input Methods: Methods, Skills and Training* (Diannao Hanzi shuru de fangfa jiqiao yu xunlian) [电脑汉字输入的方法技巧与训练]. Guangzhou: Guangdong keji chubanshe, 1993.

Caldwell, Samuel H. "Final Report on Studies Leading to Chinese and Devanagari." (December 21, 1954). Louis Rosenblum Collection. Stanford University.

Caldwell, Samuel H. Ideographic type composing machine. US Patent no. 2950800. Filed October 24, 1956, and issued August 30, 1960.

Caldwell, Samuel H. "Progress on the Chinese Studies," 1–6. In "Second Interim Report on Studies Leading to Specifications for Equipment for the Economical Composition of Chinese and Devanagari." Report by the Graphic Arts Research Foundation, Inc. Addressed to the Trustees and Officers of the Carnegie Corporation of New York. Pardee Lowe Papers. Hoover Institute Archives. Accession No. 98055–16.370/376. Box No. 276.

Caldwell, Samuel H. *Switching Circuits and Logical Design*. New York: Wiley, 1958.

Caldwell, Samuel H. "The Sinotype: A Machine for the Composition of Chinese from a Keyboard." *Journal of The Franklin Institute* 267, no. 6 (June 1959): 471–502.

Caldwell, Samuel H., and W. W. Garth Jr. "Proposal for Studies Leading to Specifications for Equipment for the Economical Composition of Chinese and Devanagari." Marked "Confidential." Graphic Arts Research Foundation. Cambridge, MA. March 25, 1953. Louis Rosenblum Collection. Stanford University.

Cameron, Deborah. *Verbal Hygiene*. London: Routledge, 1995.

Campbell-Kelly, Martin. "The History of the History of Software." *IEEE Annals of the History of Computing* (December 18, 2007): 40–51.

Canagarajah, Suresh. "Codemeshing in Academic Writing: Identifying Teachable Strategies of Translanguaging." *The Modern Language Journal* 95, no. 3 (2011): 401–417.

Cao Xuenan. "Bullet Screens (Danmu): Texting, Online Streaming, and the Spectacle of Social Inequality on Chinese Social Networks." *Theory, Culture and Society* 38, no. 3 (2019): 29–49.

Carrington, Victoria. "Txting: The End of Civilization (Again)?" *Cambridge Journal of Education* 35 (2004): 161–175.

Central Intelligence Agency Directorate of Intelligence. "China: Progress in Computers." Intelligence Memorandum. December 1972, 7.

Ceruzzi, Paul E. *A History of Modern Computing*. Cambridge, MA: MIT Press, 2003 [1998], chapter 7.

"Chan H. Yeh A Brief Biography." Personal collection of author, provided by Yeh.

Cheatham, Jr., Thomas E. Wesley A. Clark, Anatoly W. Holt, Severo M. Ornstein, Alan J. Perlis, and Herbert A. Simon. "Computing in China: A Travel Report." *Science* (October 12, 1973): 134–140.

Chen, David. "The Race is On to Design New Chinese System." *South China Morning Post*, September 1, 1987, 25.

Chen Jianwen [陈建文]. *Research Report on the Double Pinyin Code Chinese Character Information Processing System (Shuangpin bianma Hanzi xinxi chuli xitong yanzhi baogao)* [双拼编码汉字信息处理系统研制报告]. Nanjing: Nanjing University Library and the Suzhou University Mathematics Department, ca. 1986.

Chen Jianwen. *The Double Pinyin Chinese Character Encoding Method: Purpose and Usage (Hanyu shuangpin Hanzi Bianma de yongtu he yongfa)* [汉语双拼编码汉字的用途和用法]. Nanjing: Nanjing daxue tushuguan, 1986.

Chen Jianwen, Xue Shiquan [薛士权], and Qian Peide [钱培德]. *Double Pinyin Encoding Chinese Character Information Processing System Development Report (Shuangpin bianma Hanzi xinxi chuli xitong yanzhi baogao)* [双拼编码汉字信息处理系统研制报告]. Nanjing and Suzhou: Nanjing daxue tushuguan and Suzhou daxue jisuanji jiaoyanshi, ca. May 1986.

Chen, Joyce. *Joyce Chen Cook Book*. Philadelphia: J. B. Lippincott Company, 1962.

Chen Liwei [陈力为]. "Develop Chinese Information Technology, Promote Computer Use in More Situations" [发展中文信息技术, 促进计算机推广应用]. *Journal of Chinese Information Processing* 1, no.1 (1986): 5–7.

Chen Qixiu [陈启秀]. *Chinese Character Input and Word Processing: 100 Questions (Hanzi shuru yu wenzi chuli 100 wen)* [汉字输入与文字处理100问]. Nanjing: Jiangsu kexue jishu chubanshe [江苏科学技术出版社], 1996.

Chen Shu. "Symposium on Chinese Character Processing Systems Reviewed." FBIS report based on *Jisuanji shijie* 20 (October 1982): 13. In *FBIS* 83, 20–22.

Chen Shukai [陈树楷]. "Review of Academic Activities in 1982 (*1982 niandu xueshu huodong shuping*) [一九八二年度学术活动述评]. In *1983 Spring Festival Symposium*

Proceedings (*1983 chunjie zuotanhui wenji*) [1983春节座谈会文集]. Ed. Zhongguo Zhongwen xinxi yanjiuhui. n.p.: Beijing, February 1983.

Chen Shukai and Jiang Decun [姜德存]. "Overview of the Activities of the Chinese Information Processing Society of China in 1982" (Zhongguo Zhongwen xinxi yanjiuhui 1982 nian huodong zongshu) [中国中文信息研究会1982年活动综述]. In *1983 Spring Festival Symposium Proceedings* (*1983 chunjie zuotanhui wenji*) [1983春节座谈会文集]. Ed. Zhongguo Zhongwen xinxi yanjiuhui, 105–110, 109. n.p.: Beijing, February 1983.

Chen Xiangwen [陈相文], ed. *The Five-Stroke and Natural Code Chinese Character Input Methods on a Microcomputer* (*Weixing jisuanji wubi zixing ji ziranma Hanzi shurufa*) [微型计算机五笔字型及自然码汉字输入法]. Tianjin: Nankai daxue chubanshe, 1994.

Chen Yaoxing [陈耀星]. "Introduction to 'Chinese Codes for Information Interchange'" ('Xinxi jiaohuan yong Hanzi bianma zifu ji' jianjie) ['信息交换用汉字编码字符集'简介]. *Wenzi gaige* [*Writing Reform*] 4 (1983): 5–7.

Chen Zengwu [陈增武] and Jin Lianfu [金连甫]. *Chinese Language Information Processing System* (*Zhongwen xinxi chuli xitong*) [中文信息处理系统]. Beijing: Zhongguo jisuanji yonghu xiehui zonghui, 1984.

Chiang, Yee. *Chinese Calligraphy: An Introduction to Its Aesthetics and Techniques*. Cambridge, MA: Harvard University Press, 1973 [1938].

Chin, Francis. "IBM Shows Chinese PC at Information Week." *Asia Computer Weekly*, December 23, 1983, 24.

China Great Wall Computer Group (Zhongguo changcheng jisuanji jituan gongsi) [中国长城计算机集团公司]. *Great Wall 95 DOS Chinese System Great Wall Tianhui Standard Chinese Character System User's Manual* (*Changchang 95 DOS Zhongwen xitong changcheng tianhui biaozhun Hanzi xitong yonghu shouce*) [长城95DOS中文系统长城天汇标准汉字系统用户手册]. Zhongguo changcheng jisuanji jituan gongsi, n.d.

"China Will Apply for International Registration of Chinese Graphic Character Set for Computers." *Xinhua General Overseas News Service*, March 27, 1981.

Chinese Character Encoding Research Group of China (*Zhongguo Hanzi bianma yanjiuhui*), ed. *Compendium of Proposals for Chinese Character Encodings* (*Hanzi bianma fang'an huibian*) [汉字编码方案汇编]. Shanghai: Kexue jishu wenxuan chubanshe, n.d. (ca. 1979).

"Chinese Characters Have Entered the Computer" (*Hanzi jinru le jisuanji*) [汉子进入了计算机]. *Wenhuibao* (July 19, 1978): 1, 3. Housed in Graphic Arts Research Foundation (October 1976), Box "Oct 94 Sinotype '81," Folder "Sinotype Vol VI Wang Pinyin Sequences." Graphic Arts Research Foundation Materials. Museum of Printing. North Andover, MA.

"Chinese Computer." *South China Morning Post*, December 7, 1983, 33.

"Chinese Computer, 1981–1982." Wang Laboratories Corporate Papers. Harvard University Business School Special Collections. Subseries IB2 ("Horace Tsiang records, 1974–1992"). Box 97, Folder 7.

Chinese Computer (*Zhongwen diannao*) [中文电脑]. Inaugural issue, no. 1 (January 1, 1986).

"Chinese Computer Launched." *South China Morning Post*, July 17, 1984, 27.

"Chinese Computer on Way." *South China Morning Post*, August 21, 1975, 32.

"Chinese Computers: Dr. Yeh Chen-hui Won the Day in London." *Kung Sheung Evening News* (September 11, 1978). Clipping located in SCC2/24/1. Needham Research Institute.

"Chinese Divisible Type." *Chinese Repository* 14 (March 1845): 124–129.

"Chinese Engineers Meet." *New York Times*, June 30, 1946, 9.

"Chinese Ideographic VS Programmers Manual." Box 136, Folder 4. Wang Laboratories Corporate Papers. Harvard University Business School Special Collections.

Chinese Information Processing (*Hanzi xinxi chuli*) [汉字信息处理]. Beijing: Zhongguo shehui kexue chubanshe, 1979.

"Chinese Language Photocomposition Machine." May 4, 1959. Pardee Lowe Papers. Hoover Institute Archives. Accession No. 98055–16.370/376. Box No. 193.

"Chinese Machine, 1981–1982." Wang Laboratories Corporate Papers. Box 97, Folder 8. Harvard University Business School Special Collections.

"Chinese Progress in the Production of Integrated Circuits." Report by the CIA Directorate of Intelligence (March 12, 1985): 2.

"Chinese Romanized—Keyboard no. 141." Hagley Museum and Library. Accession no. 1825. Remington Rand Corporation. Records of the Advertising and Sales Promotion Department. Series I Typewriter Div. Subseries B, Remington Typewriter Company, box 3, vol. 1.

"Chinese Science and the Requirements of Economic Development: Chung-Chin Kao Creates a Chinese Typewriter" (*Zhongguo kexue yu jingji jianshe yaoxun: Gao Zhongqin chuangzhi Huawen daziji*) [中國科學與經濟建設要訊:高仲芹創制華文打字機]." *Minzhu yu kexue* [民主與科學] 1, no. 4 (1945): 64.

"Chinese Typewriter Inventor Chung-Chin Kao Brings His Invention to San Francisco Recently, Attends American Business Stationery Exhibition; Kao Accompanied by Demonstrators Chen Rujin and Liu Shulian to Demonstrate Typing on the Machine" (Huawen daziji famingren Gao Zhongqin jun, jin tui qi faming pin fu Sanfanshi canjia 'quan Mei shangye wenju zhanlanhui': dangzhong biaoyan dazi qingxing, sui Gao jun tongwang huichang biaoyanzhe you Chen Rujin, Liu Shulian liang nvshi) [華文打字機發明人高仲芹君,近攜其發明品赴三藩市參加"全美商業文具展覽

會":當眾表演打字情形,隨高君同往會場表演者有陳如金劉淑蓮兩女士]. *Chinese-American Weekly (Zhong-Mei zhoubao)* [中美周報] 230 (1947): 1.

"Chinese Typewriter, Shown to Engineers, Prints 5,400 Characters with Only 36 Keys." *New York Times*, July 1, 1946, 26.

Chu Bong-Foo [朱邦复]. *On Chinese Computing (Zhongwen diannao mantan)* [中文电脑漫谈]. Self-Published: ca. 1982.

Chu, Yaohan. "Structure of a Direct-Execution High-Level Chinese Programming Language Processor." *ACM '74: Proceedings of the 1974 Annual Conference* 1 (January 1974): 19–27.

Clark, Mary Allen. "China Diary." *Washington University Magazine* (Fall 1972), 11. Bernard Becker Medical Library Archives, Washington University School of Medicine, St. Louis, Missouri, https://digitalcommons.wustl.edu/ad_wumag/48/.

The Collected Works of Margaret C. Fung, PhD. Taipei: Showwe Information Corporation, 2004.

Conley, Rita K. "At Plant 3, Rochester." *Business Machines* 27, no. 20 (May 10, 1945): 5.

"Contract No. DA19–129–QM-458 Quarterly Progress Report of June 18, 1958, Report No. 12." Included in letter from R. G. Crockett to T. S. Bonczyk. June 18, 1958.

Cortada, James. *IBM: The Rise and Fall and Reinvention of a Global Icon.* Cambridge, MA: MIT Press, 2019.

CROMEMCO Microcomputer Software Data Compilation 1–6 (CROMEMCO weixing jisuanji ruanjian ziliao huibian 1–6) [CROMEMCO 微型计算机软件资料汇编1–6]. Beijing: Tsinghua University Computing Center, 1980. Thomas S. Mullaney East Asian Information Technology History Collection. Stanford University.

Crystal, David. *A Glossary of Netspeak and Textspeak.* Edinburgh, Edinburgh University Press, 2004.

Crystal, David. *Txtng: The Gr8 Db8.* Oxford: Oxford University Press, 2008.

Dai Qingxia [戴庆夏], Xu Shouchun [许寿椿], and Gao Xikui [高喜奎]. *Chinese Ethnic Minority Languages and Computer Information Processing (Zhongguo ge minzu wenzi yu diannao xinxi chuli)* [中国各民族文字与电脑信息处理]. Zhongyan minzu xueyuan chubanshe, 1991.

Dale, N. B. "An Overview of Computer Science in China: Research Interests and Educational Directions." *ACM SIGCSE Bulletin* 12, no. 1 (February 1980): 186–190.

Danet, Brenda. *Cyberpl@y: Communicating Online.* Oxford: Berg Publishers, 2001.

David, Paul A. "Clio and the Economics of QWERTY." *American Economic Review* 75, no. 2 (1985): 332–337.

Davies, Margery W. *Woman's Place Is at the Typewriter: Office Work and Office Workers 1870–1930.* Philadelphia: Temple University Press, 1982.

Deceased Persons: Shanghai Cultural Yearbook 1994 (Shishi renwu: Shanghai wenhua nianjian 1994) [逝世人物上海文化年鉴 1994], 297–298.

DeFrancis, John. *Nationalism and Language Reform in China*. Princeton, NJ: Princeton University Press, 1950.

DeFrancis, John. *The Chinese Language: Fact and Fantasy*. Honolulu: University of Hawai'i Press, 1984.

DeFrancis, John. *Visible Speech: The Diverse Oneness of Writing Systems*. Honolulu: University of Hawai'i Press, 1989.

Demick, Barbara. "China Worries about Losing its Character(s)." *Los Angeles Times*, July 12, 2010. http://articles.latimes.com/2010/jul/12/world/la-fg-china-characters -20100712 Accessed September 2, 2014.

Denson, Shane. *Discorrelated Images*. Durham, NC: Duke University Press, 2020.

Diamond, Jared. "The Curse of QWERTY: O typewriter? Quit your torture!" *Discover Magazine*, April 1997. https://www.discovermagazine.com/technology/the-curse-of -qwerty.

Dianma xinbian. Shanghai: Shanghai Zhonghua shuju, n.d., ca. 1920. Thomas S. Mullaney East Asian Information Technology History Collection. Stanford University.

Dictionary of Famous Chinese Women (Huaxia funü mingren cidian) [华夏妇女名人词典]. Beijing: Huaxia chubanshe, 1988.

"Difficult Oriental Languages Ready for Computer Technology." *The Telegraph-Herald*, June 4, 1978, 17.

Dikötter, Frank. *The Cultural Revolution: A People's History, 1962–1976*. London: Bloomsbury Press, 2016.

Ding Xilin [丁西林]. *Chinese Stroke-Shape Look-Up System and Encoding System (Hanzi bixing chazifa bianmafa)* [汉字笔形查字法编码法]. Beijing: Beijing shifan daxue Zhongwen xinxi keti zu, 1979.

Double Pinyin Encoding Chinese Character Information Processing System User Report (Shuangpin bianma Hanzi xinxi chuli xitong yonghu baogao) [双拼编码汉字信息处理系统用户报告] Nanjing: Nanjing daxue jisuanjichang (May 26, 1986).

Downey, Greg. "Constructing 'Computer-Compatible' Stenographers: The Transition to Real-Time Transcription in Courtroom Reporting." *Technology and Culture* 47, no. 1 (Jan 2006): 1–26.

Dreaper, James. "Geared for Export: Three Cases Histories," *Design* (1968): 30–39.

Duncan, S., T. Mukaii, and S. Kuno. "A Computer Graphics System for Non-Alphabetic Orthographies," *Computer Studies in the Humanities and Verbal Behavior* (October 1969): 5.2–5.14.

Dyer, Samuel. *A Selection of Three Thousand Characters Being the Most Important in the Chinese Language for the Purpose of Facilitating the Cutting of Punches and Casting Metal Type in Chinese.* Malacca: Anglo-Chinese College, 1834.

"Education and Culture: Chung-Chin Kao Invents an Electric Chinese Typewriter" (Jiaoyu yu wenhua: Gao Zhongqin faming diandong Zhongwen daziji) [教育與文化: 高仲芹發明電動中文打字機]. *Jiaoyu tongxun* [教育通訊] 4, no. 1 (1947): 29–30.

Edwards, Paul N. *The Closed World: Computers and the Politics of Discourse in Cold War America.* Cambridge, MA: MIT Press, 1997.

Eisenstein, Elizabeth. *The Printing Press as an Agent of Change: Communications and Cultural Transformations in Early-Modern Europe.* Cambridge: Cambridge University Press, 1980.

"Electric Chinese Typewriter" (Dian Hua daziji) [電華打字機]. January 17, 1948, 1–10. Tianjin Municipal Archives J66-3-410. Addressed to the Tianjin branch of the China Textile Industries Corporation (*Zhongguo fangzhi jianshe gongsi Tianjin fen gongsi*) [中國紡織建設公司天津分公司].

"Electric Chinese Typewriter" (Diandong Huawen daziji) [電動華文打字機]. *Tiandiren* [天地人] 2, no. 1 (1946): 36.

"Electric Chinese Typewriter" (Diandong Huawen daziji) [電動華文打字機]. *Kexue* 29, no. 12 (1947): 378.

"Electric Chinese Typewriter" (Diandong Huawen daziji) [電動華文打字機]. *Qingnian shiji* [青年世紀] 1, no. 1 (1946): 10.

"Electric Chinese Typewriter" (Diandong Huawen daziji) [電動華文打字機]. *Shizhao yuebao* [時兆月報] 42, no. 3 (1947): 33.

"Electric Chinese Typewriter" (Diandong Huawen daziji) [電動華文打字機]. *Shizheng pinglun* [市政評論] 9, no. 12 (1947): 17.

"Electric Chinese Typewriter" (Diandong Huawen daziji) [電動華文打字機]. *Zhonghua shaonian* [中華少年] 3, no. 11 (1946): 1.

"Electric Chinese Typewriter" (Diandong Zhongwen daziji) [電動中文打字機]. *Shenbao* (July 16, 1946): 3.

"Electric Chinese Typewriter: First Electric Chinese Typewriter Displayed in New York at the National Trade Show, Invented by Chung-Chin Kao, Chinese" (*Diandong Zhongwen daziji: di'yi jia diandong Zhongwen daizji ceng zai Niuyue quanguo shangye zhanlanhui zhong chenlie ke ji wei guoren Gao Zhongqin shi suo faming*) [電動中文打字機:第一架電動中文打字機曾在紐約全國商業展覽會中陳列該機為國人高仲芹氏所發明]. *Xinwen tiandi* [新聞天地] 18 (1946): cover.

"Electromechanics Multilingual Typesetters." *Electronic Age: RCA in 1962 Year-End Report by RCA* (Winter 1962–63): 27–29.

"Electronic Type-Setting Machine for Chinese Invented." November 3, 1962. Pardee Lowe Papers. Accession No. 98055–16.370/376. Box No. 276. Hoover Institute Archives.

Ensmenger, Nathan L. *The Computer Boys Take Over: Computers, Programmers, and the Politics of Technical Expertise*. Cambridge, MA: MIT Press, 2012.

"Faster Chinese." *Time*, July 15, 1946, 86.

"Final Report on Studies Leading to Chinese and Devanagari." December 21, 1954. Graphic Arts Research Foundation. Louis Rosenblum Collection. Stanford University.

"Final Round of the 2013 'National Chinese Characters Typing Competition' Takes Place in the Pingqiao District of Xinyang City" (2013 "Quanguo Hanzi shuru dasai" zongjuesai zi Xinyang Pingqiaoqu juxing) [2013"全国汉字输入大赛"总决赛在信阳平桥区举行]. *Henan xinwen* (broadcast December 12, 2013, 7:11 p.m.).

"The First Article Written by an Artificial Intelligence in the Guardian." *Web24.news*, September 15, 2020.

"First Domestically Produced Microcomputer to Enter the International Market (*Shouci jinru guoji shichang guochan weixing jisuanji*)" [首次进入国际市场国产微型计算机]. *Jiefang ribao* (October 26, 1982): 3.

Freedman, Alisa, Laura Miller, and Christine R. Yano, eds. *Modern Girls on the Go: Gender, Mobility, and Labor in Japan*. Stanford, CA: Stanford University Press, 2013.

Fuller, Matthew. *Behind the Blip: Essays on the Culture of Software*. Brooklyn, NY: Automedia, 2003.

Gaboury, Jacob. "Hidden Surface Problems: On the Digital Image as Material Object." *Journal of Visual Culture* 14, no. 1 (2015): 40–60.

Gaboury, Jacob. "Image Objects: An Archaeology of Computer Graphics, 1965–1979." PhD dissertation, New York University, 2014.

Gabrilovich, Evgeniy and Alex Gontmakher. "The Homograph Attack." December 2001. https://gabrilovich.com/publications/papers/Gabrilovich2002THA.pdf. Accessed August 10, 2022.

Galloway, Alexander R. *The Interface Effect*. Cambridge: Polity, 2012.

Gao Yongzu [高永祖]. "A Newly Invented Chinese Tele-Typewriter (*Xin faming de Zhongwen dianbao daziji*) [新發明的中文電報打字機]." *World Today* (March 16, 1962). Pardee Lowe Papers. Accession No. 98055–16.370/376. Hoover Institute Archives. Box No. 276.

"Gao Zhongqin Invented Automatic Chinese Typewriter" (Gao Zhongqin faming zidong shi Zhongwen daziji) [高仲芹發明自動式中文打字機]. *Xinan shiye tongxun* [西南實業通訊] 12, no. 3/4 (1945): 59.

"Gao Zhongqin Invents an Electric Chinese Typewriter" (Gao Zhongqin faming diandong Zhongwen daziji) [高仲芹發明電動中文打字機]. *Jiaoyu tongxun* 4, no. 1 (1947): 29–30.

Garth, William W., IV, *Entrepreneur: A Biography of William W. Garth, Jr. and the Early History of Photocomposition*. Self-Published by Author, 2002.

Geoghegan, Bernard Dionysius. "An Ecology of Operations: Vigilance, Radar, and the Birth of the Computer Screen." *Representations* 147, no. 1 (August 2019): 59–95.

Geoghegan, Bernard Dionysius. "From Information Theory to French Theory: Jakobson, Lévi-Strauss, and the Cybernetic Apparatus." *Critical Inquiry* 38, no. 1 (2011): 96–126.

George A. Kennedy Papers. Manuscripts and Archives, Yale University Library, MS 308, Box 3, Folder 39.

Gere, Charlie. "Genealogy of the Computer Screen." *Visual Communication* 5, no. 2 (2006): 141–52.

Giannoulis, Elena, and Lukas R. A. Wilde, eds. *Emoticons, Kaomoji, and Emoji: The Transformation of Communication in the Digital Age*. New York: Routledge, 2020.

"A Girl from Shanghai and Her Father Create a 'New Character Retrieval Method': To Look Up Characters One Need Only Rely on Character Radicals" (Shanghai guniang yu qi fu chuang 'jianzi xinfa' chazi zhi xu yiju bushoujian) [上海姑娘与其父创检字新法查字只需依据部首件]. *Dongfangwang*, October 20, 2013.

Gitelman, Lisa. *Scripts, Grooves, and Writing Machines: Representing Technology in the Edison Era*. Stanford, CA: Stanford University Press, 2000.

Goody, Jack. *The Interface between the Written and the Oral*. Cambridge: Cambridge University Press, 1987.

Goody, Jack. "Technologies of the Intellect: Writing and the Written Word." In *The Power of the Written Tradition*. Washington: Smithsonian Institution Press, 2000: 133–138.

Gottlieb, Nanette. *Word-Processing Technology in Japan: Kanji and the Keyboard*. Richmond, VA: Curzon, 2000.

Graphic Arts Research Foundation. "GARF 5-Year Plan 1980." Louis Rosenblum Collection. Stanford University.

Graphic Arts Research Foundation. Press Release Announcing News of Joint Protocol Signed in Shanghai. November 1979. Louis Rosenblum Collection. Stanford University.

Graphic Arts Research Foundation. "Second Interim Report on Studies Leading to Specifications for Equipment for the Economical Composition of Chinese and Devanagari." (December 1, 1953) Pardee Lowe Papers. Accession No. 98055–16.370/376. Box No. 276. Hoover Institute Archives.

Graphic Arts Research Foundation. "Sinotype II Demonstration." August 1, 1979, 1–2. Louis Rosenblum Collection. Stanford University.

Graphic Arts Research Foundation. "Technical Report." December 6, 1979. Louis Rosenblum Collection. Stanford University.

Graphic Arts Research Foundation. "Trip Report. Peking, China October 29-November 22, 1980." December 1, 1980. Louis Rosenblum Collection. Stanford University.

Grier, David Alan. *When Computers Were Human*. Princeton, NJ: Princeton University Press, 2007.

"A Guide to the OSCO-Method of Coding Chinese Characters for Entry into Olympia Chinese/English Memory Typewriters." Olympia Werke AG T1/04–10/84. Rolf Heinen Technology Collection. Drolshagen, Germany.

Gu Jingwen [顾景文]. *Driver Design for the Display of Chinese Characters and Images on a Microcomputer (Weiji Hanzi yu tuxing de xianshi jiqi jiekou chengxu sheji)* [微机汉字与图形的显示及其接口程序设计]. Shanghai: Tongji daxue chubanshe, 1995.

Handel, Zev. *Sinography: The Borrowing and Adaptation of the Chinese Script*. Leiden: Brill, 2019.

Hannas, William C. *The Writing on the Wall: How Asian Orthography Curbs Creativity*. Philadelphia: University of Pennsylvania Press, 2003.

Hansen, Mark B. N. *Feed-Forward: On the Future of Twenty-First-Century Media*. Chicago: University of Chicago Press, 2014.

Harrist, Robert E. and Wen Fong. *The Embodied Image: Chinese Calligraphy from the John B. Elliott Collection*. Princeton, NJ: Art Museum, Princeton University in association with Harry N. Abrams, 1999.

Hayashi, Hideyuki, Sheila Duncan, and Susumu Kuno. "Computational Linguistics: Graphical Input/Output of Nonstandard Characters." *Communications of the ACM* 11, no. 9 (1968): 613–618.

Hayles, N. Katherine. *How We Think: Digital Media and Contemporary Technogenesis*. Chicago: University of Chicago Press, 2012.

Hayles, N. Katherine. *Postprint: Books and Becoming Computational*. New York: Columbia University Press, 2021.

Hayles, N. Katherine. *Writing Machines*. Cambridge, MA: MIT Press, 2002.

Heijdra, Martin J. "The Development of Modern Typography in East Asia, 1850–2000." *East Asia Library Journal* 11, no. 2 (Autumn 2004): 100–168.

Hicks, Mar. *Programmed Inequality: How Britain Discarded Women Technologists and Lost Its Edge in Computing*. Cambridge, MA: MIT Press, 2018.

Hjorth, Larissa. "Cute@keitai.com." In *Japanese Cybercultures*. Ed. Nanette Gottlieb and Mark J. McLelland. London: Routledge, 2003.

Hoare, R. "Keyboard Diagram for Chinese Phonetic." Mergenthaler Linotype Collection. Museum of Printing, North Andover, Massachusetts, February 4, 1921.

Hoare, R. "Keyboard Diagram for Chinese Phonetic Amended." Mergenthaler Linotype Collection. Museum of Printing, North Andover, Massachusetts, March 3, 1921.

Hockx, Michel. *Internet Literature in China*. New York: Columbia University Press, 2015.

Hofheinz, Roy. "Conversation from 2 p.m. to 5 p.m. November 14 in the Chinchiang Hotel with Representatives of the *People's Daily* and the First Ministry of Machine Building." Memorandum. November 14, 1979. Louis Rosenblum Collection. Stanford University.

Hofheinz, Roy. "Memorandum of Conversation with Dr. Chih Ping-I." Louis Rosenblum Collection. Stanford University.

Hou, Dongchen. "Writing Sound: Stenography, Writing Technology, and National Modernity in China, 1890s." *Journal of Linguistic Anthropology* 30, no. 1 (2019): 103–122.

"How Electronics Helped Solve a Chinese Puzzle." 1977. DOC/CW/12/262. Porthcurno Telegraph Museum Collection and Cable and Wireless Archives.

How to Study Hanyu Pinyin (*Zenyang xuexi Hanyu Pinyin fang'an*) [怎样学习汉语拼音方案]. Beijing: Wenzi gaige chubanshe, 1958.

Howard, Philip. "When Chinese is a String of Two-Letter Words." *The Times*, January 16, 1978, 12.

Hsieh, Ching-chun, Kai-tung Huang, Chung-tao Chang, and Chen-chau Yang. "The Design and Application of the Chinese Character Code for Information Interchange (CCCII)." Louis Rosenblum Collection. Stanford University.

Hu Jintao [胡锦涛]. "Hold High the Great Banner of Socialism with Chinese Characteristics and Strive for New Victories in Building a Moderately Prosperous Society in All Respects." Report to the Seventeenth National Congress of the Communist Party of China (October 15, 2007). Official English-language translation: https://www.chinadaily.com.cn/china/2007-10/25/content_6204663.htm.

Huang Jinfu [黄金富]. *The Weiwu Chinese Dictionary* (*Weiwu Zhongwen zidian*) [唯物中文字典]. Beijing: Jijie gongye chubanshe, 1988.

Huang, Jack Kai-tung. "Principles of Chinese Keyboard Layout." Draft. April 25, 1981. Louis Rosenblum Collection. Stanford University.

Huang, Jack Kai-tung, and Timothy D. Huang. *An Introduction to Chinese, Japanese and Korean Computing*. Singapore: World Scientific, 1989.

Hunter, Duncan B. "Chinese Computer that's Got Character." *South China Morning Post*, January 17, 1986, 30.

Hunter, Janet. "Technology Transfer and the Gendering of Communications Work: Meiji Japan in Comparative Historical Perspective." *Social Science Japan Journal* 14, no. 1 (Winter 2011): 1–20.

Hurst, Jan et al. "Retrospectives: The Early Years in Computer Graphics at MIT, Lincoln Lab and Harvard." *SIGGRAPH '89 Panel Proceedings* (1989): 19–38.

Hutchins, John W. and Harold L. Somers. *An Introduction to Machine Translation*. Cambridge, MA: Academic Press, 1992.

Huters, Theodore. *Bringing the World Home: Appropriating the West In Late Qing and Early Republican China*. Honolulu: University of Hawai'i Press, 2005.

Hutton, Chris. "Writing and Speech in Western Views of the Chinese Language." In Q. S. Tong and D. Kerr, eds., *Critical Zone 2: A Forum of Chinese and Western Knowledge*, 83–105. Hong Kong: Hong Kong University Press, 2006.

IBM Brochure for Electric Chinese Typewriter. IBM Corporate Archives. New York.

IBM Electric Chinese Typewriter Four-Digit Code Tables. Smithsonian. Mergenthaler Papers.

"IBM Goes West: A 73-Year-Long Saga, From Punch Cards to Watson." *Fast Company*, October 28, 2016. https://www.fastcompany.com/3064902/ibm-goes-west-a-73-year-long-saga-from-punch-cards-to-watson.

IBM Multistation 5550 (Hanzi bianma shouce) [汉字编码手册]. October 1984. Thomas S. Mullaney East Asian Information Technology History Collection. Stanford University.

"IBM's Chinese Typewriter Demonstrated in New York." *Business Machines*, July 9, 1946.

"Ideographic Encoder Handbook (April 1978), 2.5.06, H1050B. Porthcurno Telegraph Museum Collection and Cable & Wireless Archives.

"Ideographic Professional Computer Database Reference Guide." Wang Laboratories Corporate Papers. Box 137, Folder 2. Harvard University Business School Special Collections.

Inoue, Miyako. "Word for Word: Verbatim as Political Technologies." *Annual Review of Anthropology* 47, no. 1 (2018): 217–32.

"Invention: Electric Chinese Typewriter" (Faming: diandong Huawen daziji) [發明:電動華文打字機]. *Qingnian wenti* [青年問題] 3, no. 6 (1946): 21.

"Invitation Card to the Computational Chinese Information Processing Systems Exhibition" (Jisuanji Zhongwen xinxi chuli xitong zhanlanhui qingjian) [计算机中文信息处理系统展览会请柬]. Thomas S. Mullaney East Asian Information Technology History Collection. Stanford University.

IPX Materials. SCC2/24/11. Needham Research Institute. Cambridge University.

"IPX Model 9600 Intelligent Keyboard User's Manual." Included inside "IPX Model 9600 Intelligent K'Board," Box 22 of 27, Folder 003047. Computer History Museum. Mountain View, CA.

Irvine, M. M. "Early Digital Computers at Bell Telephone Laboratories." *IEEE Annals of the History of Computing* (July–September 2001): 22–42.

Ishida, Haruhisa. "Chinese Character Input/Output and Transmission in Japanese Personal Computing." *IEEE* (1979): 402–409.

Ishii, Kae. "The Gendering of Workplace Culture: An Example from Japanese Telegraph Operators," *Bulletin of Health Science University* 1, no. 1 (2005): 37–48.

Jacobsen, Kurt. "A Danish Watchmaker Created the Chinese Morse System." *NIAS-nytt (Nordic Institute of Asian Studies) Nordic Newletter* 2 (July 2001): 17–21.

"Japan Sells China 'Strategic' Computer System." *New Scientist*, April 13, 1978, 69.

"Japanese Language Telegraph Printer." US Patent 2728816. Filed March 24, 1953, and issued December 27, 1955. Assignor to Trasia Corporation, NY.

"Jenny Chuang, 1986." Wang Laboratories Corporate Papers. Box 97, Folder 1. Harvard University Business School Special Collections.

Jiaotong bu. *Ming mi dianma xinbian*. 1935. Thomas S. Mullaney East Asian Information Technology History Collection, Stanford University.

Jones, Stacy V. "Telegraph Printer in Japanese with 2,300 Symbols Patented." *New York Times* (December 31, 1955): 19.

Kachru, Braj B. "The Bilinguals' Creativity." *Annual Review of Applied Linguistics* 6 (1985): 20–33.

Kachru, Braj B. "The Bilingual's Creativity and Contact Literatures." In Braj B. Kachru, ed., *The Alchemy of English: The Spread, Functions, and Models of Non-Native Englishes*, 159–170. Oxford: Pergamon Press, 1986.

Kachru, Yamuna. "Code-Mixing, Style Repertoire and Language Variation: English in Hindu Poetic Creativity." *World Englishes* 8, no. 3 (1989): 311–319.

Kao Chung-Chin. "Chinese language typewriter and the like." US Patent 2412777A. Filed June 28, 1944, and issued December 17, 1946.

Kao Chung-Chin. "Keyboard-controlled ideographic printer having permutation type selection." US Patent 2427214A. Filed December 11, 1943, and issued September 9, 1947.

Kao Chung-Chin [高仲芹]. "The Design and Applications of the Electric Chinese Typewriter (Diandong Huawen daziji zhi sheji ji qi yingyong) [電動華文打字機之設計及其應用]." *Kexue huabao* 13, no. 12 (1947): 746–748.

Kaske, Elisabeth. *The Politics of Language in Chinese Education, 1895–1919*. Leiden: Brill, 2008.

Katsuno, Hirofumi, and Christine R. Yano. "Kaomoji and Expressivity in a Japanese Housewives' Chat Room." In *The Multilingual Internet: Language, Culture, and Communication Online*. Ed. Brenda Danet and Susan C. Herring, 278–302. New York: Oxford University Press, 2007.

Kennedy, George A., ed. *Minimum Vocabularies of Written Chinese*. New Haven, CT: Far Eastern Publications, 1966.

"Kickin' the Bucket: 12 Outrageous Fake KFC Restaurants." https://weburbanist.com /2014/05/04/kickin-the-bucket-12-outrageous-fake-kfc-restaurants/.

King, Bob. "Enter the Dragon: IBM Launches Chinese Style Computer." *Financial Times*, February 8, 1984, 13.

Kirschenbaum, Matthew. *Mechanisms: New Media and the Forensic Imagination*. Cambridge, MA: MIT Press, 2012.

Kirschenbaum, Matthew G. *Track Changes: A Literary History of Word Processing*. Cambridge, MA: Harvard University Press, 2016.

Kittler, Friedrich. "Computer Graphics: A Semi-Technical Introduction." Trans. Sara Ogger. *Grey Room* 2 (Winter 2001): 30–45.

Kittler, Friedrich. *Optical Media: Berlin Lectures*. Trans. Anthony Enns. Cambridge: Polity, 2009.

Kline, Ronald. *The Cybernetics Moment: Or Why We Call Our Age the Information Age*. Baltimore, MD: Johns Hopkins University Press, 2015.

Kline, Ronald and Trevor Pinch. "Users as Agents of Technological Change: The Social Construction of the Automobile in the Rural United States." *Technology and Culture* 37 (1996): 763–795.

Kurtz, Joachim. *The Discovery of Chinese Logic*. Leiden: Brill, 2011.

Kurzon, Dennis. "Romanisation of Bengali and Other Indian Scripts." *Journal of the Royal Asiatic Society* 20 (January 2010): 61–74.

Kuzuoğlu, Uluğ. "Codebooks for the Mind: Dictionary Index Reforms in Republican China, 1912–1937." *Information & Culture* 53, no. 3/4 (2018): 337–366.

"Lab's Chris Berry Marks 40th Anniversary with IBM." *IBM News* 1, no. 4 (February 25, 1964): 4. Personal archives of Richard Foss and John O'Farrell.

Lam Man-Wah. "Now . . . a Chinese Language Computer!" *South China Morning Post*, July 21, 1980, 22.

Lanham, Richard A. *The Electronic Word*. Chicago: University of Chicago Press, 1993.

Lawrence, Halcyon M. "Siri Disciplines." In *Your Computer Is on Fire*, edited by Thomas S. Mullaney, Benjamin Peters, Mar Hicks, and Kavita Philip, 179-198. Cambridge, MA: MIT Press, 2021.

"Leaflet advertising the 'Ideo-Matic' Chinese character encoder, engineered by Robert Sloss and Peter Nancarrow of Cambridge University." June 1981. Needham Research Institute. SCC2/24/7.

Lean, Eugenia. *Vernacular Industrialism in China: Local Innovation and Translated Technologies in the Making of a Cosmetics Empire, 1900–1940.* New York: Columbia University Press, 2020.

Lee, Jennifer 8. "In China, Computer Use Erodes Traditional Handwriting, Stirring a Cultural Debate." *New York Times*, February 1, 2001.

Legrand, Marcellin. *Spécimen de caractères chinois gravés sur acier et fondus en types mobiles par Marcellin Legrand.* Paris: n.p., 1859.

Levy, Steven *Hackers: Heroes of the Computer Revolution*, 25th Anniversary Edition. Sebastopol, CA: O'Reilly Media, 2010.

Li, Jie. *Shanghai Homes: Palimpsests of Private Life.* New York: Columbia University Press, 2014.

Li, Jinying. "The Interface Affect of a Contact Zone: Danmaku on Video Streaming Platforms." *Asiascape: Digital Asia* 4, no. 3 (2017): 233–256.

Li Wei [李嵬]. "New Chinglish and the Post-Multilingualism Challenge: Translanguaging ELF in China." *Journal of English as a Lingua Franca* 5, no. 1 (2016): 1–25.

Li Wei and Zhu Hua. "Tranßcripting: Playful Subversion with Chinese Characters." *International Journal of Multilingualism* 16, no. 2 (2019): 145–161.

Light, Jennifer S. "When Computers Were Women." *Technology and Culture* 40, no. 3 (July 1999): 455–483.

Lilly, Edward P. "Memorandum for the Executive Officer: Chinese Ideograph Typesetting Machine." April 23, 1959. OCB Secretariat Series, Box 3: Ideographic Composing Machine. Dwight D. Eisenhower Presidential Library.

Lin Shuzhen. *Household Computer: Chinese Character Encoding Rapid Look-Up (Jiating diannao: Hanzi bianma sucha)* [家庭电脑: 汉字编码速查]. Fuzhou: Fujian kexue jishu chubanshe, 1994.

Lin, Ming-Yao, and Wen-Hsiang Tsai. "Removing the Ambiguity of Chinese Input by the Relaxation Technique." *Computer Processing of Chinese and Oriental Languages* 3, no. 1 (May 1987): 1–24.

Lindtner, Silvia M. *Prototype Nation: China and the Contested Promise of Innovation.* Princeton, NJ: Princeton University Press, 2020.

Ling Zhijun. *The Lenovo Affair: The Growth of China's Computer Giant and Its Takeover of IBM-PC.* Trans. Martha Avery. Malden, MA: Wiley, 2006.

Lipartito, Ken. "When Women Were Switches: Technology, Work, and Gender in the Telephone Industry, 1890–1920." *American Historical Review* 99, no. 4 (1994): 1074–1111.

Liu Fuying [刘甫迎] and He Xiqiong [何希琼]. *Chinese Language FoxBASE+ Novel Relational Database* (*Hanzi FoxBASE+ Xinying guanxi shujuku*) [汉字FoxBASE+新颖关系数据库]. Zhongguo kexueyuan changdu jisuanji yingyong yanjiusuo qingbao shi. November 1987.

Liu Kaiying [刘开瑛] and Xing Zuolin [刑作林]. *DJS-21 Type Electronic Computer Algorithmic Language* (*DJS-21 xing dianzi shuzi jisuanji suanfa yuyan*) [DJS-21型电子数字计算机算法语言]. Shanxi renmin chubanshe, 1979.

Liu, Lydia H. *The Freudian Robot Digital Media and the Future of the Unconscious*. Chicago: The University of Chicago Press, 2010.

Liu Qi [刘奇]. *Chinese Language dBASE-II Relational Database Management System* (*Hanzi xing dBASE-II guanxi shujuku guanli xitong*) [汉字型dBASE-II关系数据库管理系统]. Beijing: Beijing Yanshan shiyou huagongsi qiye guanli xiehui, 1984.

Liu Shuji [刘树吉] et al. *Impact Printers: Principles of Usage and Repair* (*Zhenshi dayinji yuanli shiyong yu weixiu*) [针式打印机原理使用与维修]. Beijing: Guoji gongye chubanshe, 1988.

Liu, Xiao. *Information Fantasies: Precarious Mediation in Postsocialist China*. Minneapolis: University of Minnesota Press, 2019.

Liu Yongquan [刘涌泉]. "A Discussion of Problems Related to Character Compound Repositories (*Tantan ciku wenti*) [谈谈词库问题]. *Journal of Chinese Information Processing* 1, no. 1 (1986): 8–11.

Loh, Shiu C. "Ideographic Character Selection." US Patent 4270022A. Filed June 18, 1979, and issued May 26, 1981.

Loy, D. Gareth. "The Systems Concepts Digital Synthesizer: An Architectural Retrospective." *Computer Music Journal* 37, no. 3 (2013): 49–67.

Lu Xun [鲁迅]. "Reply to an Interview from My Sickbed" (Bingzhong da jiumang qingbao fangyuan) [病中答救亡情报訪員] (1938), 160. In *Complete Works of Lu Xun* (*Lu Xun Quanji*) [鲁迅全集]. Vol. 6 (Beijing: Renmin Wenxue, 1981).

Lunde, Ken. *CJKV Information Processing: Chinese, Japanese, Korean & Vietnamese Computing*. Sebastopol, CA: O'Reilly Media, 2009.

Luo Yinghui [罗英辉] and Zhang Yala [张亚拉]. *Shape-Meaning Three Letter Code: A Revolution in Chinese Character Input Methods* (*Xingyi sanma: Hanzi shurufa de geming*) [形意三码: 汉字输入法的革命]. Guangzhou: Zhongshan daxue chubanshe, 1995.

Lüthi, Lorenz M. *The Sino-Soviet Split: Cold War in the Communist World*. Princeton, NJ: Princeton University Press, 2008.

MacFarquhar, Roderick, and Michael Schoenhals. *Mao's Last Revolution*. Cambridge, MA: Belknap Press, 2008.

"Machine Heralds New Printing Era." *Boston Herald*, September 16, 1949, 1.

"Machine Seen as Possible 'Breakthrough' in Chinese Printing." File No. 147 (June 22, 1959). Pardee Lowe Papers. Hoover Institute Archives. Accession No. 98055–16.370/376. Box No. 276.

MacKenzie, I. Scott, and Kumiko Tanaka-Ishii. *Text Entry Systems Mobility, Accessibility, Universality*. Amsterdam: Morgan Kaufman, 2007.

Maddox, Brenda. "Women and the Switchboard." In *The Social History of the Telephone*. Ed. Ithiel de Sola Pool (Cambridge, MA: MIT Press, 1977), 262–280.

Maier, John H. "Computer Science Education in the People's Republic of China." *Computer*, June 1986, 50–56.

Maier, John H. "Thirty Years of Computer Science Developments in the People's Republic of China: 1956–1985." *IEEE Annals of the History of Computing* 10, no. 1 (1988): 19–34.

Mair, Victor. "Character Amnesia." *Language Log*, July 22, 2010. https://languagelog.ldc.upenn.edu/nll/?p=2473.

Manovich, Lev. "An Archeology of a Computer Screen." Moscow: Soros Center for Contemporary Art. http://www.manovich.net/TEXT/digital_nature.html.

Marshall, Alan. *Du Plomb à la Lumière: La Lumitype-Photon et la Naissance des Industries Graphiques Modernes*. Paris: Maison des Sciences de L'Homme, 2003.

Martin, Michele. *"Hello, Central?": Gender, Technology and Culture in the Formation of Telephone Systems*. Montreal: McGill-Queens University Press, 1991.

Martin, W. A. *The History of the Art of Writing*. New York: Macmillan, 1920.

Materials related to IBM Electric Chinese Typewriter Demonstration in Shanghai. Shanghai Municipal Archives. Q449-1-535.

Mathias, Jim, and Thomas L. Kennedy, eds. *Computers, Language Reform, and Lexicography in China. A Report by the CETA Delegation*. Pullman: Washington State University Press, 1980.

Mauran, Cecily. "Pee and Me: The Gadget That Turns Toilets into Urine Labs. Why 2023 Could Be the Year of Personal Pee-Testing." *Medium.com*, January 3, 2023, https://mashable.com/article/withings-u-scan-urine-analysis-ces-2023.

McLuhan, Marshall. *The Gutenberg Galaxy: The Making of Typographic Man*. Toronto: The University of Toronto Press, 1962.

Medina, Eden. *Cybernetic Revolutionaries: Technology and Politics in Allende's Chile*. Cambridge, MA: MIT Press, 2014.

"Memorandum for the Operations Coordinating Board: Interim Report of the Chinese Ideographic Composing Machine (CICM)." May 18, 1959. OCB Secretariat Series, Box 3: Ideographic Composing Machine. Dwight D. Eisenhower Presidential Library.

"Memorandum for the Executive Office: Chinese Ideographic Composing Machine."
May 21, 1959. OCB Secretariat Series, Box 3: Ideographic Composing Machine.
Dwight D. Eisenhower Presidential Library.

"Memorandum for the Executive Office: Chinese Ideographic Composing Machine—
Briefing Memo on Deferral of Board Consideration." May 20, 1959. OCB Secretariat
Series, Box 3: Ideographic Composing Machine. Dwight D. Eisenhower Presidential
Library.

"Memorandum of Meeting: OCB Ad Hoc Working Group on Exploitation of the
Chinese Ideographic Composing Machine." May 7, 1959. OCB Secretariat Series,
Box 3: Ideographic Composing Machine. Dwight D. Eisenhower Presidential Library.

Mergenthaler Linotype Company. *China's Phonetic Script and the Linotype*. Brooklyn:
Mergenthaler Linotype Co., April 1922. Smithsonian National Museum of American
History Archives Center. Collection no. 666, box LIZ0589 ("History—Non-Roman
Faces"), folder "Chinese," subfolder "Chinese Typewriter."

Miller, George A. "The Magical Number Seven, Plus or Minus Two: Some Limits on
Our Capacity for Processing Information." *The Psychological Review* 63, no. 2 (March
1956): 81–97.

Mindell, David A. *Between Human and Machine: Feedback, Control, and Computing
Before Cybernetics*. Baltimore: Johns Hopkins University Press, 2004.

Mittler, Barbara. *A Continuous Revolution: Making Sense of Cultural Revolution Culture*.
Cambridge, MA: Harvard University Asia Center, 2016.

"Modern Business Machines for Writing, Duplicating, and Recording." 1947 Film
produced by Teaching Aids Exchange. Director not listed. Film housed online
at http://www.archive.org/details/modern_business_machines_for_writing (segment
begins at 12m35s).

Montfort, Nick. "Continuous Paper: The Early Materiality and Workings of Electronic Literature." MLA Annual Conference, Philadelphia. December 2004. Text of
talk here: https://nickm.com/writing/essays/continuous_paper_mla.html

Morita, Ichiko. "Japanese Character Input: Its State and Problems." *Journal of Library
Automation* 14, no. 1 (March 1981): 6–23.

Morrell, Alan. "Whatever Happened to . . . Cathay Pagoda?" *Democrat and Chronicle*
(May 6, 2017). https://www.democratandchronicle.com/story/local/rocroots/2017
/05/06/whatever-happened-cathay-pagoda/101345224/. Accessed January 2, 2019.

Moser, David. *A Billion Voices: China's Search for a Common Language*. New York:
Penguin, 2016.

Mott, Walter N. "Service Test of Ideographic Composing Machine Final Report."
Fort Bragg, NC: US Army Airborne, Electronics and Special Warfare Board (March
1968).

Mullaney, Thomas S. "Controlling the Kanjisphere: The Rise of the Sino-Japanese Typewriter and the Birth of CJK." *Journal of Asian Studies* 75, no. 3 (August 2016): 725–753.

Mullaney, Thomas S. "How to Spy on 600 Million People: The Hidden Vulnerabilities in Chinese Information Technology." *Foreign Affairs*, June 5, 2016, https://www.foreignaffairs.com/articles/china/2016-06-05/how-spy-600-million-people.

Mullaney, Thomas S. "QWERTY in China: Chinese Computing and the Radical Alphabet." *Technology and Culture* 59, no. 4 Supplement (October 2018): S34-S65.

Mullaney, Thomas S. *The Chinese Typewriter: A History*. Cambridge, MA: MIT Press, 2017.

Mullaney, Thomas S. "The Font That Never Was: Linotype and the 'Phonetic Chinese Alphabet' of 1921." *Philological Encounters/Brill* 3, no. 4 (November 2018): 550–566.

Mullaney, Thomas S. "The Movable Typewriter: How Chinese Typists Developed Predictive Text during the Height of Maoism." *Technology and Culture* 53, no. 4 (October 2012): 777–814.

Mullaney, Thomas S. "The Origins of Chinese Supercomputing and an American Delegation's Mao-Era Visit." *Foreign Affairs* (August 4, 2016).

Mullaney, Thomas S., Benjamin Peters, Mar Hicks, and Kavita Philip, eds. *Your Computer Is on Fire*. Cambridge, MA: MIT Press, 2021.

"Museum of Printing Long-Time Board Member Louis Rosenblum Passes." https://museumofprinting.org/news-and-events/long-time-board-member-louis-rosenblum-passes/.

Nancarrow, Peter, and Richard Kunst. "The Computer Generation of Character Indexes to Classical Chinese Texts." In *Sixth International Conference on Computers and the Humanities*. Ed. Sarah K. Burton and Douglas D. Short, 772–780. Rockville, MD: Computer Science Press, 1983.

National Archives and Records Administration, Washington, DC. Passenger and Crew Lists of Vessels Arriving at Seattle, Washington, NAI Number: 4449160. "Records of the Immigration and Naturalization Service, 1787–2004," Record Group Number: 85, Series Number: M1383, Roll Number: 230. Ancestry.com.

"New Invention" (Xin faming) [新發明]. *Guofang yuekan* 2, no. 1 (1947): 2.

"New Machine Sets Type on Film Instead of Metal." *Christian Science Monitor*, September 16, 1949, 1.

"News: Chung-Chin Kao Invents Electric Chinese Typewriter" (Xiao xiaoxi: Gao Zhongqin faming diandong Zhongwen daziji) [小消息:高仲芹氏發明電動中文打字機]." *Tianjia banyuekan* [田家半月報] 13, no. 1/2/3/4 (1946): 3.

Ng, K. W. "An Intelligent CRT Terminal for Chinese Characters." *Microprocessing and Microprogramming* 8 (1981): 22–31.

Nickjoo, Mahvash. "A Century of Struggle for the Reform of the Persian Script." *The Reading Teacher* 32 (May 1979): 926–929.

Norman, Jeremy. *Chinese*. Cambridge: Cambridge University Press, 1988.

Obituary of Robert Sloss. *The Darwinian: Newsletter of Darwin College* (Spring 2008): 14. https://www.darwin.cam.ac.uk/drupal7/sites/default/files/downloads/Alumni -Darwinian10-2008.pdf

Obituary of T. Kevin Mallen. February 2, 2000. http://www.almanacnews.com /morgue/2000/2000_02_02.obit02.html. Accessed December 12, 2012.

O'Connell, Brigid. "New 'Smart Chair' Designed to Improve Posture and Protect Back Health." *The West Australian*, perthnow.com, June 16, 2015, https://www .perthnow.com.au/news/new-smart-chair-designed-to-improve-posture-and-protect -back-health-ng-1233bbbe13e3f6536fca4e13445f4d6a.

"Olympia 1011." *Electronics* 54, no. 9–13 (1981): 71.

"Olympia 1011 Chinese Word Processor System." FBIS reprint from *China Computer World* (*Beijing jisuanji shijie*) no. 20 (October 20, 1982): 4. Translation in *China Report: Science and Technology* no. 189 FBIS (March 1, 1983): 12–13.

Opinions on the Use of Double Pinyin Encoding Chinese Character Information Processing System (*Shuangpin bianma Hanzi xinxi chuli xitong shiyong yijian*) [双拼编码汉字信息处理系统使用意见] Nanjing: Nanjing daxue jisuanjichang, May 28, 1986.

Ornstein, Severo. *Computing in the Middle Ages: A View from the Trenches 1955–1983*. Bloomington, IN: AuthorHouse, 2002.

"Over the Past Two Years, Tsinghua University Prototypes 18 Reliable Electronic Computers of Varying Types" (Liangnian lai Qinghua daxue shizhi le 18 tai ji zhong butong leixing de gongzuo kekao de dianzi jisuanji) [两年来清华大学试制了18台几种不同类型的工作可靠的电子计算机]. *Kexue jishu gongzuo jianbao* [科学技术工作简报] 30 (March 7, 1960): 64–70. Beijing Municipal Archives (BMA) 001-022-00494.

Packer, Jeremy and Kathleen F. Oswald. "From Windscreen to Widescreen: Screening Technologies and Mobile Communication." *The Communication Review* 13, no. 4 (2010): 309–39.

Packer, Jeremy. "Screens in the Sky: SAGE, Surveillance, and the Automation of Perceptual, Mnemonic, and Epistemological Labor." *Social Semiotics* 23, no. 2 (2013): 173–95.

Pan Defu [潘德孚] and Zhan Zhenquan [詹振权]. *Chinese Character Encoding Design* (*Hanzi bianma shejixue*) [汉字编码设计学]. ca. 1997. Thomas S. Mullaney East Asian Information Technology History Collection, Stanford University.

Parisi, David. *Archaeologies of Touch Interfacing with Haptics from Electricity to Computing*. Minneapolis: University of Minnesota Press, 2018.

PDP-1 Restoration Project, *"Spacewar!"* Computer History Museum website, www.computerhistory.org/pdp-1/spacewar/.

Peckley, Silahis O. No Title. *Asia Africa Intelligence Wire*, October 14, 2002.

Pei Jie [裴杰]. *CC-DOS V4.2 Chinese Character Operating System User Guide* (*CC-DOS V4.2 Hanzi caozuo xitong shiyong zhinan*) [CC-DOS V4.2 汉字操作系统使用指南]. Shanghai: Shanghai jiaotong daxue chubanshe, 1991.

Peiser, Jaclyn. "The Rise of the Robot Reporter." *New York Times*. February 5, 2019.

Peking University Chinese Character Information Processing Technology Research Office (*Beijing daxue Hanzi xinxi chuli jishu yanjiushi*) [北京大学汉字信息处理技术研究室], "Design Proposal for Medium-Sized Keyboard Chinese Character Information Processing and Input System" (Hanzi xinxi chuli shuru xitong zhongxing jianpan sheji fang'an) [汉字信息处理输入系统中型键盘设计方案]. In *Chinese Information Processing* (*Hanzi xinxi chuli*) [汉字信息处理]. Ed. Chinese Language Journal Editorial Department (Zhongguo yuwen bianjibu), 1–18. Beijing: Zhongguo shehui kexue chubanshe, 1979.

"Percentages of Letter Frequencies per 1000 Words." http://www.cs.trincoll.edu/~crypto/resources/LetFreq.html. Accessed August 2, 2020.

Peters, Benjamin. *Digital Keywords: A Vocabulary of Information Society and Culture*. Princeton, NJ: Princeton University Press, 2016.

Philip, Kavita. "What is a Technological Author? The Pirate Function and Intellectual Property." *Postcolonial Studies* 8, no. 2 (2005): 199-218.

Photograph of Chinese typist with "Eputima" label behind her. Thomas S. Mullaney East Asian Information Technology Collection. Stanford University.

Photograph of Harry Garland and Roger Melen (1980). Wikimedia Commons. https://commons.wikimedia.org/wiki/File:Harry_Garland_and_Roger_Melen_-_with_Cromemco_shipment_to_China_(1980).jpg. Accessed June 17, 2020.

Photographic Album from Chung-Chin Kao to J. T. Mackey. August 14, 1946. Mergenthaler Linotype Company Records, 1905–1993, Archives Center, National Museum of American History. Smithsonian Institution. Box 3628.

Poo, Gee-Swee, Beng-Cheng Lim, and Edward Tan. "An Efficient Data Structure for Hanyu Pinyin Input System." *Computer Processing of Chinese and Oriental Languages* 4, no. 1 (November 1988): 1–17.

"The Press: Peace in Chicago." *Time*, September 26, 1949. https://content.time.com/time/subscriber/article/0,33009,800775,00.html. Retrieved February 19, 2022.

"Printing without Type." *Business Week*, October 1, 1949, 57.

Proceedings of the First International Symposium on Computers and Chinese Input/Output Systems. Taipei, Taiwan: Academia Sinica, August 14–16, 1973.

"Prof. Zhi Bingyi." College of Electrical Engineering, Zhejiang University. http://ee.zju.edu.cn/english/redir.php?catalog_id=18647&object_id=19246. Accessed May 5, 2012.

Purdon, James. "Teletype." In *Writing, Medium, Machine: Modern Technographies.* Ed. S. Pryor and D. Trotter, 120–136. London: Open Humanities Press, 2016.

Qian Peide. "An Analysis of CC-DOS" (*CC-DOS fenxi*) [CC-DOS 分析]. *Microcomputer Applications (Weijisuanji yingyong)* [微计算机应用] 5 (1985): 1–10.

Qian Peide [钱培得], ed. *CC-DOS Chinese Character Operating System for Microcomputers (Weixing jisuanji Hanzi caozuo xitong CC-DOS)* [微型计算机汉字操作系统CC-DOS]. Xi'an: Shaanxi dianzi bianjibu, 1988.

Qian Yuzhi [钱玉趾]. "Prospects for the Development of a 'Unified Script for Computers' ('Diannao shu tongwen' de fazhan qianying) [电脑书同文的发展前景]." *Zhongwen xinxi* 1 (1992): 6–9.

Qin Dulie [秦笃烈]. "Principles of Developing Expert Systems of Traditional Chinese Medicine." In *Expert Systems and Decision Support in Medicine.* Ed. Otto Rienhoff, Ursula Piccolo, and Berthold Schneider, 85–93. Berlin: Springer, 1988.

Quartermaster Research and Engineering Center. "Chinese Photocomposing Machine." Natick, MA: Headquarters Quartermaster Research and Engineering Command, U.S. Army, March 1960. Louis Rosenblum Collection. Stanford University.

"Radical Machines Chinese in the Information Age." Exhibition held October 18, 2018–March 24, 2019, at the Museum of Chinese in America. Curated by the author.

RCA Advanced Technology Labs. "Language Manual for Use with Ideographic Composing Machines." Camden, NJ: RCA, February 1970.

"Revolutionary Chinese Typewriter Displayed." *The North-China Daily News* (October 21, 1947).

"Richard Solomon, Former Diplomat Who Helped Nixon Open Relations with China, Dies." *Wall Street Journal*, March 14, 2017, https://www.wsj.com/articles/richard-solomon-former-diplomat-who-helped-nixon-open-relations-with-china-dies-1489532159.

"Robot Writes *LA Times* Earthquake Breaking News Article." *BBC.com*, March 18, 2014. https://www.bbc.com/news/technology-26614051.

Romano, Frank J. *History of the Linotype Company.* Rochester, NY: RIT Press, 2014.

Romano, Frank J. *History of the Phototypesetting Era.* San Luis Obispo, CA: Graphic Communication Institute, California Polytechnic State University, 2014.

Rosenblum, Louis. "Photocomposition of Chinese: Input Systems and Output Results." n.p., n.d. Louis Rosenblum Collection. Stanford University.

Rosenblum, Louis, B. D. Rosenblum, G. P. Low, and W. W. Garth. *Computerized Typesetting of Chinese: Text Processing on the Sinotype III*. Research Triangle Park, NC: Instrument Society of America, 1983.

Rotman, Brian. *Becoming Beside Ourselves: The Alphabet, Ghosts, and Distributed Human Being*. Durham: Duke University Press, 2008.

Schlombs, Corinna. "Women, Gender and Computing: The Social Shaping of a Technical Field from Ada Lovelace's Algorithm to Anita Borg's 'Systers.'" In *The Palgrave Handbook of Women and Science: History, Culture and Practice since 1660*, 307–332. London: Palgrave MacMillan, 2022.

"Science and Technology News: Domestic: Chung-Chin Kao Invents a Chinese Character Technology Application" (Kexue jishu xiaoxi: Guonei: Gao Zhongqin faming 'Zhongguo wenzi jishu yingyong') [科學技術消息:國內:高仲芹發明'中國文字技術應用']. *Kexue yu jishu* [科學與技術] 1, no. 4 (1944): 87.

Scott, Matt [Mate Sikete 马特·斯科特]. "The Background Technology and Story of Microsoft's Engkoo Pinyin Input Method" (Weiruan yingku pinyin shurufa beihou de jishu he gushi) [微软英库拼音输入法背后的技术和故事]." http://tech.sina.com.cn/it/csj/2013-01-25/08418014659.shtml.

Shannon, Claude E., and Warren Weaver. *The Mathematical Theory of Communication*. Urbana: University of Illinois Press, 1949.

Shanxi Province Chinese Character Encoding Research Special Issue (*Shanxi sheng Hanzi bianma yanjiu zhuanji*) [山西省汉字编码研究专辑]. Shanxi sheng kexue jishu qingbao yanjiusuo, March 1979. Thomas S. Mullaney East Asian Information Technology History Collection, Stanford University.

Shashoua, Fred E. "Photocomposition Machine for the Chinese Language." *RCA, Camden, NJ, RCA Tech. Paper* 101 (1964).

Shashoua, Fred E., Warren R. Isom, and Harold E. Haynes. "Machine for Composing Ideographs." US Patent 3325786. Filed June 2, 1964, and issued June 13, 1967.

"S. H. Caldwell: 1904–1960." *The Technology Review* 63, no. 2 (December 1960): 4.

Sherblom-Woodward, Blake. "Hackers, Gamers and Lamers: The Use of l33t in the Computer Sub-Culture." Unpublished, Swarthmore College, Swarthmore, PA, 2008. https://www.swarthmore.edu/sites/default/files/assets/documents/linguistics/2003_sherblom-woodward_blake.pdf.

Shivtiel, Shlomit Shraybom. "The Question of Romanisation of the Script and the Emergence of Nationalism in the Middle East." *Mediterranean Language Review* 10 (1998): 179–196.

Shortis, Tim. "'Gr8 Txtpectations': The Creativity of Text Spelling." *English Drama Media* 8 (June 2007): 21–26.

Shortis, Tim. "Revoicing Txt: Spelling, Vernacular Orthography and 'Unregimented Writing.'" In *The Texture of Internet: Netlinguistics in Progress*. Ed. Santiago

Posteguillo, María José Esteve, and Lluïsa Gea-Valor. Cambridge: Cambridge Scholar Press, 2007.

Siegert, Bernard. *Cultural Techniques: Grids, Filters, Doors, and Other Articulations of the Real*. New York: Fordham University Press, 2015.

Simon, Denis Fred Simon and Merle Goldman, eds. *Science and Technology in Post-Mao China*. Cambridge, MA: Harvard University Asia Center, 1988.

Sloss, Robert P., and Peter H. Nancarrow. "A Binary Signal Generator for Encoding Chinese Characters into Machine-Compatible Form." Chinese Language Project. Cambridge, England. 1976. FH.410.45. Cambridge University Library Special Collections.

Sloss, Robert P., and P. H. Nancarrow. "C.L.P. Ideo-Matic 66: A Pre-Production Prototype Encoder for Chinese Characters." Chinese Language Project. Cambridge, England. May 1976. FH.410.46. Cambridge University Library Special Collections.

Smith, Eleanor. "Life after QWERTY." *The Atlantic* (November 2013).

"Software Exhibition '87." *South China Morning Post*, May 25, 1987, 47.

"Software Exhibition '88." *South China Morning Post*, August 22, 1988, 7.

"Software Exhibition '89." *South China Morning Post*, October 10, 1989, 28.

"Sogou Cloud Input: An Introduction to Cloud Computing" (Sougou yunrufa: yunjisuan jieshao) [搜狗输入法云计算介绍]." https://pinyin.sogou.com/features/cloud/.

"Sound Code Input System" (Yinma shurufa) [音码输入法]. PRC Patent Document CN 1277379A. December 20, 2000.

"Stanford Libraries Receives a Remarkable East Asian Information Technology Collection." *Stanford Libraries Newsletter* (May 26, 2021). https://library.stanford.edu/node/172367. Accessed July 2, 2022.

Stanley, Autumn. *Mothers and Daughters of Invention: Notes for a Revised History of Technology*. New Jersey: Rutgers University Press, 1995.

Stanton, Andrea. "Broken is Word." In *Your Computer Is on Fire*. Ed. Thomas S. Mullaney, Benjamin Peters, Mar Hicks, and Kavita Philip, 213–230. Cambridge, MA: MIT Press, 2021.

Stepanek, James B. "Microcomputers in China." *The Chinese Business Review*, May–June 1984, 26–29, 29.

Stern, Adi. "Aleph = X: Hebrew Type on the Edge." Presentation at the 2007 ATypI Annual Conference. Helsinki.

Strom, Sharon Hartman. *Beyond the Typewriter: Gender, Class, and the Origins of Modern American Office Work, 1900–1930*. Chicago: University of Illinois Press, 1992.

Suchenwirth, Richard. *Systeme zur Datenverarbeitung in chinesischer Schrift. Eine Markt-übersicht*. Beijing: April 15, 1985. Original housed in the Rolf Heinen Technology Collection in Drolshagen, Germany. Photocopy in author's collection.

"Suggested Outline X of License Agreement Between Chung-Chin Kao and Mergenthaler Linotype Company, N.Y." October 24, 1943. Mergenthaler Linotype Company Records, 1905–1993, Archives Center, National Museum of American History. Smithsonian Institution. Box 3628.

Su, Hsi-Yao. "The Multilingual and Multi-Orthographic Taiwan-Based Internet: Creative Uses of Writing Systems on College-Affiliated BBSs." *Journal of Computer-Mediated Communication* 9, no. 1 (2003), JCMC912, https://doi.org/10.1111/j.1083 -6101.2003.tb00357.x.

"Summary of Cable & Wireless' History in China." DOC/CW/12/54. Porthcurno Telegraph Museum Collection and Cable & Wireless Archives.

"Summary of the Discussions and Samples on the Sinotype Project." December 18, 1975. Graphic Arts Research Foundation. In Louis Rosenblum Collection. Stanford University.

Sun Qiang [孙强]. "Chinese Character Matrix Generator that Can Generate Multiple Fonts" (Yi zhong neng shengcheng duozhong ziti de Hanzi zimo fashengqi) [一种能生成多种字体的汉字字模发生器]. CN1031140A. Assignee: Beijing Sitong Group. Date of Application April 10, 1987. Patent Date February 15, 1989.

Szuprowicz, Bohdan O. "CDC's China Sale Seen Focusing Western Attention." *Computerworld*, November 29, 1976, 51.

Szuprowicz, Bohdan O. "Expanding Chinese Micro Market Triggering Frenzy." *Computerworld*, May 21, 1984, 10.

Tam, Gina Anne. *Dialect and Nationalism in China, 1860–1960*. Cambridge: Cambridge University Press, 2020.

"Technical Data: IPX 5486 Automatic Send-Receive (ASR) Telecommunications Terminal." Thomas S. Mullaney East Asian Information Technology History Collection. Stanford University.

Thornton, Tamara Plakins. *Handwriting in America*. New Haven, CT: Yale University Press, 1996.

Thurlow, Crispin. "Generation Txt? The Sociolinguistics of Young People's Text-Messaging." *Discourse Analysis Online* 1, no. 1 (2003). https://extra.shu.ac.uk/daol /articles/v1/n1/a3/thurlow2002003-paper.html.

Tian Wuzhao [田吳炤], trans., *Outline of logic* (*Lunlixue gangyao*) [論理學綱要]. Shanghai: Shangwu yinshuguan, 1903.

Tianhui ABC Chinese Input Method (*Tianhui ABC Hanzi shurufa*) [天汇ABC汉字输入法]. Thomas S. Mullaney East Asian Information Technology History Collection. Stanford University.

Tianjin Chinese Character Encoding and Information Processing Research Society (Tianjin shi Hanzi bianma xinxi chuli yanjiuhui) [天津市汉字编码信息处理研究

会]. *Chinese Character Encoding Compilation* (*Hanzi bianma huibian*) [汉字编码汇编]. August 1979. Thomas S. Mullaney East Asian Information Technology History Collection. Stanford University.

Tianjin City Zhonghuan Electronic Computer Company (Tianjin shi zhonghuan dianzi jisuanji gongsi) [天津市中环电子计算机公司]. *Chinese Character Encoding Manual* (*Hanzi bianma shouce*) [汉字编码手册]. Tianjin: Tianjin City Zhonghuan Electronic Computer Company, 1982.

Tien, H. C. "On Learning Chinese." *World Journal of Psychosynthesis* 4, no. 7 (July 1972): 4–5, *h.n.*

"'To Lift One's Brush and Forget the Character': Did Informationalization Cause the Chinese Character Crisis?" ('Tibi wangzi': shi xinxihua zaocheng le Hanzi weiji ma?) ["提笔忘字": 是信息化造成了汉字危机吗?]. *China Youth Daily* (*Zhongguo qingnian bao*) [中国青年报]. September 16, 2013. http://www.cernet.edu.cn/zhong_guo_jiao_yu /yiwujiaoyu/201309/t20130916_1017489.shtml.

"'To Lift One's Brush and Forget the Character': What Have We Forgotten?" ("Tibi wangzi" Women jiujing wangdiao le shenme?) ['提笔忘字' 我们究竟忘掉了什么?]. *New China Net* (*Xinhua wang*) [新华网]. November 4, 2013. http://edu.qq.com/a /20131104/002402.htm. Accessed March 1, 2019.

Tong, Q. S. "Inventing China: The Use of Orientalist Views on the Chinese Language." *Interventions* 2 (2000): 11–4.

"Toynbee Lectures." *VMI Cadet* (February 10, 1958): 2.

"Tsinghua University Organizes 30-Plus Professors and Students to Test Electronic Computer Designed for Statistical Purposes" (Qinghua daxue zuzhi 30 duo ge shisheng wei shijiwei shi shituo tongji yong dianzi jisuanji) [清华大学组织30多个师生为市计委试试托统计用电子计算机]. Beijing Municipal Archives (BMA) 001-022-00494 (March 11, 1960).

Tsu, Jing. *Kingdom of Characters: The Language Revolution That Made China Modern*. New York: Riverhead Books, 2022.

Turner, R. H. "Chinese Type Casting Machine." November 19, 1943, 1. Mergenthaler Linotype Company Records, 1905–1993, Archives Center, National Museum of American History. Smithsonian Institution. Box 3628.

"Two Newly Invented Chinese Typewriters" (Zhongwen daziji liang qi xin faming) [中文打字機兩起新發明]. *Kexue yuekan* 15 (1947): 23–44.

UCDOS High-Level Chinese Character System User Manual v. 2.0 (*UCDOS gaoji Hanzi xitong yonghu shouce v. 2.0*) [UCDOS高级汉字系统用户手册 v 2.0]. Beijing: Zhongguo kexueyuan xiwang gaoji dianna jishu gongsi, 1990.

Uzman, Mehmet. "Romanisation in Uzbekistan Past and Present." *Journal of the Royal Asiatic Society* 20 (January 2010): 61–74.

Vaisman, Carmel. "Performing Girlhood through Typographic Play in Hebrew Blogs." In *Digital Discourse: Language in the New Media*, eds. Crispin Thurlow and Kristine Mroczek. Oxford: Oxford University Press, 2011, 177–196.

Viguier, Septime Auguste (Weijiye [威基謁]), *Dianbao xinshu* [電報新書] (Guangxu 18). In "Extension Selskabet—Kinesisk Telegrafordbog." 1871. Arkiv nr. 10.619. In "Love og vedtægter med anordninger." GN Store Nord A/S SN China and Japan Extension Telegraf. Rigsarkivet [Danish National Archives]. Copenhagen, Denmark.

"Visit of Caryl P. Haskins to Graphic Arts Foundation. Carnegie Corporation of New York Record of Interview." February 20, 1953. Louis Rosenblum Collection. Stanford University.

Walder, Andrew. *Agents of Disorder: Inside China's Cultural Revolution*. Cambridge, MA: Belknap Press, 2019.

Wang, Chi. *Building a Better Chinese Collection for the Library of Congress. Selected Writings*. Lanham: The Scarecrow Press, 2012.

Wang Code Computer: Special Issue on Five-Stroke Technology (*Wangma diannao: Wubi zixing jishu zhuankan*) 6 (October 1991).

Wang Guanwei [王官伟], Chen Hongzhong [陈闳中] and Wang Xiaoyu [王晓宇], eds. *Natural Code Chinese Character Input Method Tutorial* (*Ziran ma Hanzi shurufa jiaocheng*) [自然码汉字输入法教程]. Shanghai: Tongji University Press (*Tongji daxue chubanshe*), 1994.

Wang Jizhi [王辑志]. "My Autobiography (25): Making a Chinese Character Card" (*Wo de zizhuan (25): zhizuo Hanka*) [我的自传25:制作汉卡]. http://blog.sina.com.cn/wangjizhi. Accessed January 2, 2018.

Wang, Maojiang. "On the Interface Between the High-Level Languages and Chinese Character Information." *Computer Standards and Interfaces* 6, no. 2 (1987): 181–186.

Wang Qihong. "PASCAL Compiler with Chinese Identifier" (Ke yong Hanzi biaoshifu de PASCAL bianyi chengxu) [可用汉字标识符的PASCAL编译程序]. *Journal of Changchun Post and Telecommunication Institute* (*Changchun youdian xueyuan xuebao*) [长春邮电学院学报]] 2 (1987): 61–71.

Wang Songping [王颂平]. *Complete Illustrated Guide to Stroke Order Code* (*Bishunma tujie quanji*) [笔顺码图解全集]. Beijing: Zhongguo funü chubanshe, 1998.

Wang Xuan [王选]. *Beijing University Founder Group Book Typesetting Technology and Application* (*Beida fangzheng shuban paiban jishu he yingyong*) [北大方正书版排版技术和应用]. Beijing: Founder Group: 1993.

Wang, Xinyuan. *Social Media in Industrial China*. London: UCL Press, 2016.

Wardrip-Fruin, Noah. *Expressive Processing: Digital Fictions, Computer Games, and Software Studies*. Cambridge, MA: MIT Press, 2009.

Wei Tang [唯唐]. "A Computational Chinese Character System with Multiple Input Methods" (Ke yong duozhong shuru fangfa de dianzi jisuanji Hanzi xitong) [可用多种输入方法的电子计算机汉字系统]. *Wenzi gaige* [*Writing Reform*] 4 (1985): 47–48.

Wen Desheng [汶德胜]. *Research on Computer Recognition of Printed Chinese Characters (Yinshua Hanzi jisuanji shibie de yanjiu)* [印刷汉字计算机识别的研究]. Zhongguo kexue jishu daxue [中国科学技术大学], 1988.

"West Germany—Co-Operation in Typewriters." *BBC Summary of World Broadcasts*, July 30, 1980.

Wiener, Norbert. *Cybernetics: Or, Control and Communication in the Animal and the Machine.* Cambridge, MA: MIT Press, 1961.

Winder, Robert. "The Qwerty Conspiracy." *The Independent*, August 12, 1995.

Witzell, Otto W. and J.K. Lee Smith. *Closing the Gap: Computer Development in the People's Republic of China.* Boulder, CO: Westview Press, 1989.

Wood, Helen M., Donald J. Reifer, and Martha Sloan. "A Tour of Computing Facilities in China." *Computer*, January 1985, 80–87.

Wu, K. T. "The Development of Typography in China During the Nineteenth Century." *Library Quarterly* 22, no. 3 (July 1952): 288–301.

Wu Qidi [吴启迪], ed. *The History of Chinese Engineers, Volume III, Innovation and Transcendence: The Rise and Engineering Achievements of Engineers in the Contemporary Period (Zhongguo gongchengshi shi di'san juan chuangxin chaoyue: dangdai gongchengshi qunti de jueti yu gongcheng chengjiu)* [中国工程师史 第三卷 创新超越: 当代工程师群体的崛起与工程成就]. Shanghai: Tongji University Press, 2017.

Wu Xiaojun [吴晓军]. *2.13 Series Chinese Character System User Manual (2.13 xilie Hanzi xitong yonghu shouce)* [2.13系列汉字系统用户手册]. Beijing: Jixie gongye chubanshe, 1993.

Wu Youshou [吴佑寿] and Ding Xiaoqing [丁晓青]. *The Principle Method and Realization of Chinese Character Recognition (Hanzi shibie yuanli fangfa yu shixian)* [汉字识别原理方法与实现]. Gaodeng jiaoyu chubanshe, 1992.

Xu, Yizhou. "The Postmodern Aesthetic of Chinese Online Comment Cultures." *Communication and the Public* 1, no. 4 (2016): 436–451.

Xue Shiquan [薛士权] and Qian Peide [钱培德]. *Guanyu shuangpin bianma Hanzi chuli xitong de shiyan baogao* [关于双拼编码汉字处理系统的实验报告]. N.p.: 1986.

Yang, Guobin. *The Power of the Internet in China: Citizen Activism Online.* New York: Colombia University Press, 2011.

Yang Huimin [杨惠民] and Jiang Zifang [蒋子放]. *Sinicized COBOL Language for Microcomputers (Programming Methods and Skills) (Weixing jisuanji Hanzihua COBOL yuyan)* [微型计算机汉字化COBOL语言]. Beijing: Dianzi gongye chubanshe, 1987.

Yang Jisheng [杨继绳]. *The World Turned Upside Down: A History of the Chinese Cultural Revolution*. New York: Farrar, Straus and Giroux, 2021.

Yang, Shou-chuan, and Charlotte W. Yang. 1969. "A Universal Graphic Character Writer. In International Conference on Computational Linguistics COLING 1969: Preprint No. 42, Sånga Säby, Sweden. See this link for info: https://aclanthology.org/C69-4201/.

Ye Naibeng [叶乃辇], Zhang Xizhong [张炘中] and Xia Ying [夏莹]. *Chinese Language Microcomputers and Chinese Character Recognition (Hanzi weixing jisuanji yu Hanzi shibie)* [汉字微型计算机与汉字识别]. Jixie gongye chubanshe [机械工业出版社], 1989.

Ye, Sang. "Computer Insects." In *The China Reader: The Reform Era*. Ed. Orville Schell and David Shambaugh, 291–296. New York: Vintage Books, 1999.

Ye, Weili. *Seeking Modernity in China's Name: Chinese Students in the United States, 1900–1927*. Stanford, CA: Stanford University Press, 2001.

Yeh, Chan-hui. "System for the Electronic Data Processing of Chinese Characters." US Patent no. 3820644A. Filed May 7, 1973 and issued June 28, 1974.

Yuan Qi [袁琦]. "Chinese Information Technology and Processing Natural Languages" (Zhongwen xinxi jishu he ziran yuyan chuli) [中文信息技术和自然语言处理]. *Journal of Chinese Information Processing* 1, no. 1 (1986): 33–36.

Zhang Guofang [张国防]. *Fifty Character Element Computational Chinese Input Usage Manual (50 ziyuan jisuan Hanzi shurufa shiyong shouce)* [五十字元计算汉字输入法使用手册]. Beijing: Zhongguo jiliang chubanshe, 1989.

Zhang Hongfan [张闳凡]. "Chinese Character Shape-Letter Encoding Method" (*Hanzi xingmu bianmafa*) [汉字形母编码法].

Zhang Shoudong [张寿董], Xu Jianyi [徐建毅], and Zhang Jiansheng [张建生]. *Fundamentals of Chinese Language Computing (Zhongwen xinxi de jisuanji chuli)* [中文信息的计算机处理]. Shanghai: Yuzhou chubanshe, 1984.

Zhang Tinghua [张廷华]. *Double Pinyin Double Radical Encoding System Table of 4000 Commonly Used Chinese Characters (Shuangpin shuangbu bianmafa siqian changyong Hanzi bianma biao)* [双拼双部编码法四千常用汉字编码表]. Shanxi: Shanxi Linyi zhongxue, 1979.

Zhang Xizhong [张炘中]. *Chinese Character Recognition Technology (Hanzi shibie jishu)* [汉字识别技术]. Qinghua daxue chubanshe, 1992.

Zhang Wei [张薇]. "Multilingual Creativity on China's Internet." *World Englishes* (May 2015): 231–246.

Zhang, Weiyu. *The Internet and New Social Formation in China: Fan Publics in the Making*. New York: Routledge, 2016.

Zheng Yi [郑邑], ed. *The Apple II Microcomputer and Chinese System (Apple II weiji ji Hanzi xitong)* [Apple II 微机及汉字系统]. Shanghai: Tongji daxue chubanshe, 1985.

Zhi Bingyi [支秉彝]. "A Cursory Discussion of On-Site Coding (*Qiantan jianzi shima*) [浅谈见字识码]. *Ziran zazhi* [自然杂志] (1978): 1.

Zhi Bingyi [支秉彝]. "An Introduction to 'On Sight Encoding of Character.'" *Nature Magazine* 1, no. 6 (October 1979): 350–353.

Zhi Bingyi [支秉彝]. "Recommendation for a New Method of Chinese Character Encoding" (*Jianyi yi zhong Hanzi bianma xin fangfa*) [建议一种汉字编码新方法]. *Diangong yiqi* [电工仪器] (1975).

Zhi Bingyi [支秉彝]. "The On-Site Coding Chinese Encoding Method and its Implementation on Computers (*Jianzi shima Hanzi bianma fangfa ji qi jisuanji shixian*) [见字识码汉字编码方法及其在计算机实现]." *Zhongguo yuwen* [中国语文] (1979).

Zhi Bingyi [支秉彝]. "The On-Site Coding Chinese Encoding System and Its Realization (*Jianzi shima Hanzi bianma fangfa jiqi zai yingyongzhong shixian*) [见字识码汉字编码方法及其应用中实现]." *Shanghai Wenhuibao* [上海文汇报] (August 19, 1978).

Zhi Bingyi [支秉彝]. *On-Site Code Chinese Encoding Method* (*Jianzi shima Hanzi bianma fangfa*) [见字识码汉字编码方法]. Shanghai: Shanghai Yiqi Yibiao Yanjiusuo [上海仪器仪表研究所], 1982.

Zhi Bingyi [支秉彝] and Qian Feng [钱锋]. "'On-Sight Coding' and its Realization (*Jianzi shima ji qi shixian*) [见字识码及其实现]."

Zhonghua renmin gongheguo guojia biaozhun. "General Word Set for Chinese Character Keyboard Input" (*Hanzi jianpan shuru yong tongyong ciyu ji*). [汉字键盘输入用通用词语集]. GB/T 15732–1995. Circulated August 1, 1995. Adopted April 1, 1996.

Zhou, Yongmin. *Historicizing Online Politics: Telegraphy, the Internet, and Political Participation in China*. Stanford, CA: Stanford University Press, 2005.

Zhou Youguang [周有光]. *Basics of Pinyin Letters* (*Pinyin zimu jichu zhishi*) [拼音字母基础知识]. Beijing: Wenzi gaige chubanshe, 1962.

Zhou Youguang. *General Theory of Chinese Character Reform* (*Hanzi gaige gailun*) [汉字改革概论]. Beijing: Wenzi gaige chubanshe, 1961.

Zhou Youguang. *Questions Regarding Pinyin-ization* (*Pinyinhua wenti*) [拼音化问题]. Zhongguo wenzi gaige weiyuanhui yanjiuzu [中国文字改革委员会研究组]. 1978.

Zhou Youguang. *The Pinyin-ization of Telegrams* (*Dianbao pinyinhua*) [電報拼音化]. Beijing: Wenzi gaige chubanshe, 1965.

Zhu Shili [朱世立]. *Common Methods in Chinese TRUE BASIC Language and System Engineering* (*Hanzi TRUE BASIC yuyan he xitong gongcheng changyong fangfa*) [汉字 TRUE BASIC 语言和系统工程常用方法]. Beijing: Dianzi gongye chubanshe, 1988.

Zimin Wu and J. D. White. "Computer Processing of Chinese Characters: An Overview of Two Decades' Research and Development." *Information Processing and Management* 26, no. 5 (1990): 681–692.

PREVIOUS PUBLICATIONS

The arguments in this book draw from the author's previous publications including the following articles:

"America Has a Rich History of Innovation by Asian Immigrants." *Quartz*, May 29, 2021.

"America's Secret Cold War Mission to Build the First Chinese Computer." *The Atlantic* (September 14, 2016).

"Behind the Painstaking Process of Creating Chinese Computer Fonts." *MIT Technology Review*, May 31, 2021.

"Chinese Is Not a Backward Language." *Foreign Policy*, May 12, 2016.

"The Engineering Daring That Led to the First Chinese Personal Computer." *TechCrunch,*June 29, 2021.

"How a Solitary Prisoner Decoded Chinese for the QWERTY Keyboard." *Psyche*, July 21, 2021.

"How to Spy on 600 Million People: The Hidden Vulnerabilities in Chinese Information Technology." *Foreign Affairs*, June 5, 2016.

"Meet the Mystery Woman Who Mastered IBM's 5,400-Character Chinese Typewriter." *Fast Company*, May 17, 2021.

"The Origins of Chinese Supercomputing, and an American Delegation's Mao-Era Visit." *Foreign Affairs*, August 4, 2016.

"'Security Is Only as Good as Your Fastest Computer': China Now Dominates Supercomputing. That Matters for US National Security." *Foreign Policy*, July 21, 2016.

"The Underground Zines That Kept Self-Expression Alive in Mao's China." *Boston Globe*, June 6, 2021.

"Why Is the World's Largest Collection on China's Modern IT History in the US?" *South China Morning Post*, June 6, 2021.

FIGURES

Figure 2.5: Diagram and photographs of Sinotype showing flashback prism matrix and "flashback window"

Figure 2.6: "Entities" developed by Caldwell

Figure 2.7: Keyboard of the Sinotype

Figure 3.1: Bitmap diagram of the character *ying* (鷹 "eagle")

Figure 3.2: Diagram of IPX keyboard

Figure 3.3: Close-Up of IPX keypad containing the character *zhong* (中 "central")

Figure 3.4: IPX promotional film, stills

Figure 3.5: Comparison of tray bed of mechanical Chinese typewriter (Double Pigeon brand) and keyboard of the IPX by Ideographix

Figure 3.6: Patent featuring machine-readable codes printed with mechanical Chinese typewriter

Figure 3.7: Photographs from 1972 delegation of American Computer Scientists

Figure 3.8: Example of "large-keyboard" approach to Chinese keyboard interface

Figure 3.9: Sample button on the Peking University "medium-sized" keyboard

Figure 3.10: Peking University keyboard and explanatory diagram, 1975

Figure 3.11: Comparison of "divisible type" Chinese printing, the Chinese typewriter prototype by Qi Xuan, the MingKwai Chinese typewriter by Lin Yutang, and the Peking University "medium-sized" keyboard

Figure 3.12: Loh's keyboard, by Loh Shiu-chang

Figure 3.13: Ideo-Matic Encoder

Figure 3.14: Comparison of the Zhou Houkun's first Chinese typewriter prototype, the Toshiba Japanese typewriter, and the Ideo-Matic 66 by Sloss and Nancarrow

Figure 4.1: Photograph of Zhi Bingyi

Figure 4.2: The Three-Corner Coding look-up table and sample encoding

Figure 4.3: "The First Radical Transalphabet Table" by H. C. Tien (samples)

Figure 4.4: Encoding 電 (*dian*, "electricity") using H. C. Tien's Chinese Transalphabet

Figure 4.5: Olympia 1011 by Olympia Werke

Figure 4.6: RCA, Fred Shashoua, and the revival of Sinotype

Figure 4.7: Photograph of An Wenyi, deputy secretary of the *People's Daily*, entering characters on the Sinotype using Standard Telegraph Code (STC) input

INDEX

Chinese (language) (cont.)
  and character crisis (*Hanzi weiji*), 1, 28
  and Character Input Code, 179,
    297n59
  and Character Internal Code, 179,
    297n59
  and character libraries, 208–210, 220,
    295n35, 296n40
  and characters-per-minute, 248n14
  and character processing, 180, 193–
    194, 269n22, 295n35
  characters, 1–3, 5–6, 8–10, 12–13, 16–
    22, 26–28, 32, 34, 36, 40, 44, 46–48,
    56, 64–70, 75–76, 79, 81–82, 85,
    87–90, 92–93, 95, 97, 101, 105–106,
    108, 110, 112–113, 121, 127–128,
    130–132, 134, 144, 146, 149, 154,
    157–158, 164, 166–167, 174, 178,
    191–192, 201, 208, 272n50, 289n99,
    294n35
  computer users, 18–19, 26, 32–33, 63
  and Devanagari, 71–72
  dictionaries, 112, 126, 131, 203
  and digitization, 2, 90–91, 112, 115,
    164–166, 217, 295n35
  and divisible type Chinese printing,
    106, 109
  and Double Pinyin, 195–196
  and electro-automation, 30
  electronic, 2, 29
  and encoding methods, 192, 201,
    305nn24–25
  and experts, 39
  and GARF, 72–74, 287n83, 289n99,
    293n30
  and Hanyu pinyin, 189–190, 196,
    200
  and homographs, 81
  and homophones, 135, 255n12
  and human-computer interaction
    (HCI), 174, 218, 227
  and hypography, 20, 26, 187, 197
  "immediate," 11

  and information technology, 38, 61,
    101, 127, 140, 206, 291n3, 292n12,
    306n27
  and inputs, 23, 44, 79, 147, 188–190,
    199, 212, 220, 231, 297n59
  and Input Method Editors (IMEs), 5,
    14–15, 123, 128–129, 135, 153–54,
    156, 175, 177, 179, 183, 185, 188,
    192, 205–206, 210–211, 215, 220,
    228, 248n9, 309n8
  and internet, 2
  and kanji, 204
  and keyboards, 103–104, 107, 218
  lexicon, 36, 200
  and mechanized textual production,
    41
  and Microsoft, 211
  and modernization, 91, 216
  and movable type, 166
  and MS-DOS, 177–178
  and operating systems, 185
  and OSCO system, 197
  and pinyin input, 212–214
  and predictive text, 188
  and programming languages, 299n66
  and script, 36, 58, 66, 87–88, 130,
    251n26, 280n24
  and Sinotype, 77, 150, 272n51, 301n9
  and Robert Sloss, 111–113
  and spelling, 70, 75, 82
  and standardization, 308n8
  and strokes, 79, 110, 130
  symbols, 21
  and telegrams, 37
  and telegraph code, 37, 46, 143, 150,
    254n12
  and telegraphy, 35, 38, 44–46, 48, 55,
    60, 219, 254n11
  and text processing, 118, 168, 178,
    207
  and translation to English, 112
  and transliteration, 189
  and typecasting, 52

Studies of the Weatherhead East Asian Institute
Columbia University

Selected Titles
(Complete list at: weai.columbia.edu/content/publications)

Territorializing Manchuria: The Transnational Frontier and Literatures of East Asia, by Miya Xie. Harvard East Asian Monographs, 2023.

Takamure Itsue, Japanese Antiquity, and Matricultural Paradigms that Address the Crisis of Modernity: A Woman from the Land of Fire, by Yasuko Sato. Palgrave Macmillan, 2023.

Rejuvenating Communism: Youth Organizations and Elite Renewal in Post-Mao China, by Jérôme Doyon. University of Michigan Press, 2023.

From Japanese Empire to American Hegemony: Koreans and Okinawans in the Resettlement of Northeast Asia, by Matthew R. Augustine. University of Hawai'i Press, 2023.

Building a Republican Nation in Vietnam, 1920-1963, edited by Nu-Anh Tran and Tuong Vu. University of Hawai'i Press, 2022.

China Urbanizing: Impacts and Transitions, edited by Weiping Wu and Qin Gao. University of Pennsylvania Press, 2022.

Common Ground: Tibetan Buddhist Expansion and Qing China's Inner Asia, by Lan Wu. Columbia University Press, 2022.

Narratives of Civic Duty: How National Stories Shape Democracy in Asia, by Aram Hur. Cornell University Press, 2022.

The Concrete Plateau: Urban Tibetans and the Chinese Civilizing Machine, by Andrew Grant. Cornell University Press, 2022.

Confluence and Conflict: Reading Transwar Japanese Literature and Thought, by Brian Hurley. Harvard East Asian Monographs, 2022.

Inglorious, Illegal Bastards: Japan's Self-Defense Force During the Cold War, by Aaron Skabelund. Cornell University Press, 2022.

Madness in the Family: Women Care, and Illness in Japan, by H. Yumi Kim. Oxford University Press, 2022.

Uncertainty in the Empire of Routine: The Administrative Revolution of the Eighteenth-Century Qing State, by Maura Dykstra. Harvard University Press, 2022.

Outsourcing Repression: Everyday State Power in Contemporary China, by Lynette H. Ong. Oxford University Press, 2022.

Diasporic Cold Warriors: Nationalist China, Anticommunism, and the Philippine Chinese, 1930s–1970s, by Chien-Wen Kung. Cornell University Press, 2022.

Dream Super-Express: A Cultural History of the World's First Bullet Train, by Jessamyn Abel. Stanford University Press, 2022.

The Sound of Salvation: Voice, Gender, and the Sufi Mediascape in China, by Guangtian Ha. Columbia University Press, 2022.

Carbon Technocracy: Energy Regimes in Modern East Asia, by Victor Seow. The University of Chicago Press, 2022.

Disunion: Anticommunist Nationalism and the Making of the Republic of Vietnam, by Nu-Anh Tran. University of Hawai'i Press, 2022.

Learning to Rule: Court Education and the Remaking of the Qing State, 1861–1912, by Daniel Barish. Columbia University Press, 2022.